Materials
Engineering, Science, Processing and Design

Michael Ashby, Hugh Shercliff and David Cebon

University of Cambridge,
UK

ELSEVIER

AMSTERDAM • BOSTON • HEIDELBERG • LONDON • NEW YORK • OXFORD
PARIS • SAN DIEGO • SAN FRANCISCO • SINGAPORE • SYDNEY • TOKYO
Butterworth-Heinemann is an imprint of Elsevier

Butterworth-Heinemann is an imprint of Elsevier
Linacre House, Jordan Hill, Oxford OX2 8DP, UK
30 Corporate Drive, Suite 400, Burlington, MA 01803, USA

First edition 2007
Reprinted 2007

Copyright © 2007, Michael Ashby, Hugh Shercliff and David Cebon.
Published by Elsevier Ltd. All rights reserved

The right of Michael Ashby, Hugh Shercliff and David Cebon to be identified
as the authors of this work has been asserted in accordance with the Copyright,
Designs and Patents Act 1988

No part of this publication may be reproduced, stored in a retrieval system
or transmitted in any form or by any means electronic, mechanical, photocopying,
recording or otherwise without the prior written permission of the publisher

Permissions may be sought directly from Elsevier's Science & Technology Rights
Department in Oxford, UK: phone (+44) (0) 1865 843830; fax (+44) (0) 1865 853333;
email: permissions@elsevier.com. Alternatively you can submit your request online by
visiting the Elsevier web site at http://elsevier.com/locate/permissions, and selecting
Obtaining permission to use Elsevier material

Notice
No responsibility is assumed by the publisher for any injury and/or damage to persons
or property as a matter of products liability, negligence or otherwise, or from any use
or operation of any methods, products, instructions or ideas contained in the material
herein. Because of rapid advances in the medical sciences, in particular, independent
verification of diagnoses and drug dosages should be made

British Library Cataloguing in Publication Data
A catalogue record for this book is available from the British Library

Library of Congress Cataloging-in-Publication Data
A catalog record for this book is available from the Library of Congress

ISBN: 978-0-7506-8391-3

For information on all Butterworth-Heinemann publications
visit our website at books.elsevier.com

Printed and bound in *Italy*

07 08 09 10 10 9 8 7 6 5 4 3 2

Working together to grow
libraries in developing countries

www.elsevier.com | www.bookaid.org | www.sabre.org

ELSEVIER   BOOK AID International   Sabre Foundation

Contents

Preface

Science-led or Design-led? Two approaches to materials teaching

Most things can be approached in more than one way. In teaching this is especially true. The way to teach a foreign language, for example, depends on the way the student wishes to use it—to read the literature, say, or to find accommodation, order meals and buy beer. So it is with the teaching of this subject.

The traditional approach to it starts with fundamentals: the electron, the atom, atomic bonding, and packing, crystallography and crystal defects. Onto this is built alloy theory, the kinetics of phase transformation and the development of microstructure on scales made visible by electron and optical microscopes. This sets the stage for the understanding and control of properties at the millimeter or centimeter scale at which they are usually measured. The approach gives little emphasis to the behavior of structures, methods for material selection, and design.

The other approach is design-led. The starting point is the need: the requirements that materials must meet if they are to perform properly in a given design. To match materials to designs requires a perspective of the range of properties they offer and the other information that will be needed about them to enable successful selection. Once the importance of a property is established there is good reason to 'drill down', so to speak, to examine the science that lies behind it—valuable because an understanding of the fundamentals itself informs material choice and usage.

There is sense in both approaches. It depends on the way the student wishes to use the information. If the intent is scientific research, the first is the logical way to go. If it is engineering design, the second makes better sense. This book follows the second.

What is different about this book?

There are many books about the science of engineering materials and many more about design. What is different about this one?

First, a *design-led approach* specifically developed to guide material selection and manipulation. The approach is systematic, leading from design requirements to a prescription for optimized material choice. The approach is illustrated by numerous case studies. Practice in using it is provided by Exercises.

Second, an emphasis on *visual communication* and a unique graphical presentation of material properties as *material property charts*. These are a central feature of the approach, helpful both in understanding the origins of properties, their manipulation and their fundamental limits, as well as providing a tool for selection and for understanding the ways in which materials are used.

Third, its *breadth*. We aim here to present the properties of materials, their origins and the way they enter engineering design. A glance at the Contents pages will show sections dealing with:

- Physical properties
- Mechanical characteristics
- Thermal behavior

- Electrical, magnetic and optical response
- Durability
- Processing and the way it influences properties
- Environmental issues

Throughout we aim for a simple, straightforward presentation, developing the materials science as far as is it helpful in guiding engineering design, avoiding detail where this does not contribute to this end.

And fourth, *synergy* with the Cambridge Engineering Selector (CES)[1]—a powerful and widely used PC-based software package that is both a source of material and process information and a tool that implements the methods developed in this book. The book is self-contained: access to the software is not a prerequisite for its use. Availability of the CES EduPack software suite enhances the learning experience. It allows realistic selection studies that properly combine multiple constraints on material and processes attributes, and it enables the user to explore the ways in which properties are manipulated.

The CES EduPack contains an additional tool to allow the science of materials to be explored in more depth. The CES Elements database stores fundamental data for the physical, crystallographic, mechanical, thermal, electrical, magnetic and optical properties of all 111 elements. It allows interrelationships between properties, developed in the text, to be explored in depth.

The approach is developed to a higher level in two further textbooks, the first relating to mechanical design[2], the second to industrial design[3].

[1] The CES EduPack 2007, Granta Design Ltd., Rustat House, 62 Clifton Court, Cambridge CB1 7EG, UK, www.grantadesign.com.

[2] Ashby, M.F. (2005), *Materials Selection in Mechanical Design*, 3rd edition, Butterworth-Heinemann, Oxford, UK, Chapter 4. ISBN 0-7506-6168-2. (*A more advanced text that develops the ideas presented here in greater depth.*)

[3] Ashby, M.F. and Johnson, K. (2002) *Materials and Design—The Art and Science of Material Selection in Product Design*, Butterworth-Heinemann, Oxford, UK. ISBN 0-7506-5554-2. (*Materials and processes from an aesthetic point of view, emphasizing product design.*)

Acknowledgements

No book of this sort is possible without advice, constructive criticism and ideas from others. Numerous colleagues have been generous with their time and thoughts. We would particularly like to recognize suggestions made by Professors Mick Brown, Archie Campbell, Dave Cardwell, Ken Wallace and Ken Johnson, all of Cambridge University, and acknowledge their willingness to help. Equally valuable has been the contribution of the team at Granta Design, Cambridge, responsible for the development of the CES software that has been used to make the material property charts that are a feature of this book.

Resources that accompany this book

Exercises

Each chapter ends with exercises of three types: the first rely only on information, diagrams and data contained in the book itself; the second makes use of the CES software in ways that use the methods developed here, and the third explores the science more deeply using the CES Elements database that is part of the CES system.

Instructor's manual

The book itself contains a comprehensive set of exercises. Worked-out solutions to the exercises are freely available to teachers and lecturers who adopt this book. To access this material online please visit http://textbooks.elsevier.com and follow the instructions on screen.

Image Bank

The Image Bank provides adopting tutors and lecturers with jpegs and gifs of the figures from the book that may be used in lecture slides and class presentations. To access this material please visit http://textbooks.elsevier.com and follow the instructions on screen.

The CES EduPack

CES EduPack is the software-based package to accompany this book, developed by Michael Ashby and Granta Design. Used together, *Materials: Engineering, Science, Processing and Design* and CES EduPack provide a complete materials, manufacturing and design course. For further information please see the last page of this book, or visit www.grantadesign.com.

Chapter 1
Introduction: materials— history and character

Professor James Stuart, the first Professor of Engineering at Cambridge.

Chapter contents

1.1 Materials, processes and choice

Engineers *make* things. They make them out of *materials*. The materials have to support loads, to insulate or conduct heat and electricity, to accept or reject magnetic flux, to transmit or reflect light, to survive in often-hostile surroundings, and to do all this without damage to the environment or costing too much.

And there is the partner in all this. To *make* something out of a material you also need a process. Not just any process—the one you choose has to be compatible with the material you plan to use. Sometimes it is the process that is the dominant partner and a material-mate must be found that is compatible with it. It is a marriage. Compatibility is not easily found—many marriages fail— and material failure can be catastrophic, with issues of liability and compensation. This sounds like food for lawyers, and sometimes it is: some specialists make their living as expert witnesses in court cases involving failed materials. But our aim here is not contention; rather, it is to give you a vision of the materials universe (since, even on the remotest planets you will find the same elements) and of the universe of processes, and to provide methods and tools for choosing them to ensure a happy, durable union.

But, you may say, engineers have been making things out of materials for centuries, and successfully so—think of Isambard Kingdom Brunel, Thomas Telford, Gustave Eiffel, Henry Ford, Karl Benz and Gottlieb Daimler, the Wright brothers. Why do we need new ways to choose them? A little history helps here. Glance at the portrait with which this chapter starts: it shows James Stuart, the first Professor of Engineering at Cambridge University from 1875 to 1890 (note the cigar). In his day the number of materials available to engineers was small—a few hundred at most. There were no synthetic polymers—there are now over 45 000 of them. There were no light alloys (aluminum was first established as an engineering material only in the 20th century)—now there are thousands. There were no high-performance composites—now there are hundreds of them. The history is developed further in Figure 1.1, the time-axis of which spans 10 000 years. It shows roughly when each of the main classes of materials first evolved. The time-scale is nonlinear—almost all the materials we use today were developed in the last 100 years. And this number is enormous: over 160 000 materials are available to today's engineer, presenting us with a problem that Professor Stuart did not have: that of optimally selecting from this huge menu. With the ever-increasing drive for performance, economy and efficiency, and the imperative to avoid damage to the environment, making the right choice becomes very important. Innovative design means the imaginative exploitation of the properties offered by materials.

These properties, today, are largely known and documented in handbooks; one such—the *ASM Materials Handbook*—runs to 22 fat volumes, and it is one of many. How are we to deal with this vast body of information? Fortunately another thing has changed since Prof. Stuart's day: we now have digital information storage and manipulation. Computer-aided design is now a standard part

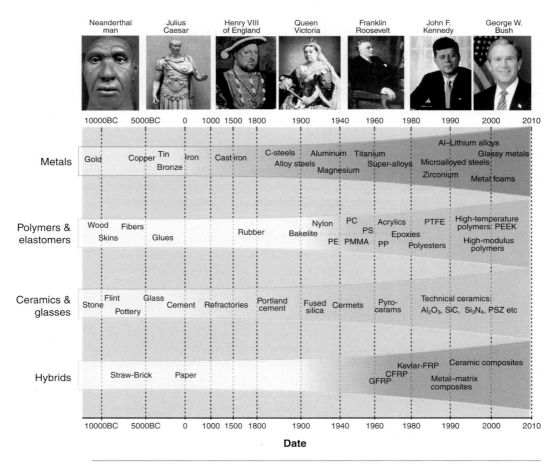

Figure 1.1   The development of materials over time. The materials of pre-history, on the left, all occur naturally; the challenge for the engineers of that era was one of shaping them. The development of thermochemistry and (later) of polymer chemistry enabled man-made materials, shown in the colored zones. Three—stone, bronze and iron—were of such importance that the era of their dominance is named after them.

of an engineer's training, and it is backed up by widely available packages for solid modeling, finite-element analysis, optimization, and for material and process selection. Software for the last of these—the selection of materials and processes—draws on databases of the attributes of materials and processes, documenting their mutual compatibility, and allows them to be searched and displayed in ways that enable selections that best meet the requirements of a design.

If you travel by foot, bicycle or car, you take a map. The materials landscape, like the terrestrial one, can be complex and confusing; maps, here, are also a good idea. This text presents a design-led approach to materials and manufacturing processes that makes use of maps: novel graphics to display the world of materials and processes in easily accessible ways. They present the properties of materials in ways that give a global view, that reveal relationships between properties and that enable selection.

1.2 Material properties

So what are these properties? Some, like density (mass per unit volume) and price (the cost per unit volume or weight) are familiar enough, but others are not, and getting them straight is essential. Think first of those that have to do with carrying load safely—the *mechanical properties*.

Mechanical properties

A steel ruler is easy to bend *elastically*—'elastic' means that it springs back when released. Its elastic stiffness (here, resistance to bending) is set partly by its shape—thin strips are easy to bend—and partly by a property of the steel itself: its *elastic modulus*, E. Materials with high E, like steel, are intrinsically stiff; those with low E, like polyethylene, are not. Figure 1.2(b) illustrates the consequences of inadequate stiffness.

The steel ruler bends elastically, but if it is a good one, it is hard to give it a permanent bend. Permanent deformation has to do with *strength*, not stiffness. The ease with which a ruler can be permanently bent depends, again, on its shape and on a different property of the steel—its *yield strength*, σ_y. Materials with large σ_y, like titanium alloys, are hard to deform permanently even though their stiffness, coming from E, may not be high; those with low σ_y, like lead, can be deformed with ease. When metals deform, they generally get stronger (this is called 'work hardening'), but there is an ultimate limit, called the *tensile strength*, σ_{ts}, beyond which the material fails (the amount it stretches before it breaks is called the *ductility*). Figure 1.2(c) gives an idea of the consequences of inadequate strength.

So far so good. One more. If the ruler were made not of steel but of glass or of PMMA (Plexiglas, Perspex), as transparent rulers are, it is not possible to bend it permanently at all. The ruler will fracture suddenly, without warning, before it acquires a permanent bend. We think of materials that break in this way as brittle, and materials that do not as tough. There is no permanent deformation here, so σ_y is not the right property. The resistance of materials to cracking and fracture is measured instead by the *fracture toughness*, K_{1c}. Steels are tough—well, most are (steels *can* be made brittle)—they have a high K_{1c}. Glass epitomizes brittleness; it has a very low K_{1c}. Figure 1.2(d) suggests consequences of inadequate fracture and toughness.

We started with the material property *density*, mass per unit volume, symbol ρ. Density, in a ruler, is irrelevant. But for almost anything that moves, weight carries a fuel penalty, modest for automobiles, greater for trucks and trains, greater still for aircraft, and enormous in space vehicles. Minimizing weight has

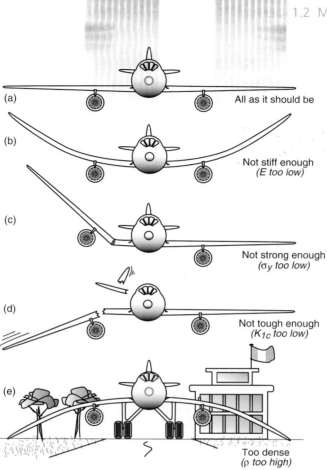

(a)   All as it should be

(b)   Not stiff enough
(E too low)

(c)   Not strong enough
(σ_y too low)

(d)   Not tough enough
(K_{1c} too low)

(e)   Too dense
(ρ too high)

Figure 1.2   Mechanical properties.

much to do with clever design—we will get to that later—but equally to choice of material. Aluminum has a low density, lead a high one. If our little aircraft were made of lead, it would never get off the ground at all (Figure 1.2(e)).

These are not the only mechanical properties, but they are the most important ones. We will meet them, and the others, in Chapters 4–11.

Thermal properties

The properties of a material change with temperature, usually for the worse. Its strength falls, it starts to 'creep' (to sag slowly over time), it may oxidize, degrade or decompose (Figure 1.3(a)). This means that there is a limiting temperature called the *maximum service temperature*, T_{max}, above which its use is impractical. Stainless steel has a high T_{max}—it can be used up to 800°C; most polymers have a low T_{max} and are seldom used above 150°C.

(a) High service temperature T_{max} Low service temperature T_{max}

(b) High expansion coefficient α Low expansion coefficient α

(c) High conductivity λ Low conductivity λ

(d) High T-diffusivity a Low T-diffusivity a

Figure 1.3   Thermal properties.

Most materials expand when they are heated, but by differing amounts depending on their thermal expansion coefficient, α. The expansion is small, but its consequences can be large. If, for instance, a rod is constrained, as in Figure 1.3(b), and then heated, expansion forces the rod against the constraints, causing it to buckle. Railroad track buckles in this way if provision is not made to cope with it.

Some materials—metals, for instance—feel cold; others—like woods—feel warm. This feel has to do with two thermal properties of the material: *thermal conductivity* and *heat capacity*. The first, thermal conductivity, λ, measures the rate at which heat flows through the material when one side is hot and the other cold. Materials with high λ are what you want if you wish to conduct heat from one place to another, as in cooking pans, radiators and heat exchangers; Figure 1.3(c) suggests consequences of high and low λ for the cooking vessel. But low λ is useful too—low λ materials insulate homes, reduce the energy consumption of refrigerators and freezers, and enable space vehicles to re-enter the earth's atmosphere.

These applications have to do with long-time, steady, heat flow. When time is limited, that other property—*heat capacity*, C_p—matters. It measures the amount of heat that it takes to make the temperature of material rise by a given amount. High heat capacity materials—copper, for instance—require a lot of heat to change their temperature; low heat capacity materials, like polymer foams, take much less. Steady heat flow has, as we have said, to do with thermal conductivity. There is a subtler property that describes what happens when heat is first applied. Think of lighting the gas under a cold slab of material with a bole of ice-cream on top (here, lime ice-cream) as in Figure 1.3(d). An instant after ignition, the bottom surface is hot but the rest is cold. After a while, the middle gets hot, then later still, the top begins to warm up and the ice-cream first starts to melt. How long does this take? For a given thickness of slab, the time is inversely proportional to the *thermal diffusivity, a,* of the material of the slab. It differs from the conductivity because materials differ in their heat capacity—in fact, it is proportional to λ/C_p.

There are other thermal properties—we'll meet them in Chapters 12 and 13—but this is enough for now. We turn now to matters electrical, magnetic and optical.

Electrical, magnetic and optical properties

We start with electrical conduction and insulation (Figure 1.4(a)). Without electrical conduction we would lack the easy access to light, heat, power, control and communication that—today—we take for granted. Metals conduct well—copper and aluminum are the best of those that are affordable. But conduction is not always a good thing. Fuse boxes, switch casings, the suspensions for transmission lines all require insulators, and in addition those that can carry some load, tolerate some heat and survive a spark if there were one. Here the property we want is *resistivity*, ρ_e, the inverse of electrical conductivity κ_e. Most plastics and glass have high resistivity (Figure 1.4(a))—they are used as insulators—though, by special treatment, they can be made to conduct a little.

Figure 1.4(b) suggests further electrical properties: the ability to allow the passage of microwave radiation, as in the radome, or to reflect them, as in the passive reflector of the boat. Both have to do with *dielectric* properties, particularly the *dielectric constant* ε_D. Materials with high ε_D respond to an electric field by shifting their electrons about, even reorienting their molecules; those with low ε_D are immune to the field and do not respond. We explore this and other electrical properties in Chapter 14.

Electricity and magnetism are closely linked. Electric currents induce magnetic fields; a moving magnet induces, in any nearby conductor, an electric current. The response of most materials to magnetic fields is too small to be of practical value. But a few—called ferromagnets and ferrimagnets—have the capacity to trap a magnetic field permanently. These are called 'hard' magnetic materials because, once magnetized, they are hard to demagnetize. They are used as permanent magnets in headphones, motors and dynamos. Here the key property is the *remanence*, a measure of the intensity of the retained magnetism. A few others—'soft'

(a) Low resistivity ρ_e High resistivity ρ_e

(b) Low dielectric response   High dielectric response

(c) 'Hard' magnetic behavior   Soft magnetic behavior

(d) Refraction   Absorption

Figure 1.4   Electrical, magnetic and optical properties.

magnet materials—are easy to magnetize and demagnetize. They are the materials of transformer cores and the deflection coils of a TV tube. They have the capacity to conduct a magnetic field, but not retain it permanently (Figure 1.4(c)). For these a key property is the *saturation magnetization*, which measures how large a field the material can conduct. These we meet again in Chapter 15.

Materials respond to light as well as to electricity and magnetism—hardly surprising, since light itself is an electromagnetic wave. Materials that are opaque *reflect* light; those that are transparent *refract* it, and some have the ability to *absorb* some wavelengths (colors) while allowing others to pass freely (Figure 1.4(d)). These are explored in more depth in Chapter 16.

Chemical properties

Products often have to function in hostile environments, exposed to corrosive fluids, to hot gases or to radiation. Damp air is corrosive, so is water; the sweat of your hand is particularly corrosive, and of course there are far more aggressive environments than these. If the product is to survive for its design life it must be made of materials—or at least coated with materials—that can tolerate the surroundings in which they operate. Figure 1.5 illustrates some of the commonest of these: fresh and salt water, acids and alkalis, organic solvents, oxidizing flames

(a) Fresh water

(b) Salt water

(c) Acids and alkalis

(d) Organic solvents

(e) Oxidation

(f) UV radiation

Figure 1.5   Chemical properties: resistance to water, acids, alkalis, organic solvents, oxidation and radiation.

and ultraviolet radiation. We regard the intrinsic resistance of a material to each of these as material properties, measured on a scale of 1 (very poor) to 5 (very good). Chapter 17 deals with the material durability.

1.3 Design-limiting properties

The performance of a component is limited by certain of the properties of the materials of which it is made. This means that, to achieve a desired level of performance, the values of the design-limiting properties must meet certain targets—those that fail to do so are not suitable. In the cartoon of Figure 1.2, stiffness, strength and toughness are design limiting—if any one of them were too low, the plane won't fly. In the design of power transmission lines electrical resistivity is design limiting; in the design of a camera lens, it is optical quality and refractive index.

Materials are chosen by identifying the design-limiting properties and applying limits to them, screening out those that do not meet the limits (Chapter 3).

Processes, too, have properties, although we have not met them yet. These too can be design limiting, leading to a parallel scheme for choosing viable processes (Chapters 18 and 19).

1.4 Summary and conclusions

Engineering design depends on *materials* that are shaped, joined and finished by *processes*. Design requirements define the performance required of the materials, expressed as target values for certain *design-limiting properties*. A material is chosen because it has properties that meet these targets and is compatible with the processes required to shape, join and finish it.

This chapter introduced some of the design-limiting properties: *physical properties* (like density), *mechanical properties* (like modulus and yield strength) and *functional properties* (those describing the thermal, electrical, magnetic and optical behavior). We examine all of these in more depth in the chapters that follow, but those just introduced are enough to be going on with. We turn now to the materials themselves: the families, the classes and the members.

1.5 Further reading

The history and evolution of materials

A History of Technology (1954–2001) (21 volumes), edited by Singer, C., Holmyard, E.J., Hall, A.R., Williams, T.I. and Hollister-Short, G. Oxford University Press, Oxford, UK. ISSN 0307-5451. (*A compilation of essays on aspects of technology, including materials.*)

Delmonte, J. (1985) *Origins of Materials and Processes*, Technomic Publishing Company, Pennsylvania, USA. ISBN 87762-420-8. (*A compendium of information about materials in engineering, documenting the history.*)

Tylecoate, R.F. (1992) *A History of Metallurgy*, 2nd edition, The Institute of Materials, London, UK. ISBN 0-904357-066. (*A total-immersion course in the history of the extraction and use of metals from 6000 BC to 1976, told by an author with forensic talent and love of detail.*)

1.6 Exercises

Exercise E1.1   Use Google to research the history and uses of one of the following materials:

- Tin
- Glass
- Cement

- Titanium
- Carbon fiber.

Present the result as a short report of about 100–200 words (roughly half a page).

Exercise E1.2   What is meant by the *design-limiting properties* of a material in a given application?

Exercise E1.3   There have been many attempts to manufacture and market plastic bicycles. All have been too flexible. Which design-limiting property is insufficiently large?

Exercise E1.4   What, in your judgement, are the design-limiting properties for the material for the blade of a knife that will be used to gut fish?

Exercise E1.5   What, in your judgement, are the design-limiting properties for the material of an oven glove?

Exercise E1.6   What, in your judgement, are the design-limiting properties for the material of an electric lamp filament?

Exercise E1.7   A material is needed for a tube to carry fuel from the fuel tank to the carburetor of a motor mower. The design requires that the tube can bend and that the fuel be visible. List what you would think to be the design-limiting properties.

Exercise E1.8   A material is required as the magnet for a magnetic soap holder. Soap is mildly alkaline. List what you would judge to be the design-limiting properties.

Exercise E1.9   The cases in which most CDs are sold have an irritating way of cracking and breaking. Which design-limiting property has been neglected in selecting the material of which they are made?

Exercise E1.10   List three applications that, in your judgement, need high stiffness and low weight.

Exercise E1.11   List three applications that, in your judgement, need optical quality glass.

Chapter 2
Family trees: organizing materials and processes

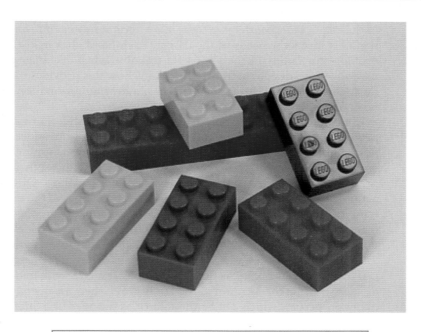

2.1 Introduction and synopsis

A successful product—one that performs well, is good value for money and gives pleasure to the user—uses the best materials for the job, and fully exploits their potential and characteristics.

The families of materials—metals, polymers, ceramics and so forth—are introduced in Section 2.2. What do we need to know about them if we are to design products using them? That is the subject of Section 2.3, in which distinctions are drawn between various types of materials information. But it is not, in the end, a *material* that we seek; it is a certain *profile of properties*—the one that best meets the needs of the design. Each family has its own characteristic profile—the 'family likeness'—useful to know when deciding which family to use for a given design. Section 2.2 explains how this provides the starting point for a classification scheme for materials, allowing information about them to be organized and manipulated.

Choosing a material is only half the story. The other half is the choice of a process route to shape, join and finish it. Section 2.3 introduces process families and their attributes. Choice of material and process are tightly coupled: a given material can be processed in some ways but not others, and a given process can be applied to some materials but not to others. On top of that, the act of processing can change, even create, the properties of the material. Process families, too, exhibit family likenesses—commonality in the materials that members of a family can handle or the shapes they can make. Section 2.3 introduces a classification for processes that parallels that for materials.

Family likenesses are most strikingly seen in *material property charts*, a central feature of this book (Section 2.5). These are charts with material properties as axes showing the location of the families and their members. Materials have many properties, which can be thought of as the axes of a 'material–property' space—one chart is a two-dimensional slice through this space. Each material family occupies a discrete part of the space, distinct from the other families. The charts give an overview of materials and their properties; they reveal aspects of the science underlying the properties, and they provide a powerful tool for materials selection. Process attributes can be treated in a similar way to create process–attribute charts—we leave these for Chapter 18.

The classification systems of Sections 2.2 and 2.3 provide a structure for computer-based information management, introduced in Section 2.6. The chapter ends with a summary, further reading and exercises.

2.2 Getting materials organized: the materials tree

Classifying materials
It is conventional to classify the materials of engineering into the six broad families shown in Figure 2.1: metals, polymers, elastomers, ceramics, glasses and

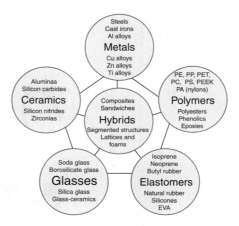

Figure 2.1   The menu of engineering materials. The basic families of metals, ceramics, glasses, polymers and elastomers can be combined in various geometries to create hybrids.

hybrids—composite materials made by combining two or more of the others. There is sense in this: the members of a family have certain features in common: similar properties, similar processing routes and, often, similar applications. Figure 2.2 shows examples of each family.

Figure 2.3 illustrates how the families are expanded to show classes, sub-classes and members, each of which is characterized by a set of *attributes*: its properties. As an example, the Materials universe contains the family '*Metals*', which in turn contains the class '*Aluminum alloys*', sub-classes such as the '*6000 series*' within which we find the particular member '*Alloy 6061*'. It, and every other member of the universe, is characterized by a set of attributes that include not only the properties mentioned in Chapter 1, but also its processing characteristics, the environmental consequences of its use and its typical applications. We call this its *property profile*. Selection involves seeking the best match between the property profiles of the materials in the universe and that required by the design. As already mentioned, the members of one family have certain characteristics in common. Here, briefly, are some of these.

Metals have relatively high stiffness, measured by the modulus, E. Most, when pure, are soft and easily deformed, meaning that σ_y is low. They can be made strong by alloying and by mechanical and heat treatment, increasing σ_y, but they remain ductile, allowing them to be formed by deformation processes. And, broadly speaking, they are tough, with a usefully high fracture toughness K_{1c}. They are good electrical and thermal conductors. But metals have weaknesses too: they are reactive; most corrode rapidly if not protected.

Ceramics are non-metallic, inorganic solids, like porcelain or alumina—the material of spark-plug insulators. They have many attractive features. They are stiff, hard and abrasion resistant, they retain their strength to high temperatures,

Figure 2.2   Examples of each material family. The arrangement follows the general pattern of Figure 2.1. The central hybrid here is a sandwich structure made by combining stiff, strong face sheets of aluminum with a low-density core of balsa wood.

and they resist corrosion well. Most are good electrical insulators. They, too, have their weaknesses: unlike metals, they are brittle, with low K_{1c}. This gives ceramics a low tolerance for stress concentrations (like holes or cracks) or for high contact stresses (at clamping points, for instance). For this reason it is more difficult to design with ceramics than with metals.

Glasses are non-crystalline ('amorphous') solids, a term explained more fully in Chapter 4. The commonest are the soda-lime and borosilicate glasses familiar as bottles and Pyrex ovenware, but there are many more. The lack of crystal structure suppresses plasticity, so, like ceramics, glasses are hard and remarkably corrosion resistant. They are excellent electrical insulators and, of course, they are transparent to light. But like ceramics, they are brittle and vulnerable to stress concentrations.

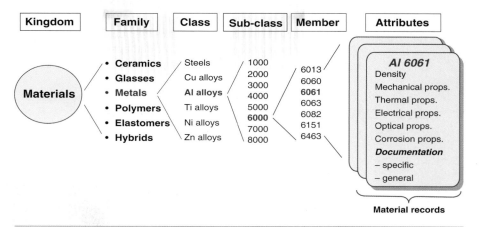

Figure 2.3 The taxonomy of the kingdom of materials and their attributes. Computer-based selection software stores data in a hierarchical structure like this.

Polymers are organic solids based on long chains of carbon (or, in a few, silicon) atoms. Polymers are light—their densities ρ are less than those of the lightest metals. Compared with other families they are floppy, with moduli E that are roughly 50 times less than those of metals. But they can be strong, and because of their low density, their strength per unit weight is comparable to that of metals. Their properties depend on temperature so that a polymer that is tough and flexible at room temperature may be brittle at the $-4°C$ of a household freezer, yet turn rubbery at the $100°C$ of boiling water. Few have useful strength above $150°C$. If these aspects are allowed for in the design, the advantages of polymers can be exploited. And there are many. They are easy to shape (that is why they are called plastics): complicated parts performing several functions can be molded from a polymer in a single operation. Their properties are well suited for components that snap together, making assembly fast and cheap. And by accurately sizing the mold and pre-coloring the polymer, no finishing operations are needed. Good design exploits these properties.

Elastomers—the material of rubber bands and running shoes—are polymers with the unique property that their stiffness, measured by E, is extremely low (500–5000 times less than those of metals) and their ability to be stretched to many times their starting length yet recover their initial shape when released. Despite their low stiffness they can be strong and tough—think of car tires.

Hybrids are combinations of two (or more) materials in an attempt to get the best of both. Glass and carbon-fiber-reinforced polymers (GFRP and CFRP) are hybrids; so, too, are sandwich structures, foams and laminates. And almost all the materials of nature (wood, bone, skin, leaf) are hybrids—bone, for instance, is a mix of collagen (a polymer) with hydroxyapatite (a mineral). Hybrid components are expensive and they are relatively difficult to form and join. So despite their attractive properties the designer will use them only when the added performance justifies the added cost. Today's growing emphasis on high performance and fuel efficiency provides increasing drivers for their use.

2.3 Organizing processes: the process tree

A *process* is a method of shaping, joining or finishing a material. *Casting, injection molding, fusion welding* and *electro-polishing* are all processes; there are hundreds of them (Figures 2.4 and 2.5). It is important to choose the right process-route at an early stage in the design before the cost-penalty of making changes becomes large. The choice, for a given component, depends on the material of which it is to be made, on its shape, dimensions and precision, and on how many are to be made—in short, on the *design requirements*.

Figure 2.4   The classes of process. The first row contains the primary shaping processes; below lie the secondary processes of machining and heat treatment, followed by the families of joining and finishing processes.

Figure 2.5   Examples of the families and classes of manufacturing processes. The arrangement follows the general pattern of Figure 2.4.

The choice of material limits the choice of process. Polymers can be molded, other materials cannot. Ductile materials can be forged, rolled and drawn but those that are brittle must be shaped in other ways. Materials that melt at modest temperatures to low-viscosity liquids can be cast; those that do not have to be processed by other routes. Shape, too, influences the choice of process. Slender shapes can be made easily by rolling or drawing but not by casting. Hollow shapes cannot be made by forging, but they can by casting or molding.

Classifying processes
Manufacturing processes are organized under the headings shown in Figure 2.4. *Primary processes* create shapes. The first row lists six primary forming processes: casting, molding, deformation, powder methods, methods for forming composites, special methods including rapid prototyping. *Secondary processes* modify shapes or properties; here they are shown as 'machining', which adds features to an already shaped body, and 'heat treatment', which enhances surface or bulk properties. Below these come *joining* and, finally, *surface treatment*. Figure 2.5 illustrates some of these; it is organized in the same way as Figure 2.4. The merit of Figure 2.4 is as a flow chart: a progression through a manufacturing route. It should not be treated too literally: the order of the steps can be varied to suit the needs of the design. The point it makes is that there are three broad process families: those of shaping, joining and finishing.

To organize information about processes, we need a hierarchical classification like that used for materials, giving each process a place. Figure 2.6 shows part of the hierarchy. The Process universe has three families: *shaping, joining* and *surface treatment*. In this figure, the shaping family is expanded to show classes: casting, deformation, molding etc. One of these—molding—is again expanded to show its members: rotation molding, blow molding, injection molding and so forth. Each process is characterized by a set of *attributes*: the materials it can handle, the shapes it can make, their size, precision and an economic batch size (the number of units that it can make most economically).

The other two families are partly expanded in Figure 2.7. There are three broad classes of joining process: adhesives, welding and fasteners. In this figure one of them—welding—is expanded to show its members. As before each member has attributes. The first is the material or materials that the process can join. After that the attribute list differs from that for shaping. Here the geometry of the joint and the way it will be loaded are important, as are requirements that the joint can or cannot be disassembled, be watertight and be electrically conducting.

The lower part of the figure expands the family of finishing processes. Some of the classes it contains are shown; one—coating—is expanded to show some of its members. Finishing adds cost: the only justification for applying a finishing process is that it hardens, or protects, or decorates the surface in ways that add value. As with joining, the material to be coated is an important attribute but the others again differ.

We return to process selection in Chapters 18 and 19.

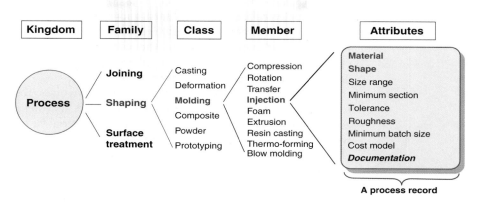

Figure 2.6 The taxonomy of the kingdom of process with part of the shaping family expanded. Each member is characterized by a set of attributes. Process selection involves matching these to the requirements of the design.

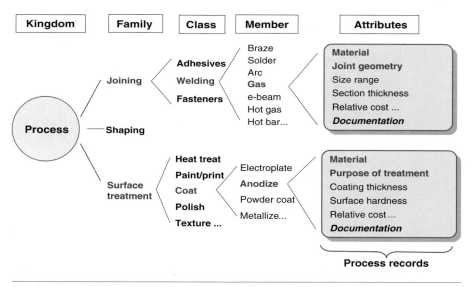

Figure 2.7   The taxonomy of the process kingdom again, with the families of joining and finishing partly expanded.

2.4 Process–property interaction

Processing can change properties. If you hammer a metal ('forging') it get harder; if you then heat it up it gets softer again ('annealing'). If polyethylene—the stuff of plastic bags—is drawn to a fiber, its strength is increased by a factor of 5. Soft, stretchy rubber is made hard and brittle by vulcanizing. Heat-treating glass in a particular way can give it enough impact resistance to withstand a projectile ('bullet-proof glass'). And composites like carbon-fiber-reinforced epoxy have no useful properties at all until processed—prior to processing they are just a soup of resin and a sheaf of fibers.

Joining, too, changes properties. Welding involves the local melting and re-solidifying of the faces of the parts to be joined. As you might expect, the weld zone has properties that differ from those of the material far from the weld—usually worse. Surface treatments, by contrast, are generally chosen to improve properties: electroplating to improve corrosion resistance, carburizing to improve wear.

Process–property interaction appears in a number of chapters. We return to it specifically in Chapter 19.

2.5 Material property charts

Data sheets for materials list their properties; they give no perspective and present no comparisons. The way to achieve these is to plot *material property charts*. They are of two types: bar charts and bubble charts.

A *bar chart* is simply a plot of one property for all the materials of the universe. Figure 2.8 shows an example: it is a bar chart for modulus, E. The largest is more than 10 million times greater than the smallest—many other properties have similar ranges—so it makes sense to plot them on logarithmic[1], not linear scales, as here. The length of each bar shows the range of the property for each material, here segregated by family. The differences between the families now become apparent. Metals and ceramics have high moduli. Those of polymers are smaller, by a factor of about 50, than those of metals; those of elastomers are some 500 times smaller still.

More information is packed into the picture if two properties are plotted to give a *bubble chart*, as in Figure 2.9, here showing modulus E and density ρ. As before, the scales are logarithmic. Now families are more distinctly separated: all metals lie in the reddish zone near the top right; all polymers lie in the dark blue envelope in the center, elastomers in the lighter blue envelope below, ceramics in the yellow envelope at the top. Each family occupies a distinct, characteristic field.

Material property charts like these are a core tool, used throughout this book.

- They give an overview of the physical, mechanical and functional properties of materials, presenting the information about them in a compact way.
- They reveal aspects of the physical origins of properties, helpful in understanding the underlying science.
- They become a tool for optimized selection of materials to meet given design requirements, and they help understand the use of materials in existing products.

---

[1] Logarithmic means that the scale goes up in constant multiples, usually of 10. We live in a logarithmic world—our senses, for instance, all respond in that way.

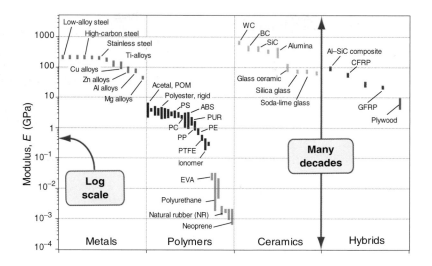

Figure 2.8 A bar chart of modulus. It reveals the difference in stiffness between the families.

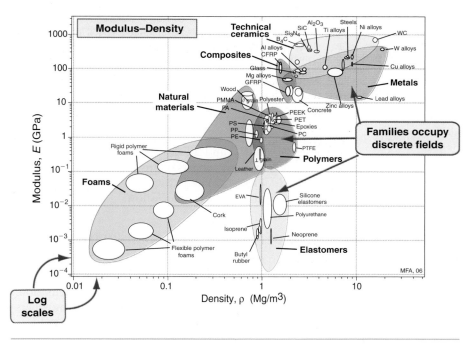

Figure 2.9 A bubble chart of modulus and density. Families occupy discrete areas of the chart.

Acrylonitrile–butadiene–styrene (ABS)

The Material
ABS (acrylonitrile–butadiene–styrene) is tough, resilient and easily molded. It is usually opaque, although some grades can now be transparent, and it can be given vivid colors. ABS–PVC alloys are tougher than standard ABS and, in self-extinguishing grades, are used for the casings of power tools.

General properties
Density	1e3	–	1.2e3	kg/m^3
Price	2	–	2.7	USD/kg

Mechanical properties
Young's modulus	1.1	–	2.9	GPa
Hardness—Vickers	5.6	–	15	HV
Elastic limit	19	–	51	MPa
Tensile strength	28	–	55	MPa
Compressive strength	31	–	86	MPa
Elongation	1.5	–	1e2	%
Endurance limit	11	–	22	MPa
Fracture toughness	1.2	–	4.3	MPa.m$^{1/2}$

Thermal properties
Thermal conductivity	0.19	–	0.34	W/m.k
Thermal expansion	85	–	230	μ strain/°C
Specific heat	1400	–	1900	J/kg.K
Glass temperature	88	–	130	°C
Max service temp.	62	–	90	°C

Electrical properties
Resistivity	2.3e21	–	3e22	μohm.cm
Dielectric constant	2.8	–	2.2	

Typical uses
Safety helmets; camper tops; automotive instrument panels and other interior components; pipe fittings; home-security devices and housings for small appliances; communications equipment; business machines; plumbing hardware; automobile grilles; wheel covers; mirror housings; refrigerator liners; luggage shells; tote trays; mower shrouds; boat hulls; large components for recreational vehicles; weather seals; glass beading; refrigerator breaker strips; conduit; pipe for drain-waste-vent (DWV) systems.

Figure 2.10   Part of a record for a material, ABS. It contains numeric data, text and image-based information.

These two charts, and all the others in the book, were made using the CES software, which allows charts of any pair of properties, or of functions of properties (like E/ρ) to be created at will. Their uses, and the operations they allow, will emerge in the chapters that follow.

2.6 Computer-aided information management for materials and processes

Classification is the first step in creating an information management system for materials and processes. In it records for the members of each universe are indexed, so to speak, by their position in the tree-like hierarchies of Figures 2.3, 2.6 and 2.7. Each record has a unique place, making retrieval easy.

The CES software is an example of an information and selection system that uses this structure. Figure 2.10 shows part of a record for a material; Figure 2.11

Injection molding

The process

No other process has changed product design more than INJECTION MOLDING. Injection molded products appear in every sector of product design: consumer products, business, industrial, computers, communication, medical and research products, toys, cosmetic packaging and sports equipment. The most common equipment for molding thermoplastics is the reciprocating screw machine, shown schematically in the figure. Polymer granules are fed into a spiral press where they mix and soften to a dough-like consistency that can be forced through one or more channels ('sprues') into the die. The polymer solidifies under pressure and the component is then ejected.

Physical attributes

Mass range	1e-3	–	25	kg
Range of section thickness	0.4	–	6.3	mm
Surface roughness (A = v. smooth)	A			

Economic attributes

Economic batch size (units)	1e4	–	1e6
Relative tooling cost	very high		
Relative equipment cost	high		
Labor intensity	low		

Shape

Circular prismatic	True
Non-circular prismatic	True
Solid 3-D	True
Hollow 3-D	True

Typical uses

Extremely varied. Housings, containers, covers, knobs, tool handles, plumbing fittings, lenses, etc.

Figure 2.11   Part of a record for a process, injection molding. The image shows how it works, and the numeric and Boolean data and text document its attributes.

shows the same for a process. A record is found by opening the tree, following the branches until the desired record is located ('*browsing*') or by locating it by name using a text-search facility ('*searching*').

Don't worry for the moment about the detailed content of the records—they are explained in later chapters. Note only that each contains data of two types. *Structured data* are numeric, Boolean (Yes/No) or discrete (e.g. Low / Medium / High), and can be stored in tables. Later chapters show how structured data are used for selection. *Unstructured data* take the form of text, images, graphs and schematics. Such information cannot so easily be used for selection but it is essential for the step we refer to in the next chapter as '*documentation*'.

2.7 Summary and conclusions

There are six broad families of materials for design: metals, ceramics, glasses, polymers, elastomers and hybrids that combine the properties of two or more of the others. Processes, similarly, can be grouped into families: those that create shape, those that join and those that modify the surface to enhance its properties or to protect or decorate it. The members of the families can be organized into a hierarchical tree-like catalog, allowing them to be 'looked up' in much the same way that you would look up a member of a company in the companies management sheet. A record for a member stores information about it: numeric and other tabular data for its properties, text, graphs and images to describe its use and applications. This structure forms the basis of computer-based selection systems of which the CES system is an example. It enables a unique way of presenting data for materials and processes as property charts, two of which appear in this chapter. They become one of the central features of the chapters that follow.

2.8 Further reading

Ashby, M.F. and Johnson, K. (2002) *Materials and Design—The Art and Science of Material Selection in Product Design*, Butterworth-Heinemann, Oxford, UK. ISBN 0-7506-5554-2. (*Materials and processes from an aesthetic point of view, emphasizing product design.*)

Bralla, J.G. (1998) *Design for Manufacturability Handbook*, 2nd edition, McGraw-Hill, New York, USA. ISBN 0-07-007139-X. (*Turgid reading, but a rich mine of information about manufacturing processes.*)

Callister, W.D. (2003) *Materials Science and Engineering, An Introduction*, 6th edition, John Wiley, New York, USA. ISBN 0-471-13576-3. (*A well-respected materials text, now in its 6th edition, widely used for materials teaching in North America.*)

Charles, J.A., Crane, F.A.A. and Furness, J.A.G. (1997) *Selection and Use of Engineering Materials*, 3rd edition, Butterworth-Heinemann, Oxford, UK. ISBN 0-7506-3277-1. (*A Materials Science approach to the selection of materials.*)

Dieter, G.E. (1991) *Engineering Design, A Materials and Processing Approach*, 2nd edition, McGraw-Hill, New York, USA. ISBN 0-07-100829-2. (*A well-balanced and respected text focusing on the place of materials and processing in technical design.*)

Farag, M.M. (1989) *Selection of Materials and Manufacturing Processes for Engineering Design*, Prentice-Hall, Englewood Cliffs, NJ, USA. ISBN 0-13-575192-6. (*A Materials Science approach to the selection of materials.*)

Kalpakjian, S. and Schmid, S.R. (2003) *Manufacturing Processes for Engineering Materials*, 4th edition, Prentice-Hall, Pearson Education, New Jersey, USA. ISBN 0-13-040871-9. (*A comprehensive and widely used text on material processing.*)

2.9 Exercises

Exercise E2.1   List the six main classes of engineering materials. Use your own experience to rank them approximately:`
(a) By stiffness (modulus, E).
(b) By thermal conductivity (λ).

Exercise E2.2	Examine the material property chart of Figure 2.9. By what factor are polymers less stiff than metals? Is wood denser or less dense than polyethylene (PE)?
Exercise E2.3	What is meant by a shaping process? Look around you and ask yourself how the things you see were shaped.
Exercise E2.4	Almost all products involve several parts that are joined. Examine the products immediately around you and list the joining methods used to assemble them.
Exercise E2.5	How many different surface treatment processes can you think of, based on your own experience? List them and annotate the list with the materials to which they are typically applied.
Exercise E2.6	How many ways can you think of for joining two sheets of a plastic like polyethylene? List each with an example of an application that might use it.
Exercise E2.7	A good classification looks simple—think, for instance, of the Periodic Table of the elements. Creating it in the first place, however, is another matter. This chapter introduced two classification schemes that work, meaning that every member of the scheme has a unique place in it, and any new member can be inserted into its proper position without disrupting the whole. Try one for yourself. Here are some scenarios. Make sure that each level of the hierarchy properly contains all those below it. There may be more than one way to do this, but one is usually better than the others. Test it by thinking how you would use it to find the information you want.

(a) You run a bike shop that stocks bikes of many types, prices and sizes. You need a classification system to allow customers to look up your bikes on the internet. How would you do it?

(b) You are asked to organize the inventory of fasteners in your company. There are several types (snap, screw, rivet) and, within each, a range of materials and sizes. Devise a classification scheme to store information about them.

2.10 Exploring design using CES

Designers need to be able to find data quickly and reliably. That is where the classifications come in. The CES system uses the classification scheme described in this chapter. Before trying these exercises, open the Materials Universe in CES and explore it. The opening screen offers options—take the Edu Level 1: Materials.

Exercise E2.8 Use the 'Browse' facility in Level 1 of the CES Software to find the record for *Copper*. What is its thermal conductivity? What is its price?

Exercise E2.9 Use the 'Browse' facility in Level 1 of the CES Software to find the record for the thermosetting polymer *Phenolic*. Are they cheaper or more expensive than *Epoxies*?

Exercise E2.10 Use the 'Browse' facility to find records for the polymer-shaping processes *Rotational molding*. What, typically, is it used to make?

Exercise E2.11 Use the 'Search' facility to find out what *Plexiglas* is. Do the same for *Pyroceram*.

Exercise E2.12 Use the 'Search' facility to find out about the process *Pultrusion*. Do the same for *TIG welding*. Remember that you need to search the Process Universe, not the Material Universe.

Exercise E2.13 Compare Young's modulus E (the stiffness property) and thermal conductivity λ (the heat transmission property) of *aluminum alloys* (a non-ferrous metal), *alumina* (a technical ceramic), *polyethylene* (a thermoplastic polymer) and *neoprene* (an elastomer) by retrieving values from CES Level 1. Which has the highest modulus? Which has the lowest thermal conductivity?

2.11 Exploring the science with CES Elements

The CES system contains a database for the Periodic Table. The records contain fundamental data for each of the elements. We will use this in the book to delve a little deeper into the science that lies behind material properties.

Exercise E2.14 Refresh your memory of the Periodic Table, perhaps the most significant classification of all time. Select *CES Elements* (*File > Change database > CES Elements*) and double-click on Periodic Table to see the table. This database, like the others described in this chapter, has a tree-like structure. Use this to find the record for Aluminum (Row 3, Atomic number 13) and explore its contents. Many of the properties won't make sense yet. We introduce them gradually throughout the book.

Chapter 3
Strategic thinking: matching material to design

Images embodying the concepts described in the text: pull, geared pull, shear and pressure. (Image courtesy of A-Best Fixture Co. 424 West Exchange Street, Akron, Ohio, 44302, USA.)

3.1 Introduction and synopsis

Our aim in this chapter is to develop a strategy for selecting materials and processes that is *design-led*; that is, the strategy uses, as inputs, the requirements of the design. To do so we must first look briefly at design itself. This chapter introduces some of the words and phrases—the vocabulary—of design, the stages in its implementation and the ways in which materials selection links with these.

Design starts with a *market need*. The need is analyzed, expressing it as a set of *design requirements*. Ways to meet these ('concepts') are sought, developed ('embodied') and refined ('detailed') to give a *product specification*. The choice of material and process evolves in parallel with this process, in the way detailed in this chapter.

With this background we can develop the selection strategy. It involves four steps: *translation, screening, ranking* and *documentation*. These steps are explained and the first, that of translation, is illustrated with examples.

3.2 The design process

Original design starts from a new concept and develops the information necessary to implement it. *Evolutionary design* (or *redesign*) starts with an existing product and seeks to change it in ways that increase its performance, reduce its cost, or both.

Original design

Original design starts from scratch. It involves a new idea or working principle (the audio tape, the compact disc and the MP3 player were all, in their day, completely new). Original design can be stimulated by new materials. Thus, high-purity silicon enabled the transistor; high-purity glass, the optical fiber; high coercive-force magnets, the miniature earphone; solid-state lasers, the compact disc. Sometimes the new material suggests the new product. Sometimes instead the new product demands the development of a new material: nuclear technology drove the development of new zirconium alloys and new stainless steels; space technology stimulated the development of beryllium alloys and lightweight composites; turbine technology today drives development of high-temperature alloys and ceramics.

The central column of Figure 3.1 shows the design process. The starting point is a *market need* or a *new idea*; the end point is the full *product specification* for a product that fills the need or embodies the idea. A need must be identified before it can be met. It is essential to define the need precisely—that is, to formulate a *need statement*, often in the form: 'a device is required to perform task X', expressed as a set of *design requirements*. Between the need statement and the product specification lie the set of stages shown in Figure 3.1: the stages of *conceptual design, embodiment design* and *detailed design*, explained in a moment.

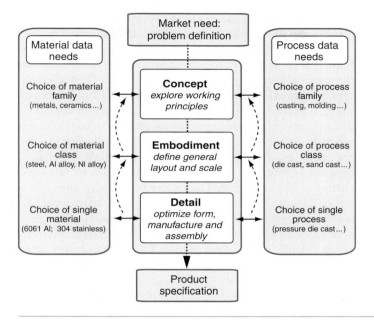

Figure 3.1   The design flow chart, showing how material and process selection enter. Information about materials is needed at each stage, but at very different levels of breadth and precision. The broken lines suggest the iterative nature of original design and the path followed in redesign.

At the conceptual design stage, all options are open: the designer considers alternative concepts and the ways in which these might be separated or combined. The next stage, embodiment, takes the promising concepts and seeks to analyze their operation at an approximate level. This involves sizing the components and a preliminary selection of materials and processes, examining the implications for performance and cost. The embodiment stage ends with a feasible layout, which becomes the input to the detailed design stage. Here specifications and dimensions for each component are drawn up. Critical components may be subjected to precise mechanical or thermal analysis. Optimization methods are applied to components and groups of components to maximize performance, minimize cost and ensure safety. A final choice of geometry and material is made, and the methods of production are analyzed and costed. The stage ends with a detailed product specification.

Redesign
Most design is not 'original' in the sense of starting from a totally new idea. It is *redesign*, starting with an existing product and correcting its shortcomings, refining it, enhancing its performance or reducing its cost, without discarding the principles on which it operates or—frequently—many of its components.

Here are some scenarios that call for redesign. First, the most obvious, the 'product recall' scenario. If a product, once released to the market, fails to meet safety standards, urgent redesign is required. Often the problem is a material failure; then an alternative must be found that retains the desirable features of the original but overcomes its weaknesses. Then there is the 'poor value for money' scenario: the product performs safely but offers performance that, at its price, is perceived to be mediocre, requiring redesign to enhance performance. There is, too, the 'inadequate profit margin' scenario in which the cost of manufacture exceeds the price that the market will bear. Much of the cost of a mass-produced product derives from the materials of which it is made and the processes chosen to make it; the response is to re-examine both, with cost-cutting as the objective. Then there is the 'stay ahead of the competition' scenario. Makers of household products and electronic gadgetry (electric kettles, mobile phones) release new editions annually; those that make cars produce new models every two or three years. In a market environment in which many almost identical products compete for the consumers' attention, it is visual and stylistic character that sets some products above others. Much creative thinking goes into this 'industrial design' and in it the choice of material, or of a change of material, is dictated mainly by aesthetics: color, texture, feel and the ability to be shaped or finished in a given way.

Much of redesign has to do with detail—the last of the three boxes in the central window of Figure 3.1. But not all. The necessary changes may require a change of configuration and layout—the embodiment phase—or even of basic concept, replacing one of the ways of performing a function by another. So the flow chart in Figure 3.1 remains a useful summary to keep in mind. It has another use, too—as a tool (one of several) for analyzing existing designs and understanding how sometimes quirky details of the final product have their origins in decisions made in the concept or embodiment stages.

Described in the abstract, these ideas are not easy to grasp. An example will help. Here it is.

Devices to open corked bottles

When you buy a bottle of wine you find, generally, that it is sealed with a cork. This creates a market need: it is the need to gain access to the wine inside. We might state it thus: 'A device is required to allow access to wine in a corked bottle' and might add, 'with convenience, at modest cost, and without contaminating the wine'.

Three concepts for doing this are shown in Figure 3.2. In order, they are: to remove the cork by axial traction (=pulling); to remove it by shear tractions; to push it out from below. In the first, a screw is threaded into the cork to which an axial pull is applied; in the second, slender elastic blades inserted down the sides of the cork apply shear tractions when pulled; and in the third, the cork is pierced by a hollow needle through which a gas is pumped to push it out.

Figure 3.3 shows embodiment sketches for devices based on concept (a), that of axial traction. The first is a direct pull; the other three use some sort of

Figure 3.2   A market need — that of gaining access to wine in corked bottles — and three concepts for meeting the need. Devices based on all three of these concepts exist and can be bought.

Figure 3.3   Embodiment sketches for the first concept: direct pull, levered pull, geared pull and spring-assisted pull. Each system is made up of components that perform a sub-function. Detailed design drawings for the lever of embodiment (b) are shown on the right.

mechanical advantage—levered pull, geared pull and spring-assisted pull. The embodiments suggest the layout, the mechanisms and the scale. In the final, detailed, stage of design, the components are dimensioned so that they carry the working loads safely, their precision and surface finish are defined, and a final choice of material and manufacturing route is made as suggested on the right of the figure. Let us examine how this is done.

3.3 Material and process information for design

Materials selection enters each stage of the design (Figure 3.1, left-hand side). The nature of the data needed in the early stages differs greatly in its level of precision and breadth from that needed later on. At the concept stage, the designer requires only approximate property values, but for the widest possible range of materials. All options are open: a polymer may be the best choice for one concept, a metal for another. The problem, at this stage, is not precision and detail, it is breadth and speed of access: how can the vast range of data be presented to give the designer the greatest freedom in considering alternatives?

At the embodiment stage the landscape has narrowed. Here we need data for a subset of materials, but at a higher level of precision and detail. These are found in more specialized handbooks and software that deal with a single class or subclass of materials—metals, or just aluminum alloys, for instance. The risk now is that of losing sight of the bigger spread of materials to which we must return if the details don't work out; it is easy to get trapped in a single line of thinking when others have potential to offer better solutions.

The final stage of detailed design requires a still higher level of precision and detail, but for only one or a very few materials. Such information is best found in the data sheets issued by the material producers themselves and in detailed databases for restricted material classes. A given material (polyethylene, for instance) has a range of properties that derive from differences in the ways different producers make it. At the detailed design stage, a supplier must be identified and the properties of his product used in the design calculations; that from another supplier may have slightly different properties. And sometimes even this is not good enough. If the component is a critical one (meaning that its failure could, in some sense or another, be disastrous) then it may be prudent to conduct in-house tests to measure the critical properties, using a sample of the material that will be used to make the product itself. The process is one of narrowing the materials search space by screening out materials that cannot meet the design requirements, ranking those that remain and identifying the most promising choice (Figure 3.4).

The materials input does not end with the establishment of production. Products fail in service and failures contain information. It is an imprudent manufacturer who does not collect and analyze data on failures. Often this points to the misuse of a material, one that redesign or re-selection can eliminate.

The selection of a material cannot be separated from that of process and of shape. To make a shape, a material is subjected to processes that, collectively, we shall call *manufacture*. Figure 2.5 of Chapter 2 introduced them. The selection of process follows a route that runs parallel to that of material (Figure 3.1, right-hand side). The starting point is a catalog of all processes, which is then narrowed by screening out those that fail to make the desired shape or are incompatible with the choice of material. Material, shape and process interact (Figure 3.5). Process choice is influenced by the material: by its formability, machinability,

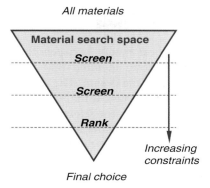

Figure 3.4   The narrowing of material search space as design constraints are applied.

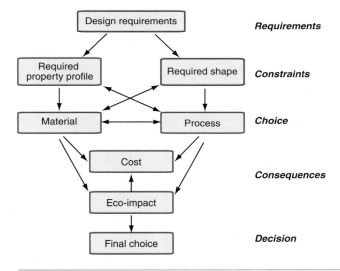

Figure 3.5   The interaction between design requirements, material, shape and process.

weldability, heat treatability and so on. Process choice is influenced by the requirements for shape—the process determines the shape, the size, the precision and, to a large extent, the cost of a component. The interactions are two-way: specification of shape restricts the choice of material and process, but equally the specification of process limits the materials you can use and the shapes they can take. The more sophisticated the design, the tighter the specifications and the greater the interactions. The interaction between material, shape and process lies at the heart of the selection process. To tackle it we need a strategy.

3.4 The strategy: translation, screening, ranking and documentation

Selection involves seeking the best match between the attribute profiles of the materials and processes—bearing in mind that these must be mutually compatible—and those required by the design. The strategy, applied to materials, is sketched in Figure 3.6. The first task is that of *translation*: converting the design requirements into a prescription for selecting a material. This proceeds by identifying the *constraints* that the material must meet and the *objectives* that the design must fulfill. These become the filters: materials that meet the constraints and rank highly in their ability to fulfill the objectives are potential candidates for the design. The second task, then, is that of *screening*: eliminating the material that cannot meet the constraints. This is followed by the *ranking* step, ordering the survivors by their ability to meet a criterion of excellence, such as that of minimizing cost. The final task is to explore the most promising candidates in depth, examining how they are used at present, how best to design with them, case histories of failures and a step we call *documentation*.

Figure 3.6   The strategy applied to materials. The same strategy is later adapted to select processes. There are four steps: translation, screening, ranking and supporting information. All can be implemented in software, allowing large populations of materials to be investigated.

Process selection follows a parallel route. In this case translation means identifying the geometric and other constraints—dimensions, shape, precision and material compatibility—that must be met, using these to screen out processes that cannot provide them. We return to process selection in Chapters 18 and 19. For now we stick to materials.

Translation

Any engineering component has one or more *functions*: to support a load, to contain a pressure, to transmit heat and so forth. This must be achieved subject to *constraints*: that certain dimensions are fixed, that the component must carry the design loads without failure, the need to insulate against or to conduct heat or electricity, that it can function in a certain range of temperature and in a given environment, and many more. In designing the component, the designer has one or more *objectives*: to make it as cheap as possible, perhaps, or as light, or as safe, or some combination of these. Certain parameters can be adjusted in order to optimize the objective—the designer is free to vary dimensions that are not constrained by design requirements and, most importantly, free to choose the material for the component and the process to shape it. We refer to these as *free variables*.

Constraints, objectives and free variables (Table 3.1) define the boundary conditions for selecting a material and—in the case of load-bearing components—a shape for its cross-section.

It is important to be clear about the distinction between constraints and objectives. A constraint is an essential condition that must be met, usually expressed as a limit on a material or process attribute. An objective is a quantity for which an extreme value (a maximum or minimum) is sought, frequently cost, mass or volume, but there are others (Table 3.2). Getting it right can take a little thought. In choosing materials for a super-light sprint bicycle, for example, the objective is to minimize mass, with an upper limit on cost, thus treating cost as a constraint. But in choosing materials for a cheap 'shopping' bike the two are reversed: now the objective is to minimize cost with a (possible) upper limit on mass, thus treating it as a constraint (Figure 3.7).

The outcome of the translation step is a list of the design-limiting properties and the constraints they must meet. The first step in relating design requirements to material properties is therefore a clear statement of function, constraints, objectives and free variables.

Table 3.1   Function, constraints, objectives and free variables

Function	• What does the component do?
Constraints	• What non-negotiable conditions must be met?
Objective	• What is to be maximized or minimized?
Free variables	• What parameters of the problem is the designer free to change?

Table 3.2   Common constraints and objectives

Common constraints	Common objectives
Meet a target value of • Stiffness • Strength • Fracture toughness • Thermal conductivity • Electrical resistivity • Magnetic remanence • Optical transparency • Cost • Mass	Minimize • Cost • Mass • Volume • Impact on the environment • Heat loss Maximize • Energy storage • Heat flow

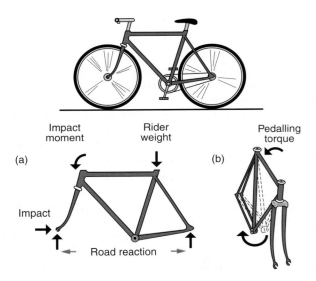

Figure 3.7   A bicycle. The forks are loaded in bending.

Screening

Constraints are gates: meet the constraint and you pass through the gate, fail to meet it and you are out. Screening (Figure 3.6) does just that: it eliminates candidates that cannot do the job at all because one or more of their attributes lies outside the limits set by the constraints. As examples, the requirement that 'the component must function in boiling water' or that 'the component must be transparent' imposes obvious limits on the attributes of *maximum service temperature* and *optical transparency* that successful candidates must meet. We refer to these as *attribute limits*.

Ranking

To rank the materials that survive the screening step we need a criterion of excellence. They are found in the *material indices*, developed below and in later chapters, which measure how well a candidate that has passed the screening step can do the job (Figure 3.6 again). Performance is sometimes limited by a single property, sometimes by a combination of them. Thus, the best materials for buoyancy are those with the lowest density, ρ; those best for thermal insulation are the ones with the smallest values of the thermal conductivity, λ—provided, of course, that they also meet all other constraints imposed by the design. Here maximizing or minimizing a single property maximizes performance. Often, though, it is not one but a group of properties that are relevant. Thus, the best materials for a light stiff tie-rod are those with the greatest value of the *specific stiffness*, E/ρ, where E is Young's modulus. The best materials for a spring are those with the greatest value of σ_y^2/E, where σ_y is the yield strength. The property or property group that maximizes performance for a given design is called its *material index*. There are many such indices, each associated with maximizing some aspect of performance. They provide criteria of excellence that allow ranking of materials by their ability to perform well in the given application.

To summarize: *screening* isolates candidates that are capable of doing the job; *ranking* identifies those among them that can do the job best.

Documentation

The outcome of the steps so far is a ranked short-list of candidates that meet the constraints and that maximize or minimize the criterion of excellence, whichever is required. You could just choose the top-ranked candidate, but what hidden weaknesses might it have? What is its reputation? Has it a good track record? To proceed further we seek a detailed profile of each: its *documentation* (Figure 3.6, bottom).

What form does documentation take? Typically, it is descriptive, graphical or pictorial: case studies of previous uses of the material, details of its corrosion behavior in particular environments, of its availability and pricing, warnings of its environmental impact or toxicity. Such information is found in handbooks, suppliers' data sheets, CD-based data sources and high-quality Websites. Documentation helps narrow the short-list to a final choice, allowing a definitive match to be made between design requirements and material and process attributes.

Why are all these steps necessary? Without screening and ranking, the candidate pool is enormous and the volume of documentation is overwhelming. Dipping into it, hoping to stumble on a good material, gets you nowhere. But once a small number of potential candidates have been identified by the screening–ranking steps, detailed documentation can be sought for these few alone, and the task becomes viable.

3.5 Examples of translation

The following examples illustrate the translation step for a number of problems, starting with the lever for the corkscrew of Figure 3.3, then an example of redesign.

Table 3.3   Translation for the corkscrew lever

Function	• Lever (beam loaded in bending)	
Constraints	• Stiff enough	
	• Strong enough	
	• Some toughness	} Functional constraints
	• Resist corrosion in wine and water	
	• Length *L* specified	A geometric constraint
Objective	• Minimize cost	
Free variables	• Choice of material	
	• Choice of cross-section area	

A corkscrew lever

Figure 3.3 shows the lever for one of the corkscrews in the design case study. In use it is loaded in bending. It must carry the bending moment without deflecting to an awkward degree, it must not yield (though some cheap corkscrews do) and it must be tough enough to withstand misuse. Finally, it must not corrode in wine or water. The length of the lever is specified, but the cross-section is not— we are free to choose a section that is sufficient to bear the use-loads. Given all these, the lever should be as cheap as possible. Table 3.3 lists the translation.

The *design-limiting properties* are those directly relating to the constraints: modulus E, strength σ_y, fracture toughness K_{1c} and corrosion resistance.

Redesign of a CD case

Music lovers will affirm that CDs—the best of them—are divine. But the cases they come in are the work of the devil (Figure 3.8). They are—for reasons of their optical clarity—called 'jewel' cases, but in performance they are far from jewels. They are usually made of polystyrene (PS), chosen for its low cost and water-clear transparency, and they are made by injection molding and that, too, is cheap if you are making millions. Polystyrene can, at least in principle, be recycled. But PS jewel cases crack easily, they jam shut, the hinges break, and the corners of the case are hard and sharp enough to inflict terminal damage on a CD. So there you have it. Decide on the features you think really matter, and formulate constraints, objective and free variables for the redesign of a CD case.

The way to tackle the problem is to seek a replacement material that retains the good properties of the old one, but without the bad. Thus, we seek a material that is optically transparent to allow the label to be read, is able to be injection molded because this is the most economic way to make large numbers, and is recyclable. But it must be tougher than polystyrene. Of the materials that meet these constraints, we want the cheapest. Table 3.4 summarizes the translation.

Potential design-limiting properties are optical transparency, fracture toughness, K_{1c} (must be better than PS), and the ability to be injection molded and recycled.

Figure 3.8   A polystyrene CD case. It is cheap, but it is brittle and cracks easily.

Table 3.4   Translation for the redesigned CD case

Function	• Contain and protect a CD	
Constraints	• Optically clear	
	• Able to be injection molded	
	• Recyclable	Functional constraints
	• Tougher than polystyrene	
	• Dimensions identical with PS case	A geometric constraint
Objective	• Minimize cost	
Free variable	• Choice of material	

Heat sinks for microchips

A microchip may only consume milliwatts, but this power is dissipated in a tiny volume, making the *power density* high. As chips shrink and clock speeds grow, overheating becomes a problem. The chip in your PC already reaches 85°C, requiring forced cooling. Multiple-chip modules (MCMs) pack as many as 130 chips on to a single substrate, and they get even hotter—up to 180°C. Heating is kept under control by attaching the chips to a heat sink (Figure 3.9), taking pains to ensure good thermal contact between chip and sink. The heat sink now becomes a critical component, limiting further development of the electronics. How can its performance be maximized?

To prevent electrical coupling and stray capacitance between chip and heat sink, the heat sink must be a good electrical insulator. If it is to work with one surface at 180°C, it must have a maximum service temperature (the temperature at which it can operate continuously without damage) that is at least as great as 180°C. These define the constraints. To drain heat away from the chip as fast as possible, it must also have the highest possible thermal conductivity, λ, defining the objective. The translation step is summarized in Table 3.5, where we assume that all dimensions are constrained by other aspects of the design.

Figure 3.9   A heat sink. It must conduct heat well, but be electrically insulating.

Table 3.5   Translation for the heat sink

Function	• Heat sink	
Constraints	• Material must be good electrical insulator	} Functional constraints
	• Maximum operating temperature >200°C	
	• All dimensions are specified	Geometric constraints
Objective	• Maximize thermal conductivity	
Free variable	• Choice of material	

120 volts   240 volts

Figure 3.10   A transformer. The core must be a soft magnetic material, and if this is a high-frequency transformer, it must be an electrical insulator.

The design-limiting properties, clearly, are maximum service temperature T_{max}, electrical resistivity ρ_e and thermal conductivity λ.

HF transformer cores
An electrical transformer uses electromagnetic induction to convert one AC voltage to another (Figure 3.10). To minimize energy loss the material must be a soft magnet—one that is easy to magnetize and demagnetize (Chapter 15). And to avoid eddy current losses at high frequencies it must also be an electrical insulator. The constraints of 'soft magnetic material' and 'electrical insulator' are very restrictive—they will screen out all but a small number of candidates. If the transformer is for an everyday product, the objective would be to minimize the cost. Table 3.6 lists the translation.

Table 3.6   Translation for the transformer core

Function	• HF transformer core	
Constraints	• Soft magnetic material	
	• Electrical insulator	} Functional constraints
	• All dimensions are specified	Geometric constraints
Objective	• Minimize cost	
Free variable	• Choice of material	

These translations are the first step in selection. In them we have identified the constraints; they will be used for screening. We have also identified the objective; it will be used for ranking. We will return to all four of these examples in later chapters when we know how to screen and rank.

3.6 Summary and conclusions

The starting point of a design is a *market need* captured in a set of *design requirements*. *Concepts* for a product that meet the need are devised. If initial estimates and exploration of alternatives suggest that the concept is viable, the design proceeds to the *embodiment* stage: working principles are selected, size and layout are decided, and initial estimates of performance and cost are made. If the outcome is successful, the designer proceeds to the *detailed design* stage: optimization of performance, full analysis of critical components, preparation of detailed production drawings (usually as a CAD file), showing dimensions, specifying precision, and identifying material and manufacturing path. But design is not a linear process, as Figure 3.1 might suggest. Some routes lead to a dead end, requiring reiteration of earlier steps. And, frequently, the task is one of redesign, requiring that constraints be rethought and objectives realigned.

The selection of material and process runs parallel to this set of stages. Initially the search space for both is wide, encompassing all possible candidates. As the design requirements are formulated in increasing detail, constraints emerge that both must meet, and one or more objectives is formulated. The constraints narrow the search space and the objective(s) allow ranking of those that remain. Identifying the constraints, the objectives and free variables (the process we called 'translation') is the first step in selection. This chapter ended with examples of translation when the task was that of choosing a material; the exercises suggest more. When the task is the choice of process, a similar translation is needed; we return to this in Chapter 18. The other steps—screening, ranking and documentation—are discussed in chapters that follow.

3.7 Further reading

Ashby, M.F. (2005) *Materials Selection in Mechanical Design*, 3rd edition, Butterworth-Heinemann, Oxford, UK, Chapter 4. ISBN 0-7506-6168-2. *(A more advanced text that develops the ideas presented here in greater depth.)*

Cross, N. (2000) *Engineering Design Methods*, 3rd edition, Wiley, Chichester, UK. ISBN 0-471-87250-3. *(A durable text describing the design process, with emphasis on developing and evaluating alternative solutions.)*

French, M.J. (1985) *Conceptual Design for Engineers*, The Design Council, London, UK and Springer, Berlin, Germany. ISBN 0-85072-155-5 and 3-540-15175-3. *(The origin of the 'Concept—Embodiment—Detail' block diagram of the design process. The book focuses on the concept stage, demonstrating how simple physical principles guide the development of solutions to design problems.)*

Pahl, G. and Beitz, W. (1997) *Engineering design*, 2nd edition, translated by K. Wallace and L. Blessing, The Design Council, London, UK and Springer, Berlin, Germany. ISBN 0-85072-124-5 and 3-540-13601-0. *(The Bible—or perhaps more exactly the Old Testament—of the technical design field, developing formal methods in the rigorous German tradition.)*

Ullman, D.G. (2003) *The Mechanical Design Process*, 3rd edition, McGraw-Hill, New York, USA. ISBN 0-07-112281-8. *(An American view of design, developing ways in which an initially ill-defined problem is tackled in a series of steps, much in the way suggested by Figure 3.1 of the present text.)*

Ulrich, K.T. and Eppinger, S.D. (1995) *Product Design and Development*, McGraw-Hill, New York, USA. ISBN 0-07-065811-0. *(A readable, comprehensible text on product design, as taught at MIT. Many helpful examples but almost no mention of materials.)*

3.8 Exercises

Exercise E3.1	What are the steps in developing an original design?
Exercise E3.2	Describe and illustrate the 'translation' step of the material selection strategy.
Exercise E3.3	What is meant by an objective and what by a constraint in the requirements for a design? How do they differ?
Exercise E3.4	You are asked to design a fuel-saving cooking pan with the goal of wasting as little heat as possible while cooking. What objective would you choose, and what constraints would you think must be met?
Exercise E3.5	Bikes come in many forms, each aimed at a particular sector of the market:

Exercise E3.5 (continued):
- Sprint bikes.
- Touring bikes.
- Mountain bikes.
- Shopping bikes.
- Children's bikes.
- Folding bikes.

Use your judgement to identify the primary objective and the constraints that must be met for each of these.

Exercise E3.6   A material is required for the windings of an electric air-furnace capable of temperatures up to 1000°C. Think out what attributes a material must have if it is to be made into windings and function properly in a furnace. List the function and the constraints; set the objective to 'minimize material price' and the free variable to 'choice of material'.

Exercise E3.7   A material is required to manufacture office scissors. Paper is an abrasive material and scissors sometimes encounter hard obstacles like staples. List function and constraints; set the objective to 'minimize material price' and the free variable to 'choice of material'.

Exercise E3.8   A material is required for a heat exchanger to extract heat from geo-thermally heated, saline, water at 120°C (and thus under pressure). List function and constraints; set the objective to 'minimize material price' and the free variable to 'choice of material'.

Exercise E3.9   A material is required for a disposable fork for a fast-food chain. List the objective and the constraints that you would see as important in this application.

Exercise E3.10   Formulate the constraints and objective you would associate with the choice of material to make the forks of a racing bicycle.

Exercise E3.11   Cheap coat-hangers used to be made of wood—now it is only expensive ones that use this material. Most coat-hangers are now metal or plastic, and both differ in shape from the wooden ones, and from each other. Examine wood, metal and plastic coat-hangers, comparing the designs, and comment on the ways in which the choice of material has influenced them.

Exercise E3.12   Cyclists carry water in bottles that slot into bottle holders on their bicycles. Examine metal and plastic bottle holders, comparing the designs, and comment on the ways in which the choice of material has influenced them.

3.9 Exploring design using CES

The 'Search' facility of CES allows a full-text search of all records, identifying those that contain the search string. The default setting applies the search to the Materials universe only. The pull-down menu headed 'Look in table:' allows this to be reset to the Process universe or to All tables.

Exercise E3.13 A company wishes to enhance its image by replacing oil-based plastics in its products by polymers based on natural materials. Use the 'Search' facility in CES to find *biopolymers* (search on 'Biopolymer'). List the material you find.

Exercise E3.14 A maker of garden furniture is concerned that the competition is stealing part of his market with furniture made by RTM, a term with which he is unfamiliar. Use the 'Search' facility in CES to find out what *RTM* is, and whether it is used to make things like garden furniture. (Remember that it is the Process universe that must be searched, since this is a process.)

Exercise E3.15 Use the 'Search' facility in CES to find materials for *furnace windings*.

Exercise E3.16 Use the 'Search' facility in CES to find materials for *scissors* and *knife blades*.

Exercise E3.17 Use the 'Search' facility in CES to find materials for *heat exchangers*.

Exercise E3.18 Use the 'Search' facility in CES to find materials for *flooring*.

Chapter 4
Stiffness and weight: density and elastic moduli

Modes of loading. (Image of Stansted Airport courtesy of Norman Foster and Partners, London, UK)

Chapter contents

4.1 Introduction and synopsis

Stress causes strain. If you are human, the ability to cope with stress without undue strain is called resilience. If you are a material, it is called elastic modulus.

Stress is something that is applied to a material by loading it. Strain—a change of shape—is its response; it depends on the magnitude of the stress and the way it is applied—the *mode of loading*. The cover picture illustrates the common ones. Ties carry tension—often, they are cables. Columns carry compression—tubes are more efficient as columns than solid rods because they don't buckle as easily. Beams carry bending moments, like the wing spar of the plane or the horizontal roof beams of the airport. Shafts carry torsion, as in the drive shaft of cars or the propeller shaft of the plane. Pressure vessels contain a pressure, as in the tires of the plane. Often they are shells: curved, thin-walled structures.

Stiffness is the resistance to change of shape that is *elastic*, meaning that the material returns to its original shape when the stress is removed. *Strength* (Chapter 6) is its resistance to permanent distortion or total failure. Stress and strain are not material properties; they describe a stimulus and a response. Stiffness (measured by the elastic modulus E, defined in a moment) and strength (measured by the elastic limit σ_y or tensile strength σ_{ts}) *are* material properties. Stiffness and strength are central to mechanical design, often in combination with the density, ρ. This chapter introduces stress and strain and the elastic moduli that relate them. These properties are neatly summarized in a *material property chart*—the modulus–density chart—the first of many that we shall explore in this book.

Density and elastic moduli reflect the mass of the atoms, the way they are packed in a material and the stiffness of the bonds that hold them together. There is not much you can do to change any of these, so the density and moduli of pure materials cannot be manipulated at all. If you want to control these properties you can either mix materials together, making composites, or disperse space within them, making foams. Property charts are a good way to show how this works.

4.2 Density, stress, strain and moduli

Density
Many applications (e.g. sports equipment, transport systems) require low weight and this depends in part on the density of the materials of which they are made. Density is mass per unit volume. It is measured in kg/m^3 or sometimes, for convenience, Mg/m^3 ($1\,Mg/m^3 = 1000\,kg/m^3$).

The density of samples with regular shapes can be determined using precision mass balance and accurate measurements of the dimensions (to give the volume), but this is not the best way. Better is the 'double weighing' method: the sample is first weighed in air and then when fully immersed in a liquid of known density. When immersed, the sample feels an upwards force equal to the weight of

$$\text{Density } \rho = \rho_{fl} \ \frac{m_1}{m_1 - m_2}$$

Figure 4.1   Measuring density by Archimedes' method.

liquid it displaces (Archimedes' principle[1]). The density is then calculated as shown in Figure 4.1.

Modes of loading

Most engineering components carry loads. Their elastic response depends on the way the loads are applied. As explained earlier, the components in both structures shown on the cover are designed to withstand different modes of loading: tension, compression, bending, torsion and internal pressure. Usually one mode dominates, and the component can be idealized as one of the simply loaded cases in Figure 4.2—*tie, column, beam, shaft* or *shell*. Ties carry simple axial tension, shown in (a); columns do the same in simple compression, as in (b). Bending of a beam (c) creates simple axial tension in elements on one side the neutral axis (the center-line, for a beam with a symmetric cross-section) and simple compression in those on the other. Shafts carry twisting or torsion (d), which generates shear rather than axial load. Pressure difference applied to a shell, like the cylindrical tube shown in (e), generates bi-axial tension or compression.

Stress

Consider a force F applied normal to the face of an element of material, as in Figure 4.3 on the left of row (a). The force is transmitted through the element and balanced by an equal but opposite force on the other side, so that it is in equilibrium (it does not move). Every plane normal to F carries the force. If the area of such a plane is A, the *tensile stress* σ in the element (neglecting its own weight) is

$$\sigma = \frac{F}{A} \tag{4.1}$$

[1] Archimedes (287—212 BC), Greek mathematician, engineer, astronomer and philosopher, designer of war machines, the Archimedean screw for lifting water, evaluator of π (as $3 + 1/7$) and conceiver, whilst taking a bath, of the principle that bears his name.

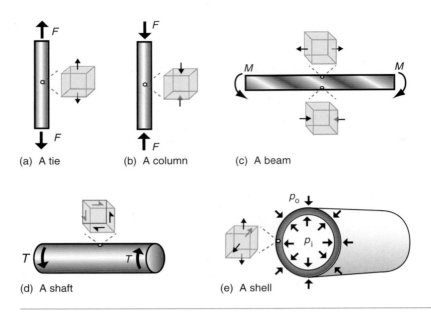

(a) A tie   (b) A column   (c) A beam

(d) A shaft   (e) A shell

Figure 4.2   Modes of loading and states of stress.

If the sign of F is reversed, the stress is compressive and given a negative sign. Forces[2] are measured in newtons (N), so stress has the dimensions of N/m². But a stress of $1\,\text{N/m}^2$ is tiny—atmospheric pressure is $10^5\,\text{N/m}^2$—so the usual unit is MN/m^2 ($10^6\,\text{N/m}^2$), called megapascals, symbol MPa[3].

If, instead, the force lies parallel to the face of the element, three other forces are needed to maintain equilibrium (Figure 4.3, row (b)). They create a state of shear in the element. The shaded plane, for instance, carries the *shear stress* τ of

$$\tau = \frac{F_s}{A} \tag{4.2}$$

The units, as before, are MPa.

One further state of multi-axial stress is useful in defining the elastic response of materials: that produced by applying equal tensile or compressive forces to all six faces of a cubic element, as in Figure 4.3, row (c). *Any* plane in the cube now carries the same state of stress—it is equal to the force on a cube face divided by its area. The state of stress is one of *hydrostatic pressure*, symbol p,

---

[2] Isaac Newton (1642–1727), scientific genius and alchemist, formulator of the laws of motion, the inverse-square law of gravity (though there is some controversy about this), laws of optics, the differential calculus, and much more.

[3] Blaise Pascal (1623–1662), philosopher, mathematician and scientist, who took a certain pleasure in publishing his results without explaining how he reached them. Almost all, however, proved to be correct.

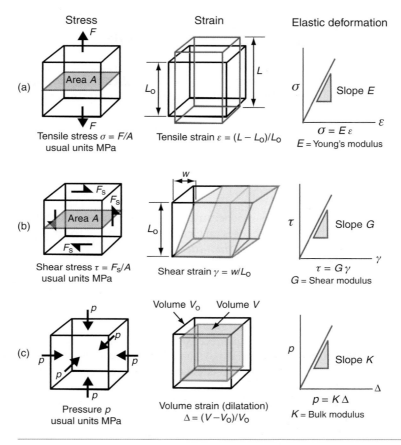

Figure 4.3   The definitions of stress, strain and elastic moduli.

again with the units of MPa. There is an unfortunate convention here. Pressures are positive when they push—the reverse of the convention for simple tension and compression.

Engineering components can have complex shapes and can be loaded in many ways, creating complex distributions of stress. But no matter how complex, the stresses in any small element within the component can always be described by a combination of tension, compression and shear. Commonly the simple cases of Figure 4.3 suffice, using superposition of two cases to capture, for example, bending plus compression.

Strain

Strain is the response of materials to stress (second column of Figure 4.3). A tensile stress σ applied to an element causes the element to stretch. If the element in Figure 4.3(a), originally of side L_o, stretches by $\delta L = L - L_o$, the nominal *tensile strain* is

$$\varepsilon = \frac{\delta L}{L_{o}} \tag{4.3}$$

A compressive stress shortens the element; the nominal compressive strain (negative) is defined in the same way. Since strain is the ratio of two lengths, it is dimensionless.

A shear stress causes a *shear strain* γ (Figure 4.3(b)). If the element shears by a distance w, the shear strain

$$\tan(\gamma) = \frac{w}{L_{o}} \approx \gamma \tag{4.4}$$

In practice $\tan \gamma \approx \gamma$ because strains are almost always small. Finally, a hydrostatic pressure p causes an element of volume V to change in volume by δV. The volumetric strain, or *dilatation* (Figure 4.3(c)), is

$$\Delta = \frac{\delta V}{V} \tag{4.5}$$

Stress–strain curves and moduli

Figure 4.4 shows typical tensile stress–strain curves for a ceramic, a metal and a polymer. The initial part, up to the elastic limit σ_{el}, is approximately linear (Hooke's[4] law), and it is elastic, meaning that the strain is recoverable—the material returns to its original shape when the stress is removed. Stresses above the elastic limit cause permanent deformation (ductile behavior) or brittle fracture.

Within the linear elastic regime, strain is proportional to stress (Figure 4.3, third column). The tensile strain is proportional to the tensile stress:

$$\sigma = E\varepsilon \tag{4.6}$$

and the same is true in compression. The constant of proportionality, E, is called *Young's[5] modulus*. Similarly, the shear strain γ is proportional to the shear stress τ:

$$\tau = G\gamma \tag{4.7}$$

and the dilatation Δ is proportional to the pressure p:

$$p = K\Delta \tag{4.8}$$

---

[4] Robert Hooke (1635–1703), able but miserable man, inventor of the microscope, and perhaps, too, of the idea of the inverse-square law of gravity. He didn't get along with Newton.

[5] Thomas Young (1773–1829), English scientist, expert on optics and deciphering ancient Egyptian hieroglyphs (among them, the Rosetta stone). It seems a little unfair that the modulus carries his name, not that of Hooke.

Figure 4.4   Tensile stress–strain curves for ceramics, metals and polymers.

where G is the *shear modulus* and K the *bulk modulus*, as illustrated in the third column of Figure 4.3. All three of these moduli have the same dimensions as stress, that of force per unit area (N/m^2 or Pa). As with stress it is convenient to use a larger unit, this time an even bigger one, that of 10^9 N/m^2, gigapascals, or GPa.

Young's modulus, the shear modulus and the bulk modulus are related, but to relate them we need one more quantity, *Poisson's*[6] *ratio*. When stretched in one direction, the element of Figure 4.3(a) generally contracts in the other two directions, as it is shown doing here. Poisson's ratio, ν, is the negative of the ratio of the lateral or transverse strain, ε_t, to the axial strain, ε, in tensile loading:

$$\nu = -\frac{\varepsilon_t}{\varepsilon} \tag{4.9}$$

Since the transverse strain itself is negative, ν is positive—it is typically about 1/3.

In an isotropic material (one for which the moduli do not depend on the direction in which the load is applied) the moduli are related in the following ways:

$$G = \frac{E}{2(1 + \nu)}; \qquad K = \frac{E}{3(1 - 2\nu)} \tag{4.10}$$

Commonly $\nu \approx 1/3$ when

$$G \approx \frac{3}{8}E \quad \text{and} \quad K \approx E \tag{4.11a}$$

Elastomers are exceptional. For these $\nu \approx 1/2$ when

$$G \approx \frac{1}{3}E \quad \text{and} \quad K \gg E \tag{4.11b}$$

This means that rubber (an elastomer) is easy to stretch in tension (low E), but if constrained from changing shape, or loaded hydrostatically, it is very stiff (large K)—a feature designers of shoes have to allow for.

Data sources like CES list values for all four moduli. In this book we examine data for E; approximate values for the others can be derived from equations (4.11) when needed.

Elastic energy
If you stretch an elastic band, energy is stored in it. The energy can be considerable: catapults can kill people. The super-weapon of the Roman arsenal at one time was a wind-up mechanism that stored enough elastic energy to hurl a 10 kg stone projectile 100 yards or more.

---

[6] Siméon Denis Poisson (1781–1840), French mathematician, known both for his constant and his distribution. He was famously uncoordinated, failed geometry at University because he could not draw, and had to abandon experimentation because of the disasters resulting from his clumsiness.

How do you calculate this energy? A force F acting through a displacement dL does work $F \, dL$. A stress $\sigma = F/A$ acting through a strain increment $d\varepsilon = dL/L$ does work per unit volume

$$dW = \frac{F \, dL}{AL} = \sigma \, d\varepsilon \qquad (4.12)$$

with units of J/m^3. If the stress is acting on an elastic material, this work is stored as elastic energy. The elastic part of all three stress–strain curves of Figure 4.4—the part of the curve before the elastic limit—is linear; in it $\sigma = E\varepsilon$. The work done per unit volume as the stress is raised from zero to a final value σ^* is the area under the stress–strain curve:

$$W = \int_0^{\sigma^*} \sigma \, d\varepsilon = \int_0^{\sigma^*} \frac{\sigma \, d\sigma}{E} = \frac{1}{2} \frac{(\sigma^*)^2}{E} \qquad (4.13)$$

This is the energy that is stored, per unit volume, in an elastically strained material. The energy is released when the stress is relaxed.

Measurement of Young's modulus

You might think that the way to measure the elastic modulus of a material would be to apply a small stress (to be sure to remain in the linear elastic region of the stress–strain curve), measure the strain and divide one by the other. In reality, moduli measured as slopes of stress–strain curves are inaccurate, often by a factor of 2 or more, because of contributions to the strain from material creep or deflection of the test machine. Accurate moduli are measured dynamically: by measuring the frequency of natural vibrations of a beam or wire, or by measuring the velocity of sound waves in the material. Both depend on $\sqrt{E/\rho}$, so if you know the density ρ you can calculate E.

Stress-free strain

Stress is not the only stimulus that causes strain. Certain materials respond to a magnetic field by undergoing strain—an effect known as magneto-striction. Others respond to an electrostatic field in the same way—they are known as piezo-electric materials. In each case a material property relates the magnitude of the strain to the intensity of the stimulus (Figure 4.5). The strains are small but can be controlled with great accuracy and, in the case of magneto-striction and piezo-electric strain, can be changed with a very high frequency. This is exploited in precision positioning devices, acoustic generators and sensors—applications we return to in Chapters 14 and 15.

A more familiar effect is that of thermal expansion: strain caused by change of temperature. The thermal strain ε_T is linearly related to the temperature change ΔT by the expansion coefficient, α:

$$\varepsilon_T = \alpha \, \Delta T \qquad (4.14)$$

where the subscript 'T' is a reminder that the strain is caused by temperature change, not stress.

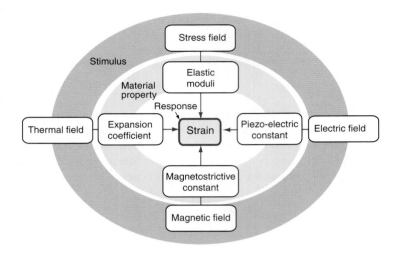

Figure 4.5 Stimuli leading to strain.

The term 'stress-free strain' is a little misleading. It correctly conveys the idea that the strain is not *caused* by stress but by something else. But these strains can none the less give rise to stresses if the body suffering the strain is constrained. Thermal stress—stress arising from thermal expansion—particularly, can be a problem, causing mechanisms to jam and railway tracks to buckle. We analyze it in Chapter 12.

4.3 The big picture: material property charts

We met the idea of material property charts in Section 2.5. Now is the time to use them. If we want materials that are stiff and light, we first need an overview of what's available. What moduli do materials offer? What are their densities? The modulus–density chart shows them.

The modulus–density chart
Figure 4.6 shows that the modulus E of engineering materials spans seven decades[7], from 0.0001 to nearly 1000 GPa; the density ρ spans a factor of 2000, from less than 0.01 to 20 Mg/m^3. The members of the ceramics and metals families have high moduli and densities; none have a modulus less than 10 GPa or a density less than 1.7 Mg/m^3. Polymers, by contrast, all have moduli below 10 GPa and densities that are lower than those of any metal or ceramic—most are close to 1 Mg/m^3. Elastomers have roughly the same density

---

[7] Very low density foams and gels (which can be thought of as molecular-scale, fluid-filled, foams) can have lower moduli than this. As an example, gelatine (as in Jello) has a modulus of about 10^{-5} GPa.

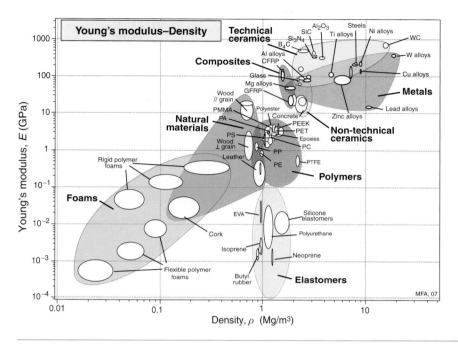

Figure 4.6   The modulus–density chart.

as other polymers but their moduli are lower by a further factor of 100 or more. Materials with a lower density than polymers are porous: man-made foams and natural cellular structures like wood and cork.

This property chart gives an overview, showing where families and their members lie in $E-\rho$ space. It helps in the common problem of material selection for stiffness-limited applications in which weight must be minimized. More on this in Chapter 5.

The modulus–relative cost chart

Often it is minimizing cost, not weight, that is the overriding objective of a design. The chart of Figure 4.7 shows, on the *x*-axis, the relative prices per unit volume of materials, normalized to that of the metal used in larger quantities than any other: mild steel. Concrete and wood are among the cheapest; polymers, steels and aluminum alloys come next; special metals like titanium, most technical ceramics and a few polymers like PTFE and PEEK are expensive. The chart allows the selection of materials that are stiff and cheap. Chapter 5 gives examples.

Anisotropy

Glasses and most polymers have disordered structures with no particular directionality about the way the atoms are arranged. They have properties that are *isotropic*, meaning the same no matter which direction they are measured. Most materials are crystalline—made up of ordered arrays of atoms. Metals

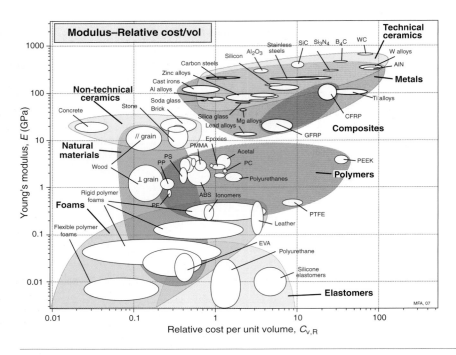

Figure 4.7 The modulus–relative cost chart. (The CES software contains material prices, regularly updated.)

and ceramics are usually polycrystalline—made up of many tiny, randomly oriented, crystals. This averages out the directionality in properties, so a single value is enough. Occasionally, though, anisotropy is important. Single crystals, drawn polymers and fibers are *anisotropic*; their properties depend on the direction in the material in which they are measured. Woods, for instance, are much stiffer along the grain than across it. Figures 4.6 and 4.7 have separate property bubbles for each of the two loading directions. Fiber composites are yet more extreme: the modulus parallel to the fibers can be larger by a factor of 20 than that perpendicular to them. Anisotropy must therefore be considered when wood and composite materials are selected.

4.4 The science: what determines density and stiffness?

Density

Atoms differ greatly in weight but little in size. Among solids, the heaviest stable atom, uranium (atomic weight 238), is about 35 times heavier than the lightest, lithium (atomic weight 6.9), yet when packed to form solids their diameters are almost exactly the same (0.32 nm). The largest atom, cesium, is only 2.5 times larger than the smallest, beryllium. Thus, the density is mainly

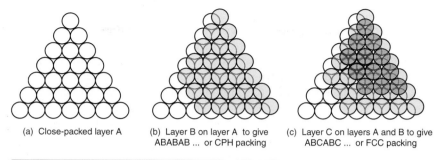

(a) Close-packed layer A

(b) Layer B on layer A to give
ABABAB ... or CPH packing

(c) Layer C on layers A and B to give
ABCABC ... or FCC packing

Figure 4.8 (a) A close-packed layer of spheres, layer A; atoms often behave as if hard and spherical. (b) A second layer, B, nesting in the first; repeating this sequence gives ABAB . . . or CPH stacking. (c) A third layer, C, can be nested so that it does not lie above A or B; if repeated this gives ABCABC . . . or FCC stacking.

determined by the atomic weight and is influenced to a lesser degree by the atom size and the way in which they are packed. Metals are dense because they are made of heavy atoms, packed densely together (iron, for instance, has an atomic weight of 56). Polymers have low densities because they are largely made of light carbon (atomic weight: 12) and hydrogen (atomic weight: 1) in low-density amorphous or semi-crystalline packings. Ceramics, for the most part, have lower densities than metals because they contain light Si, O, N or C atoms. Even the lightest atoms, packed in the most open way, give solids with a density of around 1 Mg/m^3—the same as that of water (see Figure 4.6). Materials with lower densities than this are *foams*, made up of cells containing a large fraction of pore space.

Atom packing in metals and the unit cell
Atoms often behave as if they were hard, spherical balls. The balls on a pool table, when set, are arranged as a close-packed layer, as in Figure 4.8(a). The atoms of many metals pack in this way, forming layers that are far more extensive. There is no way to pack atoms more closely than this, so this particular arrangements is called 'close packed'. Atomic structures are close packed not just in two dimensions but in three. Surprisingly, there are two ways to do this. The depressions where three atoms meet in the first layer, layer A, allow the closest nesting for a second layer, B. A third layer can be added such that its atoms are exactly above those in the first layer, so that it, too, is in the A orientation, and the sequence repeated to give a crystal with ABABAB . . . stacking, as in Figure 4.8(b); it is called *close packed hexagonal*, or *CPH* (or sometimes *HCP*) for short, for reasons explained in a moment. There is also an alternative. In placing the second layer, layer B, there are two choices of position. If the third layer, C, is nested onto B so that it lies in the alternative position, the stacking becomes (on repeating) ABCABCABC . . . as shown in (c) in the figure; it is called *face-centered cubic* or *FCC* for short. Many metals, such as copper, silver, aluminum and nickel, have the FCC structure; many others, such as magnesium, zinc and titanium, have the CPH structure. The two alternative structures have

exactly the same packing fraction, 0.74, meaning that the spheres occupy 74% of the available space. But the small difference in layout influences properties, particularly those to do with plastic deformation (Chapter 6).

Not all structures are close packed. Figure 4.9 shows one of these, made by stacking square-packed layers with a lower packing density than the hexagonal layers of the FCC and HCP structures. An ABABAB... stacking of these layers builds the *body-centered cubic* structure, *BCC* for short, with a packing fraction of 0.68. Iron and most steels have this structure. There are many other crystal structures, but for now these three are enough.

Any regular packing of atoms that repeats itself is called a *crystal*. It is possible to pack atoms in a non-crystallographic way to give what is called an *amorphous* structure, sketched in Figure 4.10. This is not such an efficient way to fill space with spheres: the packing fraction is 0.64 at best.

The characterizing unit of a crystal structure is called its *unit cell*. Figure 4.11 shows three; the red lines define the cell (the atoms have been shrunk to reveal it more clearly). In the first, shown in (a), the cell is a hexagonal prism. The atoms in the top, bottom and central planes form close-packed layers like that

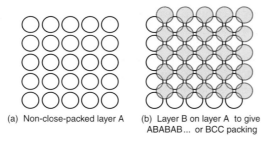

(a) Non-close-packed layer A   (b) Layer B on layer A to give
                                       ABABAB... or BCC packing

Figure 4.9   (a) A square grid of spheres; it is a less efficient packing than that of the previous figure. (b) A second layer, B, nesting in the first, A; repeating this sequence gives ABAB... packing. If the sphere spacing is adjusted so that the gray spheres lie on the corners of a cube, the result is the non-close-packed BCC structure.

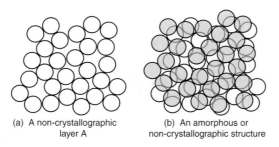

(a) A non-crystallographic   (b) An amorphous or
      layer A   non-crystallographic structure

Figure 4.10   (a) An irregular arrangement of spheres. (b) Extending this in three dimensions gives a random or amorphous structure.

of Figure 4.8(b), with ABAB... stacking. For these reasons, the structure is called close-packed hexagonal (CPH). The second, shown in (b), is also made up of close-packed layers, though this is harder to see: the shaded triangular plane is one of them. If we think of this as an A plane, atoms in the plane above it nest in the B position and those in the plane above that, in the C, giving ABCABC... stacking, as in Figure 4.8(c). The unit cell itself is a cube with an atom at each corner and one at the center of each face—for this reason it is called face-centered cubic (FCC). The final cell, shown in (c), is the characterizing unit of the square-layer structure of Figure 4.9; it is a cube with an atom at each corner and one in the middle, and is called, appropriately, body-centered cubic (BCC).

Unit cells pack to fill space as in Figure 4.12; the resulting array is called the *crystal lattice*; the points at which cell edges meet are called *lattice points*. The

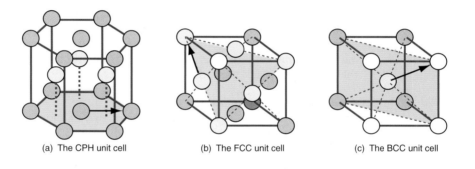

(a) The CPH unit cell   (b) The FCC unit cell   (c) The BCC unit cell

Figure 4.11   Unit cells. All the atoms are of the same type, but are shaded differently to emphasize their positions. (a) The close-packed hexagonal (CPH) structure. (b) The close-packed face-centered cubic (FCC) structure. (c) The non-close-packed body-centered cubic (BCC) structure. Arrows show nearest neighbors.

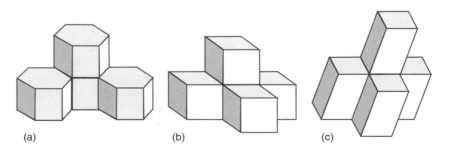

(a)   (b)   (c)

Figure 4.12   Unit cells stacked to fill space. (a) The hexagonal cell. (b) The cubic cell. (c) A cell with edges of differing length that do not meet at right angles.

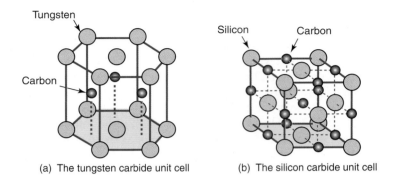

(a) The tungsten carbide unit cell   (b) The silicon carbide unit cell

Figure 4.13   Unit cells of compounds. (a) Tungsten carbide. (b) One form of silicon carbide.

crystal itself is generated by attaching one or a group of atoms to each lattice point so that they form a regular, three-dimensional, repeating pattern. The cubic and hexagonal cells are among the simplest; there are many others with edges of differing lengths meeting at differing angles. The one thing they have in common is their ability to stack with identical cells to completely fill space.

Atom packing in ceramics
Most ceramics are compounds, made up of two or more atom types. They too have characteristic unit cells. Figure 4.13 shows those of two materials that appear on the charts: tungsten carbide (WC), and silicon carbide (SiC). The cell of the first is hexagonal, that of the second is cubic, but now a pair of different atoms is associated with each lattice point: a W–C pair in the first structure and an Si–C pair in the second.

Atom packing in glasses
The crystalline state is the lowest energy state for elements and compounds. Melting disrupts the crystallinity, scrambling the atoms and destroying the regular order. The atoms in a molten metal look very like the amorphous structure of Figure 4.10. On cooling through the melting point most metals crystallize, though by cooling them exceedingly quickly it is sometimes possible to trap the molten structure to give an amorphous metallic 'glass'. With compounds it is easier to do this, and with one in particular, silica—SiO_2—crystallization is so sluggish that its usual state is the amorphous one. Figure 4.14 shows, on the left, the atom arrangement in crystalline silica: identical hexagonal Si–O rings, regularly arranged. On the right is the more usual amorphous state. Now some rings have seven sides, some have six, some five, and there is no order—the next ring could be any one of these. Amorphous silica is the basis of almost all glasses; it is mixed with Na_2O to make soda glass (windows, bottles) and with B_2O_5 to make borosilicate glasses (Pyrex), but it is the silica that gives the structure. It is for this reason than the structure itself is called 'glassy', a term used interchangeably with 'amorphous'.

(a) Crystalline   (b) Amorphous

Figure 4.14 Two alternative structures for silica, the basis of most glasses. (a) Crystalline silica. (b) Glassy or amorphous silica.

Figure 4.15 Polymer chains have a carbon–carbon backbone with hydrogen or other side groups. The figure shows three alternative representations of the polyethylene molecule.

Atom packing in polymers

Polymer structures are quite different. The backbone of a 'high' polymer ('high' means high molecular weight) is a long chain of carbon atoms, to which side groups are attached. Figure 4.15 shows a segment of the simplest: polyethylene, PE, $(-CH_2-)_n$. The chains have ends; the ends of this one are capped with a CH_3 group. PE is made by the polymerization (snapping together) of ethylene molecules, $CH_2=CH_2$; the $=$ symbol is a double bond, broken by polymerization to give the links to more carbon neighbors to the left and right. Figure 4.16 shows the chain structure of five of the most widely used linear polymers.

Polymer molecules bond together to form solids. The chains of a linear polymer, 10^3 to 10^6 $-CH_2-$ units in length, are already strongly bonded along the chain itself. Separated chains attract each other, but weakly so (Figure 4.17(a))—they are sticky, but not very sticky; the 'hydrogen' bonds that make them stick are easily broken or rearranged. The resulting structure is sketched in Figure 4.18(a): a dense spaghetti-like tangle of molecules with no order or crystallinity; it is amorphous. This is the structure of *thermoplastics* like those of Figure 4.16; the weak bonds melt easily, allowing the polymer to be molded, retaining its new shape on cooling.

H H H H H H H H
– C – C – C – C – C – C – C – C –   Polyethylene, **PE**
H H H H H H H H

H H H CH₃ H H H CH₃
– C – C – C – C – C – C – C – C –   Polypropylene, **PP**
H CH₃ H H H CH₃ H H

H H H C₆H₅ H H H C₆H₅
– C – C – C – C – C – C – C – C –   Polystyrene, **PS**
H C₆H₅ H H H C₆H₅ H H

H H H Cl H H H Cl
– C – C – C – C – C – C – C – C –   Polyvinyl chloride, **PVC**
H Cl H H H Cl H H

F F F F F F F F
– C – C – C – C – C – C – C – C –   Polytetrafluoroethylene, **PTFE**
F F F F F F F F

Figure 4.16   Five common polymers, showing the chemical make-up. The strong carbon–carbon bonds are shown in red.

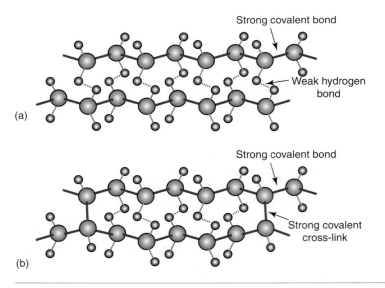

Figure 4.17   (a) Polymer chains have strong covalent 'backbones', but bond to each other only with weak hydrogen bonds unless they become cross-linked. (b) Cross-links bond the chains tightly together. The strong carbon–carbon bonds are shown as solid red lines.

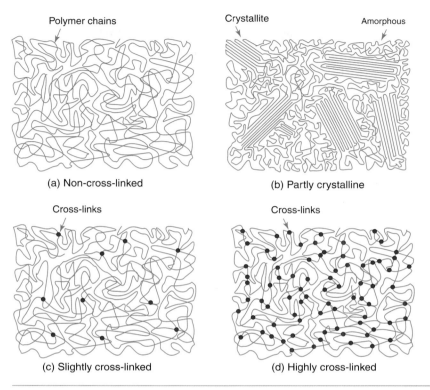

(a) Non-cross-linked

(b) Partly crystalline

(c) Slightly cross-linked

(d) Highly cross-linked

Figure 4.18   (a) Chains in polymers like polypropylene form spaghetti-like tangles with no regular repeating pattern—that structure is amorphous or 'glassy'. (b) Some polymers have the ability to form regions in which the chains line up and register, giving crystalline patches. The sketch shows a partly crystalline polymer structure. (c) Elastomers have occasional cross-links between chains, but these are far apart, allowing the chains between them to stretch. (d) Heavily cross-linked polymers like epoxy inhibit chain sliding.

The weak bonds of thermoplastics, do, however, try to line molecules up, as they are shown in Figure 4.17(a). The molecules are so long that total alignment is not possible, but segments of molecules manage it, giving small crystalline regions, as in Figure 4.18(b). These *crystallites*, as they are called, are small—often between 1 and 10 μm across, just the right size to scatter light. So amorphous polymers with no crystallites can be transparent—polycarbonate (PC), polymethyl-methacrylate (PMMA; Plexiglas) and polystyrene (PS) are examples. Those with some crystallinity, like polyethylene (PE) and nylon (PA), scatter light and are translucent.

The real change comes when chains are cross-linked by replacing some of the weak hydrogen bonds by much more muscular covalent C—C bonds, as in Figure 4.17(b), making the whole array into one huge, multiply connected network. *Elastomers* (rubbery polymers) have relatively few cross-links, as in Figure 4.18(c).

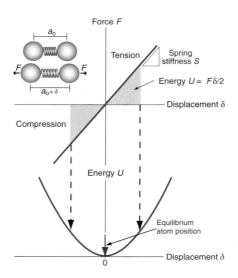

Figure 4.19   Atoms in crystals and glasses are linked by atomic bonds that behave like springs. The bond stiffness is $S = F/\delta$. Stretching or compressing the bond by a displacement δ stores energy $U = F\delta/2$. The equilibrium (no force) atom separation is at the bottom of the energy well.

Thermosets like epoxies and phenolics have many cross-links, as in Figure 4.18(d), making them stiffer and stronger than thermoplastics. The cross-links are not broken by heating, so once the links have formed thermosets cannot be thermally molded or (for that reason) recycled.

Cohesive energy and elastic moduli: crystals and glasses

Atoms bond together, some weakly, some strongly. The *cohesive energy* measures the strength of this bonding. It is defined as the energy per mol (a mol is 6×10^{23} atoms) required to separate the atoms of a solid completely, giving neutral atoms at infinity. Equally it is the energy released if the neutral, widely spaced atoms are brought together to form the solid.

The greater the cohesive energy, the stronger are the bonds between the atoms and the higher is the modulus. Think of the bonds as little springs (Figure 4.19). The atoms have equilibrium spacing a_o; a force F pulls them apart a little, to $a_o + \delta$, but when it is released they jump back to their original spacing. The same happens in compression because the potential energy of the bond increases no matter which direction the force is applied, as the curve in the figure suggests. The bond energy is a minimum at the equilibrium spacing. A spring that stretches by δ under a force F has a stiffness, S, of

$$S = \frac{F}{\delta} \tag{4.15}$$

and this is the same in compression as in tension.

Table 4.1   Bond stiffnesses, S

Bond type	Examples	Bond stiffness, S (N/m)	Young's modulus, E (GPa)
Covalent	Carbon–carbon bond	50–180	200–1000
Metallic	All metals	15–75	60–300
Ionic	Sodium chloride	8–24	32–96
Hydrogen bond	Polyethylene	3–6	2–12
Van der Waals	Waxes	0.5–1	1–4

Table 4.1 lists the stiffnesses of the different bond types; these stiffnesses largely determine the value of the modulus, E. The covalent bond is particularly stiff ($S = 20$–200 N/m); diamond has a very high modulus because the carbon atom is small (giving a high bond density) and its atoms are linked by very strong springs ($S = 200$ N/m). The metallic bond is a little less stiff ($S = 15$–100 N/m) and metal atoms are often close packed, giving them high moduli too, though not as high as that of diamond. Ionic bonds, found in many ceramics, have stiffnesses comparable with those of metals, giving them high moduli too. Polymers contain both strong diamond-like covalent bonds along the polymer chain and weak hydrogen or Van der Waals bonds ($S = 0.5$–2 N/m) between the chains; it is the weak bonds that stretch when the polymer is deformed, giving them low moduli.

How is the modulus related to the bond stiffness? When a force F is applied to a pair of atoms, they stretch apart by δ. A force F applied to an atom of diameter a_o corresponds to a stress $\sigma = F/a_o^2$, assuming each atom occupies a cube of side a_o. A stretch of δ between two atoms separated by a distance a_o corresponds to a strain $\varepsilon = \delta/a_o$. Substituting these into equation (4.15) gives

$$\sigma = \frac{S}{a_o} \varepsilon \qquad (4.16)$$

Comparing this with equation (4.6) reveals that Young's modulus, E, is approximately

$$E = \frac{S}{a_o} \qquad (4.17)$$

The largest atoms ($a_o = 4 \times 10^{-10}$ m) bonded with the weakest bonds ($S = 0.5$ N/m) will have a modulus of roughly

$$E = \frac{0.5}{4 \times 10^{-10}} \approx 1\,\text{GPa} \qquad (4.18)$$

This is the *lower limit* for true solids. Many polymers have moduli of about this value. Metals and ceramics have values 50–1000 times larger because, as Table 4.1

shows, their bonds are stiffer. But as the $E-\rho$ chart shows, materials exist that have moduli that are much lower than this limit. They are either *foams* or *elastomers*. Foams have low moduli because the cell walls bend easily (allowing large displacements) when the material is loaded. The origin of the moduli of elastomers takes a little more explaining.

The elastic moduli of elastomers

Think of it this way. An elastomer is a tangle of long-chain molecules with occasional cross-links, as on the left of Figure 4.20. The bonds between the molecules, apart from the cross-links, are weak—so weak that, at room temperature, they have melted. We describe this by saying that the glass temperature T_g of the elastomer—the temperature at which these weak bonds start to melt—is below room temperature. Segments are free to slide over each other, and were it not for the cross-links, the material would have no stiffness at all; it would be a viscous liquid.

Temperature favors randomness. That is why crystals melt into disordered fluids at their melting point and evaporate into even more random gases at the boiling point. The tangle of Figure 4.20(a) has high randomness or, expressed in the terms of thermodynamics, its entropy is high. Stretching it, as on the right of the figure, aligns the molecules—some parts of it now begin to resemble the crystallites of Figure 4.18(b). Crystals are ordered, the opposite of randomness; their entropy is low. The effect of temperature is to try to restore disorder, making the

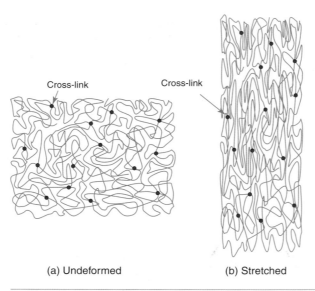

(a) Undeformed   (b) Stretched

Figure 4.20   The stretching of an elastomer. Here the structure has been stretched to twice its original length. The stretching causes alignment, producing crystal-like regions. Thermal vibration drives the structure back to the one on the left, restoring its shape.

material revert to a random tangle, and the cross-links give it a 'memory' of the disordered shape it had to start with. So there is a resistance to stretching—a stiffness—that has nothing to do with bond stretching, but with strain-induced ordering. A full theory is complicated—it involves the statistical mechanics of long-chain tangles—so it is not easy to calculate the value of the modulus. The main thing to know is that the moduli of elastomers are low because they have this strange origin and that they increase with temperature (because of the increasing tendency to randomness), whereas those of true solids decrease (because of thermal expansion).

Mixtures of atoms

Most engineering materials are not pure but contain two or more different elements. Often they dissolve in each other, like sugar in tea, but as the material is solid we call it a *solid solution*—examples are brass (a solution of zinc in copper), solder (a solution of tin in lead) and stainless steel (a solution of nickel and chromium in iron).

As we shall see later, some material properties are changed a great deal by making solid solutions. Modulus and density are not. As a general rule the density $\tilde{\rho}$ of a solid solution lies between the densities ρ_A and ρ_B of the materials that make it up, following a *rule of mixtures* (an arithmetic mean, weighted by volume fraction) known, in this instance, as Vegard's law:

$$\tilde{\rho} = f\rho_A + (1 - f)\rho_B \qquad (4.19)$$

where f is the fraction of A atoms. Modulus is a bit more complicated—pure materials have only one kind of bond, A—A say; mixtures of A and B atoms have three: A—A, B—B and A—B. Within each bond type of Table 4.1 the stiffness ranges are not large, so mixtures of bonds again average out to values between those of the pure elements.

Alloying is not therefore a route to manipulating the modulus and density very much. To do this, mixtures must be made at a more macroscopic scale to make a *hybrid* material. We can mix two discrete solids together to make *composites*, or we can mix in some space to make *foams*. The effects on modulus and density are illustrated via the property chart in the next section.

4.5 Manipulating the modulus and density

Composites

Composites are made by embedding fibers or particles in a continuous matrix of a polymer (polymer matrix composites, PMCs), a metal (MMCs) or a ceramic (CMCs), as in Figure 4.21. The development of high-performance composites is one of the great material developments of the last 40 years, now reaching maturity. Composites have high stiffness and strength per unit weight, and—in the case of MMCs and CMCs—high temperature performance. Here, then, we are interested in stiffness and weight.

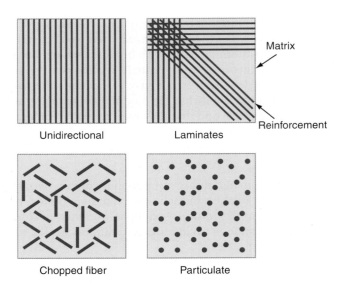

Figure 4.21   Manipulating the modulus by making composites, mixing stiff fibers or particles into a less-stiff matrix.

When a volume fraction f of a reinforcement r (density ρ_r) is mixed with a volume fraction $(1 - f)$ of a matrix m (density ρ_m) to form a composite with no residual porosity, the composite density $\widetilde{\rho}$ is given exactly by the rule of mixtures:

$$\widetilde{\rho} = f\,\rho_r + (1 - f)\rho_m \qquad (4.20)$$

The geometry or shape of the reinforcement does not matter except in determining the maximum packing fraction of reinforcement and thus the upper limit for f (typically 50%).

The modulus of a composite is bracketed by two *bounds*—limits between which the modulus must lie. The upper bound, $\widetilde{E}_U$, is found by assuming that, on loading, the two components strain by the same amount, like springs in parallel. The stress is then the average of the stresses in the matrix and the stiffer reinforcement, giving, once more, a rule of mixtures:

$$\widetilde{E}_U = fE_r + (1 - f)E_m \qquad (4.21)$$

where E_r is the Young's modulus of the reinforcement and E_m that of the matrix. To calculate the lower bound, $\widetilde{E}_L$, we assume that the two components carry the same stress, like springs in series. The strain is the average of the local strains and the composite modulus is

$$\widetilde{E}_L = \frac{E_m E_r}{fE_m + (1 - f)\,E_r} \qquad (4.22)$$

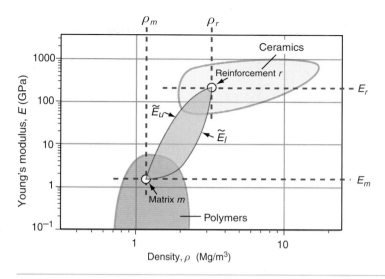

Figure 4.22 Composites made from a matrix *m* with a reinforcement *r* have moduli and densities, depending on the volume fraction and form of the reinforcement, that lie within the gray shaded lozenge bracketed by equations (4.21) and (4.22). Here the matrix is a polymer and the reinforcement a ceramic, but the same argument holds for any combination.

Figure 4.22 shows the range of composite properties that could, in principle, be obtained by mixing two materials together—the boundaries are calculated from equations (4.20)–(4.22). Composites can therefore fill in some of the otherwise empty spaces on the E–ρ chart, opening up new possibilities in design. There are practical limits of course—the matrix and reinforcement must be chemically compatible and available in the right form, and processing the mixture must be achievable at a sensible cost. Fiber-reinforced polymers are well-established examples—Figure 4.6 shows GFRP and CFRP, sitting in the composites bubble between the polymers bubble and the ceramics bubble. They are as stiff as metals, but lighter.

Foams

Foams are made much as you make bread: by mixing a matrix material (the dough) with a foaming agent (the yeast), and controlling what then happens in such a way as to trap the bubbles. Polymer foams are familiar as insulation and flotation and as the filler in cushions and packaging. You can, however, make foams from other materials: metals, ceramics and even glass. They are light for the obvious reason that they are, typically, 90% space, and they have low moduli. This might make them sound as though they are of little use, but that is mistaken: if you want to cushion or to protect a delicate object (such as yourself) what you need is a material with a low, controlled, stiffness and strength. Foams provide it.

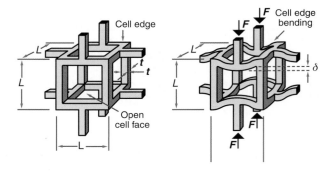

Figure 4.23 Manipulating the modulus by making a foam—a lattice of material with cell edges that bend when the foam is loaded.

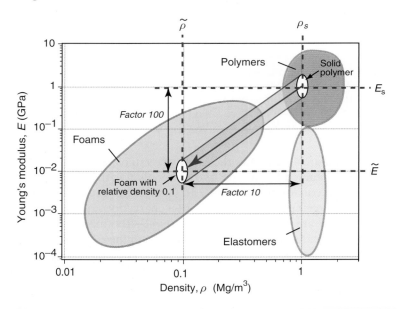

Figure 4.24 Foaming creates new materials with lower modulus and density. Low modulus is good for making packaging and protective shielding; low density is good for lightweight design and for flotation. The red arrow is a plot of equation (4.24).

Figure 4.23 shows an idealized cell of a low-density foam. It consists of solid cell walls or edges surrounding a void containing a gas. Cellular solids are characterized by their *relative density*, the fraction of the foam occupied by the solid. For the structure shown here (with $t \ll L$) it is

$$\frac{\widetilde{\rho}}{\rho_\mathrm{s}} = \left(\frac{t}{L}\right)^2 \tag{4.23}$$

where $\widetilde{\rho}$ is the density of the foam, ρ_s is the density of the solid of which it is made, L is the cell size and t is the thickness of the cell edges.

When the foam is loaded, the cell walls bend, as shown on the right of the figure. This behavior can be modeled (we will be able to do so by the end of the next chapter), giving the foam modulus $\widetilde{E}$:

$$\frac{\widetilde{E}}{E_s} = \left(\frac{\widetilde{\rho}}{\rho_s} \right)^2 \tag{4.24}$$

where E_s is the modulus of the solid from which the foam is made. This gives us a second way of manipulating the modulus, in this case, decreasing it. At a relative density of 0.1 (meaning that 90% of the material is empty space), the modulus of the foam is only 1% of that of the material in the cell wall, as sketched in Figure 4.24. The range of modulus and density for real foams is illustrated in the chart of Figure 4.6—as expected, they fall below and to the left of the polymers of which they are made.

4.6 Summary and conclusions

When a solid is loaded, it initially deforms elastically. 'Elastic' means that, when the load is removed, the solid springs back to its original shape. The material property that measures stiffness is the elastic modulus—and because solids can be loaded in different ways, we need three of them:

- *Young's modulus*, E, measuring resistance to stretching;
- the *shear modulus*, G, measuring resistance to twisting; and
- the *bulk modulus*, K, measuring the resistance to hydrostatic compression.

Many applications require stiffness at low weight, particularly ground, air and space vehicles, and that means a high modulus and a low density.

The density of a material is the weight of its atoms divided by the volume they occupy. Atoms do not differ much in volume, but they differ a great deal in weight. Thus, the density is principally set by the atomic weight; the further down the Periodic Table we go the greater becomes the density.

The moduli have their origins in the stiffness of the bonds between atoms in a solid and in the number of atoms (and thus bonds) per unit volume. The atomic volume, as we have said, does not vary much from one solid to another, so the moduli mainly reflect the stiffness of the bonds. Bonding can take several forms, depending on how the electrons of the atoms interact. The metallic, covalent and ionic bonds are generally stiff, hydrogen and Van der Waals bonds are much less so—that is why steel has high moduli and polyethylene has low.

There is very little that can be done to change the bond stiffness or atomic weight of a solid, so at first sight we are stuck with the moduli and densities of the materials we already have. But there are two ways to manipulate them: by mixing two materials to make composites, or by mixing a material with space to make foams. Both are powerful ways of creating 'new' materials that occupy regions of the E–ρ map that were previously empty.

4.7 Further reading

Ashby, M.F. and Jones, D.R.H. (2006) *Engineering Materials 1*, 3rd edition, Elsevier Butterworth-Heinemann, Oxford, UK. ISBN 0-7506-6380-4. (*One of a pair of introductory texts dealing with the engineering properties and processing of materials.*)

Askeland, D.R. and Phule, P.P. (2006) *The Science of Engineering Materials*, 5th edition, Thompson Publishing, Toronto, Canada. (*A mature text dealing with the science of materials, and taking a science-led rather than a design-led approach.*)

Callister, W.D. Jr (2007) *Materials Science and Engineering, An Introduction*, 7th edition, Wiley, New York, USA. ISBN 0-471-73696-1. (*A long established and highly respected introduction to materials, taking the science-based approach.*)

4.8 Exercises

Exercise E4.1	Identify which of the five modes of loading (Figure 4.2) is dominant in the following components:

- Fizzy drinks container.
- Overhead electric cable.
- Shoe soles.
- Wind turbine blade.
- Climbing rope.
- Bicycle forks.
- Aircraft fuselage.

Can you think of another example for each mode of loading?

Exercise E4.2 The cable of a hoist has a cross-section of 80 mm^2. The hoist is used to lift a crate weighing 500 kg. What is the stress in the cable? The free length of the cable is 3 m. How much will it extend if it is made of steel (modulus 200 GPa)? How much if it is made of polypropylene, PP (modulus 1.2 GPa)?

Exercise E4.3 Water has a density of 1000 kg/m^3. What is the hydrostatic pressure at a depth of 100 m?

Exercise E4.4 A catapult has two rubber arms, each with a square cross-section with a width 4 mm and length 300 mm. In use its arms are stretched to three times their original length before release. Assume the modulus of rubber is 10^{-3} GPa and that it does not change when the rubber is stretched. How much energy is stored in the catapult just before release?

Exercise E4.5 Use the modulus–density chart of Figure 4.6 to find, from among the materials that appear on it:

(a) The material with the highest density.
(b) The metal with the lowest modulus.
(c) The polymer with the highest density.
(d) The approximate ratio of the modulus of woods measured parallel to the grain and perpendicular to the grain.
(e) The approximate range of modulus of elastomers.

Exercise E4.6 Use the modulus–relative cost chart of Figure 4.7 to find, from among the materials that appear on it:

(a) The cheapest material with a modulus greater than 1 GPa.
(b) The cheapest metal.
(c) The cheapest polymer.
(d) Whether magnesium alloys are more or less expensive than aluminum alloys.
(e) Whether PEEK (a high-performance engineering polymer) is more or less expensive than PTFE.

Exercise E4.7 What is meant by:

- A crystalline solid?
- An amorphous solid?
- A thermoplastic?
- A thermoset?
- An elastomer?

Exercise E4.8 The stiffness S of an atomic bond in a particular material is $50 \, \text{N/m}$ and its center-to-center atom spacing is $0.3 \, \text{nm}$. What, approximately, is its elastic modulus?

Exercise E4.9 Derive the upper and lower bounds for the modulus of a composite quoted as in equations (4.21) and (4.22) of the text. To derive the first, assume that the matrix and reinforcement behave like two springs in parallel (so that each must strain by the same amount), each with a stiffness equal to its modulus E multiplied by its volume fraction, f. To derive the second, assume that the matrix and reinforcement behave like two springs in series (so that both are stressed by the same amount), again giving each a stiffness equal to its modulus E multiplied by its volume fraction, f.

Exercise E4.10 A volume fraction $f = 0.2$ of silicon carbide (SiC) particles is combined with an aluminum matrix to make a metal matrix composite. The modulus and density of the two materials are listed in the table. The modulus of particle-reinforced composites lies very close to the lower bound, equation (4.22), discussed in the text. Calculate the density and approximate modulus of the composite. Is the specific modulus, E/ρ, of the composite greater than that of unreinforced aluminum? How much larger is the specific modulus if the same volume fraction of SiC in the form of continuous fibers is used instead? For continuous fibers the modulus lies very close to the upper bound, equation (4.21).

	Density (Mg/m^3)	Modulus (GPa)
Aluminum	2.70	70
Silicon carbide	3.15	420

Exercise E4.11 Read the modulus E for polypropylene (PP) from the $E-\rho$ chart of Figure 4.6. Estimate the modulus of a PP foam with a relative density $\tilde{\rho}/\rho_s$ of 0.2.

4.9 Exploring design with CES (use Level 2, Materials, throughout)

Exercise E4.12	Make an E–ρ chart using the CES software. Use a box selection to find three materials with densities between 1000 and 3000 kg/m^3 and the highest possible modulus.
Exercise E4.13	Data estimation. The modulus E is approximately proportional to the melting point T_m in Kelvin (because strong inter-atomic bonds give both stiffness and resistance to thermal disruption). Use CES to make an E–T_m chart for metals and estimate a line of slope 1 through the data for materials. Use this line to estimate the modulus of cobalt, given that it has a melting point of 1760 K.
Exercise E4.14	Sanity checks for data. A text reports that nickel, with a melting point of 1720 K, has a modulus of 5500 GPa. Use the E–T_m correlation of the previous question to check the sanity of this claim. What would you expect it to be?
Exercise E4.15	Explore the potential of PP–SiC (polypropylene–silicon carbide) fiber composites in the following way. Make a modulus–density (E–ρ) chart and change the axis ranges so that they span the range $1 < E < 1000$ GPa and $500 < \rho < 5000$ kg/m^3. Find and label PP and SiC, then print it. Retrieve values for the modulus and density of PP and of SiC from the records for these materials (use the means of the ranges). Calculate the density $\tilde{\rho}$ and upper and lower bounds for the modulus $\tilde{E}$ at a volume fraction f of SiC of 0.5 and plot this information on the chart. Sketch by eye two arcs starting from (E, ρ) for PP, passing through each of the ($\tilde{E}$, $\tilde{\rho}$) points you have plotted and ending at the (E, ρ) point for SiC. PP–SiC composites can populate the area between the arcs roughly up to $f = 0.5$ because it is not possible to insert more than this.
Exercise E4.16	Explore the region that can be populated by making PP foams. Expand an E–ρ plot so that it spans the range $10^{-4} < E < 10$ GPa and $10 < \rho < 2000$ kg/m^3. Find and label PP, then print the chart. Construct a band starting with the PP bubble by drawing lines corresponding to the scaling law for foam modulus $\tilde{E} \propto \tilde{\rho}^2$ (equation 4.24) touching the top and the bottom of the PP bubble. The zone between these lines can be populated by PP foams.

4.10 Exploring the science with CES Elements

Exercise E4.17   The text cited the following approximate relationships between the elastic constants Young's modulus, E, the shear modulus, G, the bulk modulus, K and Poisson's ratio, ν:

$$G = \frac{E}{2(1+\nu)}; \qquad K = \frac{E}{3(1-2\nu)}$$

Use CES to make plots with the bit on the left-hand side of each equation on one axis and the bit on the right on the other. To do this you will need to use the 'Advanced' facility in the dialog box for choosing the axes to create functions on the right of the two equations. How good an approximation are they?

Exercise E4.18   The cohesive energy H_c is the energy that binds atoms together in a solid. Young's modulus E measures the force needed to stretch the atomic bonds and the melting point, T_m, is a measure of the thermal energy needed to disrupt them. Both derive from the cohesion, so you might expect E and T_m to be related. Use CES to plot one against the other to see (use absolute melting point, not centigrade or fahrenheit).

Exercise E4.19   The force required to stretch an atomic bond is

$$F = \frac{dH}{da}$$

where dH is the change in energy of the bond when it is stretched by da. This force corresponds to a stress

$$\sigma = \frac{F}{a_o^2} = \frac{1}{a_o^2}\frac{dH}{da} = \frac{1}{a_o^3}\frac{dH}{d(a/a_o)}$$

The modulus E is

$$E = \frac{d\sigma}{d\varepsilon} = \frac{d\sigma}{d(a/a_o)} = \frac{1}{a_o^3}\frac{d^2H}{d(a/a_o)^2}$$

The binding energy per atom in a crystal, H_a, is

$$H_a = \frac{H_c}{N_A}$$

where H_c is the cohesive energy and N_A is Avogadro's number (6.022×10^{23}/mol). If we assume that a stretch of 2% is enough to break the bond, we can make the approximation:

$$\frac{d^2 H}{d(a/a_o)^2} \approx \frac{H_c}{(0.02)^2}$$

giving

$$E = \frac{1}{a_o^3} \left(\frac{H_c}{0.0004 N_A} \right)$$

Make a plot of Young's modulus E against the quantity on the right of the equation (using the 'Advanced' facility in the dialog box for choosing the axes) to see how good this is. (You will need to multiply the right by 10^{-9} to convert it from pascals to GPa.)

Chapter 5
Flex, sag and wobble: stiffness-limited design

A pole vaulter – the pole stores elastic energy. (Image courtesy of Gill Athletics, 2808 Gemini Court, Champaign, IL 61822-9648, USA)

5.1 Introduction and synopsis

A few years back, with the millennium approaching, countries and cites around the world turned their minds to iconic building projects. In Britain there were several. One was—well, is—a new pedestrian bridge spanning the river Thames, linking St Paul's Cathedral to the Museum of Modern Art facing it across the river. The design was—oops, is—daring: a suspension bridge with suspension cables that barely rise above the level of the deck instead of the usual great upward sweep. The result was visually striking: a sleek, slender, span like a 'shaft of light' (the architect's words). Just one problem: it wasn't stiff enough. The bridge opened but when people walked on it, it swayed and wobbled so alarmingly that it was promptly closed. A year and \$5 000 000 later it reopened, much modified, and now it is fine.

The first thing you tend to think of with structures, bridges included, is *strength*: they must not fall down. *Stiffness*, often, is taken for granted. But, as the bridge story relates, that can be a mistake—stiffness *is* important. Here we explore stiffness-limited design and the choice of materials to achieve it. This involves the modeling of material response in a given application. The models can be simple because the selection criteria that emerge are insensitive to the details of shape and loading. The key steps are those of identifying the constraints that the material must meet and the objective by which the excellence of choice will be measured.

We saw at the beginning of Chapter 4 that there are certain common modes of loading: *tension*, *compression*, *bending* and *torsion*. The loading on any real component can be decomposed into some combination of these. So it makes sense to have a catalog of solutions for the standard modes, relating the component response to the loading. In this chapter, the response is elastic deflection; in later chapters it will be yielding or fracture. You don't need to know how to derive all of these—they are standard results—but you *do* need to know where to find them. You will find the most useful of them here; the sources listed under 'Further reading' give more. And, most important, you need to know how to use them. That needs practice—you get some of that here too.

The first section of this chapter, then, is about standard solutions to elastic problems. The second is about their use to derive *material limits* and *indices*. The third explains how to plot them onto material property charts. The last illustrates their use via case studies.

5.2 Standard solutions to elastic problems

Modeling is a key part of design. In the early, conceptual stage, approximate models establish whether a concept will work at all and identify the combination of material properties that maximize performance. At the embodiment stage, more accurate modeling brackets values for the design parameters: forces, displacements, velocities, heat fluxes and component dimensions. And in the final stage, modeling gives precise values for stresses, strains and failure probabilities in key components allowing optimized material selection and sizing.

Many common geometries and load patterns have been modeled already. A component can often be modeled approximately by idealizing it as one of these. There is no need to reanalyze the beam or the column or the pressure vessel; their behavior under all common types of loading has already been analyzed. The important thing is to know that the results exist, where to find them and how to use them.

This section is a bit tedious (as books on the strength of materials tend to be) but the results are really useful. Here they are listed, defining the quantities that enter and the components to which they apply.

Elastic extension or compression

A tensile or compressive stress $\sigma = F/A$ applied axially to a tie or strut of length L_o and constant cross-section area A suffers a strain $\varepsilon = \sigma/E$. The strain ε is related to the extension δ by $\varepsilon = \delta/L_o$ (Figure 5.1(a)). Thus, the relation between the load F and deflection δ is

$$\delta = \frac{L_o F}{AE} \tag{5.1}$$

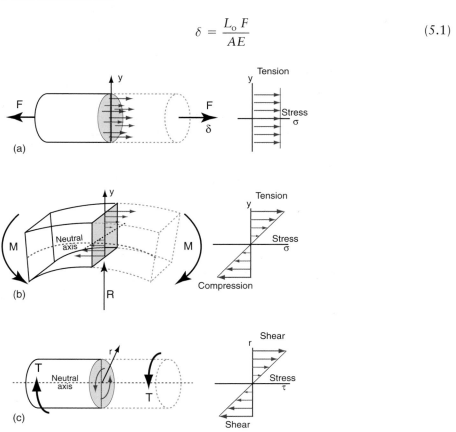

Figure 5.1   (a) A tie with a cross-section A loaded in tension. Its stiffness is $S = F/\delta$. (b) A beam of rectangular cross-section loaded in bending. The stress σ varies linearly from tension to compression, changing sign at the neutral axis, resulting in a bending moment M. (c) A shaft of circular cross-section loaded in torsion.

The stiffness S is defined as

$$S = \frac{F}{\delta} = \frac{AE}{L_o} \qquad (5.2)$$

Note that the shape of the cross-section area does not matter because the stress is uniform over the section.

Elastic bending of beams

When a beam is loaded by a bending moment M, its initially straight axis is deformed to a curvature κ (Figure 5.1(b))

$$\kappa = \frac{d^2 u}{dx^2}$$

where u is the displacement parallel to the y-axis. The curvature generates a linear variation of axial strain ε (and thus stress σ) across the section, with tension on one side and compression on the other—the position of zero stress being the *neutral axis*. The stress increases with distance y from the neutral axis. Material is more effective at resisting bending the further it is from that axis, so the shape of the cross-section is important. Elastic beam theory gives the stress σ caused by a moment M in a beam made of material of Young's modulus E, as

$$\frac{\sigma}{y} = \frac{M}{I} = E\kappa$$

$$= E\frac{d^2 u}{dx^2} \qquad (5.3)$$

where I is the second moment of area, defined as

$$I = \int_{\text{section}} y^2 b(y)\, dy \qquad (5.4)$$

The distance y is measured vertically from the neutral axis and $b(y)$ is the width of the section at y. The moment I characterizes the resistance of the section to bending—it includes the effect of both size and shape. Examples for four common sections are listed in Figure 5.2 with expressions for the cross-section area A and the second moment of area, I.

Section shape	Area A m^2	Moment I m^4	Moment K m^4
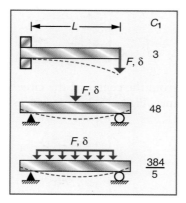 (rectangle, h, b)	bh	$\dfrac{bh^3}{12}$	$\dfrac{bh^3}{3}(1-0.58\dfrac{b}{h})$ $(h>b)$
(circle, $2r$)	πr^2	$\dfrac{\pi}{4}r^4$	$\dfrac{\pi}{2}r^4$
(tube, t, $2r_i$, $2r_0$)	$\pi(r_0^2 - r_i^2)$ $\approx 2\pi r t$	$\dfrac{\pi}{4}(r_0^4 - r_i^4)$ $\approx \pi r^3 t$	$\dfrac{\pi}{2}(r_0^4 - r_i^4)$ $\approx 2\pi r^3 t$
(box, t, h, b)	$2t(h+b)$ $(h,b >> t)$	$\dfrac{1}{6}h^3 t \left(1 + 3\dfrac{b}{h}\right)$	$\dfrac{2tb^2 h^2}{(b+h)}(1-\dfrac{t}{h})^4$

Figure 5.2 Cross-section area and second moments of sections for four section shapes.

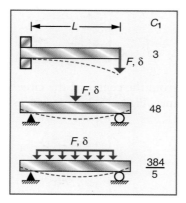

Figure 5.3 Elastic deflection of beams. The deflection δ of a span L under a force F depends on the flexural stiffness EI of the cross-section and the way the force is distributed. The constant C_1 is defined in equation (5.5).

The ratio of moment to curvature, M/κ, is called the *flexural rigidity*, EI. Figure 5.3 shows three possible distributions of load F, each creating a distribution of moment $M(x)$. The maximum deflection, δ, is found by integrating equation (5.3) twice for a particular $M(x)$. For a beam of length L with a transverse load F, the

stiffness is

$$S = \frac{F}{\delta} = \frac{C_1 EI}{L^3} \tag{5.5}$$

The result is the same for all simple distributions of load; the only thing that depends on the distribution is the value of the constant C_1; the figure lists values for the three distributions. We will find in Section 5.3 that the best choice of material is independent of the value of C_1 with the happy result that the best choice for a complex distribution of loads is the same as that for a simple one.

Torsion of shafts

A torque, T, applied to the ends of an isotropic bar of uniform section generates a shear stress τ (Figures 5.1(c) and 5.4). For circular sections, the shear stress varies with radial distance r from the axis of symmetry is

$$\frac{\tau}{r} = \frac{T}{K} \tag{5.6}$$

where K measures the resistance of the section to twisting (the torsional equivalent to I, for bending). K is easiest to calculate for circular sections, when it is equal to the polar second moment of area:

$$J = \int_{\text{section}} 2\pi r^3 \, dr \tag{5.7}$$

where r is measured radially from the center of the circular section. For non-circular sections, K is less than J. Figure 5.2 gives expressions for K for four standard shapes.

Figure 5.4   Elastic torsion of circular shafts. The stress in the shaft and the twist per unit length depend on the torque T and the torsional rigidity GK.

The shear stress acts in the plane normal to the axis of the bar. It causes the bar, with length L, to twist through an angle θ. The twist per unit length, θ/L, is related to the shear stress and the torque by

$$\frac{\tau}{r} = \frac{T}{K} = \frac{G\theta}{L} \qquad (5.8)$$

where G is the shear modulus. The ratio of torque to twist, T/θ, per unit length, is equal to GK, called the *torsional rigidity*.

Buckling of columns and plate

If sufficiently slender, an elastic column or plate, loaded in compression, fails by elastic buckling at a critical load, F_{crit}. Beam theory shows that the critical buckling load depends on the length L and flexural rigidity, EI:

$$F_{crit} = \frac{n^2 \pi^2 EI}{L^2} \qquad (5.9)$$

where n is a constant that depends on the end constraints: clamped, or free to rotate, or free also to translate (Figure 5.5). The value of n is just the number of half-wavelengths of the buckled shape (the half-wavelength is the distance between inflection points).

Some of the great engineering disasters have been caused by buckling. It can occur without warning and slight misalignment is enough to reduce the load at which it happens. When dealing with compressive loads on columns or in-plane compression of plates it is advisable to check that you are well away from the buckling load.

Figure 5.5   The buckling load of a column of length L depends on the flexural rigidity EI and on the end constraints; three are shown here, together with the value of n.

Figure 5.6   The natural vibration modes of beams clamped in different ways.

Vibrating beams and plates

Any undamped system vibrating at one of its natural frequencies can be reduced to the simple problem of a mass m attached to a spring of stiffness k—the restoring force per unit displacement. The lowest natural frequency of such a system is

$$f = \frac{1}{2\pi}\sqrt{\frac{k}{m}} \tag{5.10}$$

Different geometries require appropriate estimates of the effective k and m—often these can be estimated with sufficient accuracy by approximate modeling. The higher natural frequencies of rods and beams are simple multiples of the lowest.

Figure 5.6 shows the lowest natural frequencies of the flexural modes of uniform beams or plates of length L with various end constraints. The spring stiffness, k, is that for bending, given by equation (5.5), so the natural frequencies can be written:

$$f = \frac{C_2}{2\pi}\sqrt{\frac{EI}{m_o L^4}} \tag{5.11}$$

where C_2, listed in the figure, depends on the end constraints and m_o is the mass of the beam per unit length. The mass per unit length is just the area times the density, $A\rho$, so the natural frequency becomes

$$f = \frac{C_2}{2\pi}\sqrt{\frac{I}{AL^4}}\sqrt{\frac{E}{\rho}} \tag{5.12}$$

Thus, frequencies scale as $\sqrt{E/\rho}$.

5.3 Material indices for elastic design

We can now start to implement the steps outlined in Chapter 3 and summarized in Figure 3.6:

- *Translation*
- *Screening*, based on constraints
- *Ranking*, based on objectives, and
- *Documentation* to give greater depth.

The first two were fully described in Chapter 3. The third—ranking based on objectives—requires simple modeling to identify the *material index*.

An *objective*, it will be remembered, is a criterion of excellence for the design as a whole, something to be minimized (like cost, weight or volume) or maximized (like energy storage). Here we explore those for elastic design.

Minimizing weight: a light, stiff tie-rod

Think first of choosing a material for a cylindrical tie-rod like one of those in the cover picture of Chapter 4. Its length L_o is specified and it must carry a tensile force F without extending elastically by more than δ. Its stiffness must be at least $S^* = F/\delta$ (Figure 5.7(a)). This is a load-carrying component, so it will need to have some toughness. The objective is to make it as light as possible. The

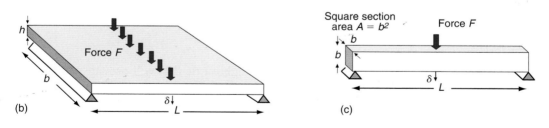

Figure 5.7   (a) A tie with cross-section area A, loaded in tension. Its stiffness is $S = F/\delta$ where F is the load and δ is the extension. (b) A panel loaded in bending. Its stiffness is $S = F/\delta$, where F is the total load and δ is the bending deflection. (c) A beam of square section, loaded in bending. Its stiffness is $S = F/\delta$, where F is the load and δ is the bending deflection.

Table 5.1   Design requirements for the light stiff tie

Function	• Tie-rod	
Constraints	• Stiffness S^* specified • Some toughness	} Functional constraints
	• Length L_o specified	Geometric constraint
Objective	• Minimize mass	
Free variables	• Choice of material • Choice of cross-section area A	

cross-section area A is free. The design requirements, translated, are listed in Table 5.1.

We first seek an equation that describes the quantity to be maximized or minimized, here the mass m of the tie. This equation, called the *objective function*, is

$$m = AL_o\rho \qquad (5.13)$$

where A is the area of the cross-section and ρ is the density of the material of which it is made. We can reduce the mass by reducing the cross-section, but there is a constraint: the section area A must be sufficient to provide a stiffness of S^*, which, for a tie, is given by equation (5.2):

$$S^* = \frac{AE}{L_o} \qquad (5.14)$$

If the material has a low modulus, a large A is needed to give the necessary stiffness; if E is high, a smaller A is needed. But which gives the lower mass? To find out, we eliminate the free variable A between these two equations, giving

$$m = S^* L_o^2 \left(\frac{\rho}{E}\right) \qquad (5.15)$$

Both S^* and L_o are specified. The lightest tie that will provide a stiffness S^* is that made of the material with the smallest value of ρ/E. We could define this as the material index of the problem, seeking the material with a minimum value, but it is more usual to express indices in a form for which a maximum is sought. We therefore invert the material properties in equation (5.15) and define the material index M_t (subscript 't' for tie), as:

$$M_t = \frac{E}{\rho} \qquad (5.16)$$

Table 5.2   Design requirements for the light stiff panel

Function	• Panel	
Constraints	• Stiffness S* specified	Functional constraint
	• Length L and width b specified	Geometric constraint
Objective	• Minimize mass	
Free variables	• Choice of material	
	• Choice of panel thickness h	

It is called the *specific stiffness*. Materials with a high value of M_t are the best choice, provided that they also meet any other constraints of the design, in this case the need for some toughness.

The mode of loading that most commonly dominates in engineering is not tension, but bending—think of floor joists, of wing spars, of golf club shafts. The index for bending differs from that for tension, and this (significantly) changes the optimal choice of material. We start by modeling panels and beams, specifying stiffness and seeking to minimize weight.

Minimizing weight: a light, stiff panel

A panel is a flat slab, like a table top. Its length L and width b are specified but its thickness h is free. It is loaded in bending by a central load F (Figure 5.7(b)). The stiffness constraint requires that it must not deflect more than δ under the load F and the objective is again to make the panel as light as possible. Table 5.2 summarizes the design requirements.

The objective function for the mass of the panel is the same as that for the tie:

$$m = AL\rho = bhL\rho \tag{5.17}$$

Its bending stiffness S is given by equation (5.5). It must be at least:

$$S^* = \frac{C_1 EI}{L^3} \tag{5.18}$$

The second moment of area, I, for a rectangular section (Table 5.2) is

$$I = \frac{bh^3}{12} \tag{5.19}$$

The stiffness S^*, the length L and the width b are specified; only the thickness h is free. We can reduce the mass by reducing h, but only so far that the stiffness constraint is still met. Using the last two equations to eliminate h in the objective function for the mass gives

$$m = \left(\frac{12 S^*}{C_1 b}\right)^{1/3} (bL^2)\left(\frac{\rho}{E^{1/3}}\right) \tag{5.20}$$

The quantities S^*, L, b and C_1 are all specified; the only freedom of choice left is that of the material. The best materials for a light, stiff panel are those with the smallest values of $\rho/E^{1/3}$ (again, so long as they meet any other constraints). As before, we will invert this, seeking instead large values of the material index M_p for the panel:

$$M_p = \frac{E^{1/3}}{\rho} \tag{5.21}$$

This doesn't look much different from the previous index, E/ρ, but it is. It leads to a different choice of material, as we shall see in a moment. For now, note the procedure. The length of the panel was specified but we were free to vary the section area. The objective is to minimize its mass, m. Use the stiffness constraint to eliminate the free variable, here h. Then read off the combination of material properties that appears in the objective function—the equation for the mass. It is the index for the problem.

It sounds easy, and it is—so long as you are clear from the start what the constraints are, what you are trying to maximize or minimize, and which parameters are specified and which are free.

Now for another bending problem, in which the freedom to choose shape is rather greater than for the panel.

Minimizing weight: a light, stiff beam

Beams come in many shapes: solid rectangles, cylindrical tubes, I-beams and more. Some of these have too many free geometric variables to apply the method above directly. However, if we constrain the shape to be *self-similar* (such that all dimensions change in proportion as we vary the overall size), the problem becomes tractable again. We therefore consider beams in two stages: first, to identify the optimum materials for a light, stiff beam of a prescribed simple shape (such as a square section); then, second, we explore how much lighter it could be made, for the same stiffness, by using a more efficient shape.

Consider a beam of square section $A = b \times b$ that may vary in size but with the square shape retained. It is loaded in bending over a span of fixed length L with a central load F (Figure 5.7(c)). The stiffness constraint is again that it must not deflect more than δ under the load F, with the objective that the beam should again be as light as possible. Table 5.3 summarizes the design requirements.

Proceeding as before, the objective function for the mass is:

$$m = AL\rho = b^2 L\rho \tag{5.22}$$

The beam bending stiffness S (equation (5.5)) is

$$S^* = \frac{C_1 EI}{L^3} \tag{5.23}$$

Table 5.3   Design requirements for the light stiff beam

Function	• Beam	
Constraints	• Stiffness S^* specified	Functional constraint
	• Length L	
	• Section shape square	} Geometric constraints
Objective	• Minimize mass	
Free variables	• Choice of material	
	• Area A of cross-section	

The second moment of area, I, for a square section beam is

$$I = \frac{b^4}{12} = \frac{A^2}{12} \tag{5.24}$$

For a given length L, the stiffness S^* is achieved by adjusting the size of the square section. Now eliminating b (or A) in the objective function for the mass gives

$$m = \left(\frac{12\, S^*\, L^3}{C_1} \right)^{1/2} (L) \left(\frac{\rho}{E^{1/2}} \right) \tag{5.25}$$

The quantities S^*, L and C_1 are all specified—the best materials for a light, stiff beam are those with the smallest values of $\rho/E^{1/2}$. Inverting this, we require large values of the material index M_b for the beam:

$$M_b = \frac{E^{1/2}}{\rho} \tag{5.26}$$

This analysis was for a square beam, but the result in fact holds for any shape, so long as the shape is held constant. This is a consequence of equation (5.24)—for a given shape, the second moment of area I can always be expressed as a constant times A^2, *so changing the shape just changes the constant C_1 in equation (5.25)*, not the resulting index.

As noted above, real beams have section shapes that improve their efficiency in bending, requiring less material to get the same stiffness. By shaping the cross-section it is possible to increase I without changing A. This is achieved by locating the material of the beam as far from the neutral axis as possible, as in thin-walled tubes or I-beams. Some materials are more amenable than others to being made into efficient shapes. Comparing materials on the basis of the index in equation (5.26) therefore requires some caution—materials with lower values

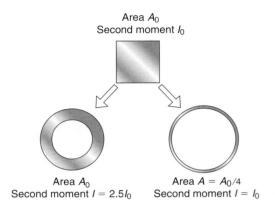

Area A_0
Second moment I_0

Area A_0
Second moment $I = 2.5I_0$

Area $A = A_0/4$
Second moment $I = I_0$

Figure 5.8   The effect of section shape on bending stiffness EI: a square-section beam compared, left, with a tube of the same area (but 2.5 times stiffer) and, right, a tube with the same stiffness (but four times lighter).

of the index may 'catch up' by being made into more efficient shapes. So we need to get an idea of the effect of shape on bending performance.

Figure 5.8 shows a solid square beam, of cross-section area A. If we turned the same area into a tube, as shown in the left of the figure, the mass of the beam is unchanged (equation (5.22)). The second moment of area, I, however, is now much greater—and so is the stiffness (equation 5.23). We define the ratio of I for the shaped section to that for a solid square section with the same area (and thus mass) as the *shape factor* Φ. The more slender the shape the larger is Φ, but there is a limit—make it too thin and the tube will buckle—so there is a maximum shape factor for each material that depends on its properties. Table 5.4 lists some typical values.

Shaping is also used to make structures lighter: it is a way to get the same stiffness with less material (Figure 5.8, right). The mass ratio is given by the reciprocal of the square root of the maximum shape factor, $\Phi^{-1/2}$ (because C_1, which is proportional to the shape factor, appears as $(C_1)^{-1/2}$ in equation (5.25)). Table 5.4 lists the factor by which a beam can be made lighter, for the same stiffness, by shaping. Metals and composites can all be improved significantly (though the metals do a little better), but wood has more limited potential because it is more difficult to shape it into efficient, thin-walled shapes. So, when comparing materials for light, stiff beams using the index in equation (5.26), the performance of wood is not as good as it looks because other materials can be made into more efficient shapes. As we will see, composites (particularly CFRP) have very high values of all three indices M_t, M_p and M_b, but this advantage relative to metals is reduced a little by the effect of shape.

Minimizing material cost
When the objective is to minimize cost rather than weight, the indices change. If the material price is C_m \$/kg, the cost of the material to make a component of mass

Table 5.4 The effect of shaping on stiffness and mass of beams in different structural materials

Material	Typical maximum shape factor (stiffness relative to that of a solid square beam)	Typical mass ratio by shaping (relative to that of a solid square beam)
Steels	64	1/8
Al alloys	49	1/7
Composites (GFRP, CFRP)	36	1/6
Wood	9	1/3

m is just mC_m. The objective function for the material cost C of the tie, panel or beam then becomes

$$C = mC_m = ALC_m\rho \qquad (5.27)$$

Proceeding as in the three previous examples then leads to indices which are just those of equations (5.16), (5.21) and (5.26), with ρ replaced by $C_m\rho$.

The material cost is only part of the cost of a shaped component; there is also the manufacturing cost—the cost to shape, join and finish it. We leave these to a later chapter.

5.4 Plotting limits and indices on charts

Screening: attribute limits on charts

Any design imposes certain non-negotiable demands ('constraints') on the material of which it is made. These limits can be plotted as horizontal or vertical lines on material property charts, as illustrated in Figure 5.9, which shows a schematic of the *E–Relative cost* chart of Figure 4.7. We suppose that the design imposes limits on these of $E > 10$ GPa and *Relative cost* < 3, shown on the figure. All materials in the window defined by the limits, labeled 'Search region', meet both constraints.

Later chapters of this book show charts for many other properties. They allow limits to be imposed on other properties.

Ranking: indices on charts

The next step is to seek, from the subset of materials that meet the property limits, those that maximize the performance of the component. We will use the design of light, stiff components as examples; the other material indices are used in a similar way.

Figure 5.10 shows a schematic of the $E–\rho$ chart of Figure 4.6. The logarithmic scales allow all three of the indices E/ρ, $E^{1/3}/\rho$ and $E^{1/2}/\rho$, derived in the last section, to be plotted onto it. Consider the condition

$$M = \frac{E}{\rho} = \text{constant}, C$$

Figure 5.9   A schematic *E–Relative cost* chart showing a lower limit for *E* and an upper one for *Relative cost*.

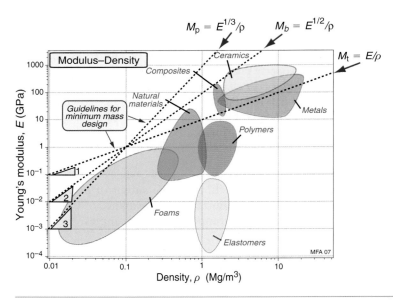

Figure 5.10   A schematic *E–ρ* chart showing guidelines for three material indices for stiff, lightweight structures.

i.e. a particular value of the specific stiffness. Taking logs,

$$\log(E) = \log(\rho) + \log(C) \tag{5.28}$$

This is the equation of a straight line of slope 1 on a plot of log(E) against log(ρ), as shown in the figure. Similarly, the condition

$$M = \frac{E^{1/3}}{\rho} = \text{constant}, C$$

becomes, on taking logs,

$$\log(E) = 3\log(\rho) + 3\log(C) \tag{5.29}$$

This is another straight line, this time with a slope of 3, also shown. And by inspection, the third index $E^{1/2}/\rho$ will plot as a line of slope 2. We refer to these lines as *selection guidelines*. They give the slope of the family of parallel lines belonging to that index.

It is now easy to read off the subset of materials that maximize performance for each loading geometry. For example, all the materials that lie on a line of constant $M = E^{1/3}/\rho$ perform equally well as a light, stiff panel; those above the line perform better, those below less well. Figure 5.11 shows a grid of lines

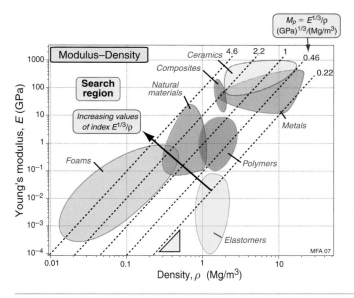

Figure 5.11   A schematic E–ρ chart showing a grid of lines for the index $E^{1/3}/\rho$. The units are (GPa)$^{1/3}$/(Mg/m^3).

corresponding to values of $M = E^{1/3}/\rho$ from $M = 0.22$ to $M = 4.6$ in units of $\text{GPa}^{1/3}/(\text{Mg/m}^3)$. A material with $M = 3$ in these units gives a panel that has one-tenth the weight of one with $M = 0.3$. The case studies in the next section give practical examples.

Computer-aided selection

The charts give an overview, but the number of materials that can be shown on any one of them is obviously limited. Selection using them is practical when there are very few constraints, but when there are many—as there usually are—checking that a given material meets them all is cumbersome. Both problems are overcome by a computer implementation of the method.

The *CES* material and process selection software[1] is an example of such an implementation. Its database contains records for materials, organized in the hierarchical manner shown in Figure 2.3 in Chapter 2. Each record contains property data for a material, each property stored as a range spanning its typical (or, often, permitted) values. It also contains limited documentation in the form of text, images and references to sources of information about the material. The data are interrogated by a search engine that offers the search interfaces shown schematically in Figure 5.12. On the left is a simple query interface for

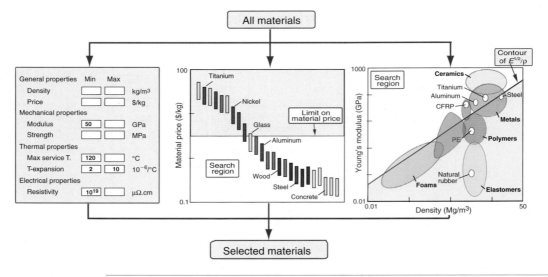

Figure 5.12   Computer-aided selection using the CES software. The schematic shows the three types of selection window. They can be used in any order and any combination. The selection engine isolates the subset of materials that pass all the selection stages.

---

[1] Granta Design Ltd, Cambridge, UK (www.grantadesign.com).

screening on single attributes. The desired upper or lower limits for constrained properties are entered; the search engine rejects all materials with attributes that lie outside the limits. In the center is shown a second way of interrogating the data: a bar chart, constructed by the software, for any numeric property in the database. It, and the bubble chart shown on the right, are ways both of applying constraints and of ranking. For screening, a selection line or box is superimposed on the charts with edges that lie at the constrained values of the property (bar chart) or properties (bubble chart). This eliminates the materials in the shaded areas and retains the materials that meet the constraints. If, instead, ranking is sought (having already applied all necessary constraints) an index-line like that shown in Figure 5.11 is positioned so that a small number—say, 10—materials are left in the selected area; these are the top-ranked candidates. The software delivers a list of the top-ranked materials that meet all the constraints.

5.5 Case studies

Here we have case studies using the two charts of Chapter 4. They are deliberately simplified to avoid obscuring the method under layers of detail. In most cases little is lost by this: the best choice of material for the simple example is the same as that for the more complex.

Light levers for corkscrews

The lever of the corkscrew (Figure 5.13) is loaded in bending: the force F creates a bending moment $M = FL$. The lever needs to be stiff enough that the bending displacement, δ, when extracting a cork, is acceptably small. If the corkscrew is intended for travelers, it should also be light. The section is rectangular. We make the assumption of self-similarity, meaning that we are free to change the scale of the section but not its shape. The material index we want was derived earlier as equation (5.26). It is that for a light, stiff beam:

$$M = \frac{E^{1/2}}{\rho} \qquad (5.30)$$

Figure 5.13   The corkscrew lever from Chapter 3. It must be adequately stiff and, for traveling, as light as possible.

Table 5.5 Design requirements for the corkscrew lever

Function	• Lightweight lever, meaning light, stiff beam
Constraints	• Stiffness $S*$ specified
	• Length L
	• Section shape rectangular
Objective	• Minimize mass
Free variables	• Choice of material
	• Area A of cross-section

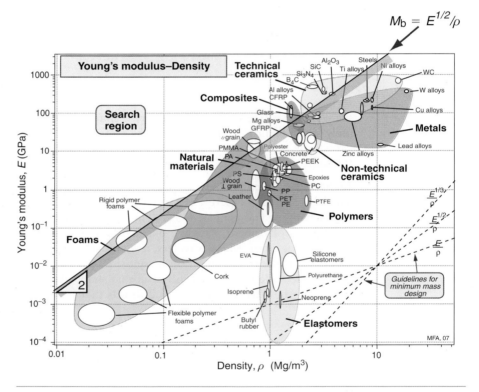

Figure 5.14 Selection of materials for the lever. The objective is to make it as light as possible while meeting a stiffness constraint.

where E is Young's modulus and ρ is the density. There are other obvious constraints. Corkscrews get dropped and must survive impacts of other kinds, so brittle materials like glass or ceramic are unacceptable. Given these requirements, summarized in Table 5.5, what materials would you choose?

Figure 5.14 shows the appropriate chart: that in which Young's modulus, E, is plotted against density, ρ. The selection line for the index M has a slope of 2, as

Floor
joists

Figure 5.15   The materials of a building are chosen to perform three different roles. Those
for the structure are chosen to carry loads. Those for the cladding provide
protection from the environment. Those for the interior control heat, light and
sound. Here we explore structural materials.

explained in Section 5.3; it is positioned so that a small group of materials is left
above it. They are the materials with the largest values of M, and it is these that
are the best choice, provided they satisfy the other constraints. Three classes of
materials lie above the line: woods, carbon-fiber reinforced polymers (CFRPs)
and a number of ceramics. Ceramics are brittle and expensive, ruling them out.
The recommendation is clear. Make your the lever out of wood or—better—out
of CFRP.

Cost: structural materials for buildings

The most expensive thing that most people ever buy is the house they live in.
Roughly half the cost of a house is the cost of the materials of which it is made,
and these are used in large quantities (family house: around 200 tonnes; large
apartment block: around 20 000 tonnes). The materials are used in three ways:
structurally to hold the building up; as cladding, to keep the weather out; and
as 'internals', to insulate against heat and sound, and to provide comfort and
decoration.

Consider the selection of materials for the structure (Figure 5.15). They must
be stiff, strong and cheap. Stiff, so that the building does not flex too much
under wind loads or internal loading. Strong, so that there is no risk of it col-
lapsing. And cheap, because such a lot of material is used. The structural frame
of a building is rarely exposed to the environment, and is not, in general, visible, so criteria of corrosion resistance or appearance are not important here.
The design goal is simple: stiffness and strength at minimum cost. To be more
specific: consider the selection of material for floor joists, focusing on stiffness.
Table 5.6 summarizes the requirements.

Table 5.6   Design requirements for floor beams

Function	• Floor beam
Constraints	• Stiffness S^* specified
	• Length L specified
	• Section shape square
Objective	• Minimize material cost
Free variables	• Choice of material
	• Area A of cross-section

The material index for a stiff beam of minimum mass, m, was developed earlier. The cost C of the beam is just its mass, m, times the cost per kg, C_m, of the material of which it is made:

$$C = mC_m = AL\rho C_m$$

Proceeding as in Section 5.3, we find the index for a stiff beam of minimum cost to be:

$$M = \frac{E^{1/2}}{\rho C_m} \tag{5.31}$$

Figure 5.16 shows the relevant chart: modulus E against relative cost per unit volume, $C_m\,\rho$ (the chart uses a *relative* cost $C_{V,R}$, defined in Chapter 4, in place of C_m but this makes no difference to the selection). The shaded band has the appropriate slope for M; it isolates concrete, stone, brick, woods, cast irons and carbon steels.

Concrete, stone and brick have strength only in compression; the form of the building must use them in this way (walls, columns, arches). Wood, steel and reinforced concrete have strength both in tension and compression, and steel, additionally, can be given efficient shapes (I-sections, box sections, tubes) that can carry bending and tensile loads as well as compression, allowing greater freedom of the form of the building.

Cushions and padding: the modulus of foams

One way of manipulating the modulus is to make a material into a foam. Figure 4.23 showed an idealized foam structure: a network of struts of length L and thickness t, connected at their mid-span to neighboring cells. Cellular solids like this one are characterized by their *relative density*, which for the structure shown here (with $t << L$) is

$$\frac{\tilde{\rho}}{\rho_s} = \left(\frac{t}{L}\right)^2 \tag{5.32}$$

$$M_b = E^{1/2}/C_{V,R}\rho$$

Figure 5.16   The selection of materials for stiff floor beams. The objective is to make them as cheap as possible while meeting a stiffness constraint.

where $\bar{\rho}$ is the density of the foam and ρ_s is the density of the solid of which it is made.

If a compressive stress σ is applied to a block of foam containing many cells, it is transmitted through the foam as forces F pushing on edges that lie parallel to the direction of σ. The area of one cell face is L^2 so the force on one strut is $F = \sigma L^2$. This force bends the cell edge to which it connects, as on the right of Figure 4.23. Thus, the cell edge is just a beam, built-in at both ends, carrying a central force F. The bending deflection is given by equation (5.5):

$$\delta = \frac{FL^3}{C_1 E_s I} \qquad (5.33)$$

where E_s is the modulus of the solid of which the foam is made and $I = t^4/12$ is the second moment of area of the cell edge of square cross-section, $t \times t$.

The compressive strain suffered by the cell as a whole is then $\varepsilon = 2\delta/L$. Assembling these results gives the modulus $\tilde{E} = \sigma/\varepsilon$ of the foam as

$$\frac{\tilde{E}}{E_s} \propto \left(\frac{\tilde{\rho}}{\rho_s} \right)^2 \tag{5.34}$$

Since $\tilde{E} = E_s$ when $\tilde{\rho} = \rho_s$, the constant of proportionality is 1, giving the result plotted earlier in Figure 4.24.

Vibration: avoiding resonance when changing material

Vibration, as the story at the start of this chapter relates, can be a big problem. Bridges are designed with sufficient stiffness to prevent wind-loads exciting their natural vibration frequencies (one, the Tacoma Straits bridge in the state of Washington, wasn't; its oscillations destroyed it). Auto-makers invest massively in computer simulation of new models to be sure that door and roof panels don't start thumping because the engine vibration hits a natural frequency. Even musical instruments, which rely on exciting natural frequencies, have problems with 'rogue' tones: notes that excite frequencies you didn't want as well as those you did.

Suppose you redesign your bridge, or car, or cello and, in a creative moment, decide to make it out of a new material. What will it do to the natural frequencies? Simple. Natural frequencies, f, as explained in Section 5.2, are proportional to $\sqrt{E/\rho}$. If the old material has a modulus E_o and density ρ_o and the new one has E_n, ρ_n, the change in frequency, Δf

$$\Delta f = \sqrt{\frac{E_n \rho_o}{E_o \rho_n}} \tag{5.35}$$

Provided this shift still leaves natural frequencies remote from those of the excitation, all is well. But if this is not so, a rethink would be prudent.

Bendy design: part-stiff, part-flexible structures

The examples thus far aimed at the design of components that did not bend too much—that is, they met a stiffness constraint. Elasticity can be used in another way: to design components that are strong but *not* stiff, arranging that they bend easily in a certain direction. Think, for instance, of a windscreen wiper blade. The frame to which the rubber squeegees are attached must adapt to the changing profile of the windscreen as it sweeps across it. It does so by flexing, maintaining an even pressure on the blades. This is a deliberately bendy structure.

Figure 5.17 shows two ways of making a spring-loaded plunger. The one on the left with a plunger and a spring involves sliding surfaces. The one on the right has none: it uses elastic bending to both locate and guide the plunger and to give the restoring force provided by the spring in the first design.

Figure 5.17   A sliding mechanism replaced by an elastic mechanism.

Figure 5.18   Elastic or 'natural' hinges allowing flexure with no sliding parts.

Exploiting elasticity in this way has many attractions. There is no friction, no wear, no need for lubrication, no precise clearances between moving parts. And in design with polymers there is another bonus; since there are no sliding surfaces it is often possible to mold the entire device as a single unit, reducing the part-count and doing away with the need for assembly. Reducing part-count is music to the ears of production engineers: it is cost-effective.

Figure 5.18 shows examples you will recognize: the 'living hinge' used on toothpaste tubes, molded plastic boxes, clips and clothes pegs. Here the function of a rotational hinge is replaced by an elastic connecting strip, wide but thin. The width gives lateral registration; the thinness allows easy flexure. The traditional three-part hinge has been reduced to a single molding.

There is much scope for imaginative design here. Figure 5.19 shows three section shapes, each allowing one or more degrees of elastic freedom, while retaining stiffness and strength in the other directions. By incorporating these into structures, parts can be allowed to move relative to the rest in controlled ways.

How do we choose materials for such elastic mechanisms? It involves a balance between stiffness and strength. Strength keeps appearing here—that means waiting until Chapter 7 for a full answer.

Figure 5.19   The flexural degrees of freedom of three alternative section shapes. (a) Thin plates are flexible about any axis in the plane of the plate, but are otherwise stiff. (b) Ribbed plates are flexible about one in-plane axis but not in others. (c) Cruciform beams are stiff in bending but can be twisted easily.

5.6 Summary and conclusions

Adequate stiffness is central to the design of structures that are *deflection limited*. The wingspar of an aircraft is an example: too much deflection compromises aerodynamic performance. Sports equipment is another: the feel of golf clubs, tennis rackets, skis and snowboards has much to do with stiffness. The sudden buckling of a drinking straw when you bend it is a stiffness-related problem—one that occurs, more disastrously, in larger structures. And when, as the turbine of an aircraft revs up, it passes through a speed at which vibration suddenly peaks, the cause is resonance linked to the stiffness of the turbine blades. Stiffness is influenced by the size and shape of the cross-section and the material of which it is made. The property that matters here is the elastic modulus E.

In selecting materials, adequate stiffness is frequently a *constraint*. This chapter explained how to meet it for various modes of loading and for differing *objectives*: minimizing mass, or volume, or cost. Simple modeling delivers expressions for the objective: for the mass, or for the material cost. These expressions contain material properties, either singly or in combination. It is these that we call the *material index*. Material indices measure the excellence of a material in a given application. They are used to rank materials that meet the other constraints.

Stiffness is useful, but lack of stiffness can be useful too. The ability to bend or twist allows *elastic mechanisms*: single components that behave as if they had moving parts with bearings. Elastic mechanisms have limitations, but—where practical—they require no assembly, they have no maintenance requirements and they are cheap.

The chapter ended by illustrating how indices are plotted onto material property charts to find the best selection. The method is a general one that we apply in later chapters to strength thermal, electrical, magnetic and optical properties.

5.7 Further reading

Ashby, M.F. (2005) *Materials Selection in Mechanical Design*, 3rd edition, Butterworth-Heinemann, Oxford, UK, Chapter 4. ISBN 0-7506-6168-2. *(A more advanced text that develops the ideas presented here, including a much fuller discussion of shape factors and an expanded catalog of simple solutions to standard problems.)*

Gere, J.M. (2006) *Mechanics of Materials*, 6th edition, Thompson Publishing, Toronto, Canada. ISBN 0-534-41793-0. *(An intermediate level text on statics of structures by one of the fathers of the field; his books with Timoshenko introduced an entire generation to the subject.)*

Hosford, W.F. (2005) *Mechanical Behavior of Materials*, Cambridge University Press, Cambridge, UK. ISBN 0-521-84670-6. *(A text that nicely links stress–strain behavior to the micromechanics of materials.)*

Jenkins, C.H.M. and Khanna, S.K. (2006) *Mechanics of Materials*, Elsevier Academic, Boston, MA, USA. ISBN 0-12-383852-5. *(A simple introduction to mechanics, emphasizing design.)*

Riley, W.F., Sturges, L.D. and Morris, D.H. (2003) *Statics and Mechanics of Materials*, 2nd edition, McGraw-Hill, Hoboken, NJ, USA. ISBN 0-471-43446-9. *(An intermediate level text on the stress, strain and the relationships between them for many modes of loading. No discussion of micromechanics—response of materials to stress at the microscopic level.)*

Vable, M. (2002) *Mechanics of Materials*, Oxford University Press, Oxford, UK. ISBN 0-19-513337-4. *(An introduction to stress–strain relations, but without discussion of the micromechanics of materials.)*

Young W.C. (1989) *Roark's Formulas for Stress and Strain*, 6th edition, McGraw-Hill, New York, USA. ISBN 0-07-100373-8. *(This is the 'Yellow Pages' of formulae for elastic problems—if the solution is not here, it doesn't exist.)*

5.8 Exercises

Exercise E5.1	Distinguish between tension, torsion, bending and buckling.
Exercise E5.2	What is meant by a *material index*?
Exercise E5.3	Plot the index for a light, stiff panel on a copy of the modulus–density chart, positioning the line such that six materials are left above it. What classes do they belong to?

Exercise E5.4   The objective in selecting a material for a panel of given in-plane dimensions for the casing of a portable computer is that of minimizing the panel thickness h while meeting a constraint on bending stiffness, S^*. What is the appropriate material index?

Exercise E5.5   Derive the material index for a torsion bar with a solid circular section. The length L and the stiffness S^* are specified, and the torsion bar is to be as light as possible. Follow the steps used in the text for the beam, but replace the bending stiffness $S^* = F/\delta$ by the torsional stiffness $S^* = T/(\theta/L)$ (equation (5.8)), using the expression for K given in Figure 5.2.

Exercise E5.6   The speed of longitudinal waves in a material is proportional to $\sqrt{E/\rho}$. Plot contours of this quantity onto a copy of an E–ρ chart allowing you to read off approximate values for any material on the chart. Which metals have about the same sound velocity as steel? Does sound move faster in titanium or glass?

Exercise E5.7   A material is required for a cheap column with a solid circular cross-section that must support a load F_{crit} without buckling. It is to have a height L. Write down an equation for the material cost of the column in terms of its dimensions, the price per kg of the material, C_m, and the material density ρ. The cross-section area A is a free variable—eliminate it by using the constraint that the buckling load must not be less than F_{crit} (equation (5.9)). Hence read off the index for finding the cheapest tie. Plot the index on a copy of the appropriate chart and identify three possible candidates.

Exercise E5.8   Devise an elastic mechanism that, when compressed, shears in a direction at right angles to the axis of compression.

Exercise E5.9   Universal joints usually have sliding bearings. Devise a universal joint that could be molded as a single elastic unit, using a polymer.

5.9 Exploring design with CES (use Level 2, Materials, throughout)

Exercise E5.10   Use a 'Limit' stage to find materials with modulus $E > 180\,\text{GPa}$ and price $C_\text{m} < 3$ \$/kg.

Exercise E5.11   Use a 'Limit' stage to find materials with modulus $E > 2\,\text{GPa}$, density $\rho < 1000\,\text{kg/m}^3$ and Price < 3/kg.

Exercise E5.12   Make a bar chart of modulus, E. Add a tree stage to limit the selection to polymers alone. Which three polymers have the highest modulus?

Exercise E5.13   Make a chart showing modulus E and density ρ. Apply a selection line of slope 1, corresponding to the index E/ρ positioning the line such that six materials are left above it. Which are they and what families do they belong to?

Exercise E5.14   A material is required for a tensile tie to link the front and back walls of a barn to stabilize both. It must meet a constraint on stiffness and be as cheap as possible. To be safe the material of the tie must have a fracture toughness $K_{1c} > 18\,\text{MPa.m}^{1/2}$ (defined in Chapter 8). The relevant index is

$$M = \frac{E}{C_\text{m}\rho}$$

Construct a chart of E plotted against $C_\text{m}\rho$. Add the constraint of adequate fracture toughness, meaning $K_{1c} > 18\,\text{MPa.m}^{1/2}$, using a 'Limit' stage. Then plot an appropriate selection line on the chart and report the three materials that are the best choices for the tie.

5.10 Exploring the science with CES Elements

Exercise E5.15   There is nothing that we can do to change the modulus or the density of the building blocks of all materials: the elements. We have to live with the ones we have got. Make a chart of modulus E plotted against the atomic number A_n to explore the modulus across the Periodic Table. (Use a linear scale for A_n. To do so, change the default log scale to linear by double-clicking on the axis name to reveal the axis-choice dialog box and choose 'Linear'.) Which element has the highest modulus? Which has the lowest?

Exercise E5.16   Repeat Exercise E5.15, exploring instead the density ρ. Which solid element has the lowest density? Which has the highest?

Exercise E5.17   Make a chart of the sound velocity $(E/\rho)^{1/2}$, for the elements. To do so, construct the quantity $(E/\rho)^{1/2}$ on the y-axis using the 'Advanced' facility in the axis-choice dialog box, and plot it against atomic number A_n. Use a linear scale for A_n as explained in Exercise E5.15. (Multiply E by 10^9 to give the velocity in m/s.)

Chapter 6
Beyond elasticity: plasticity, yielding and ductility

Dislocations in the intermetallic compound, Ni_3Al. (Image courtesy of C. Rentenberger and H.P. Karnthaler, Institute of Materials Physics, University of Vienna, Austria.)

6.1 Introduction and synopsis

The verb 'to yield' has two, seemingly contradictory, meanings. To yield under force is to submit to it, to surrender. To yield a profit has a different, more comfortable, connotation: to bear fruit, to be useful. The *yield strength*, when speaking of a material, is the stress beyond which it becomes plastic. The term is well chosen: yield and the plasticity that follows can be profitable—it allows metals to be shaped and it allows structures to tolerate impact and absorb energy. But the unplanned yield of the span of a bridge or of the wing-spar of an aircraft or of the forks of your bicycle spells disaster.

This chapter is about yield and plasticity. For that reason it is mainly (but not wholly) about metals: it is the plasticity of iron and steel that made them the structural materials on which the Industrial Revolution was built, enabling the engineering achievements of the likes of Telford[1] and Brunel[2]. The dominance of metals in engineering, even today, derives from their ability to be rolled, forged, drawn and stamped.

6.2 Strength, plastic work and ductility: definition and measurement

Yield properties and ductility are measured using the standard tensile tests introduced in Chapter 4, with the materials taken to failure. Figures 6.1–6.3 show the types of stress–strain behavior observed in different material classes. The *yield strength* σ_y (or *elastic limit* σ_{el})—units: MPa or MN/m^2—requires careful definition. For metals, the onset of plasticity is not always distinct so we identify σ_y with the *0.2% proof stress*—that is, the stress at which the stress–strain curve for axial loading deviates by a strain of 0.2% from the linear elastic line as shown in Figure 6.1. It is the same in tension and compression. When strained beyond the yield point, most metals *work harden*, causing the rising part of the curve, until a maximum, the *tensile strength*, is reached. This is followed in tension by non-uniform deformation (*necking*) and fracture.

For polymers, σ_y is identified as the stress at which the stress–strain curve becomes markedly nonlinear: typically, a strain of 1% (Figure 6.2). The behavior

---

[1] Thomas Telford (1757–1834), Scottish engineer, brilliant proponent of the suspension bridge at a time when its safety was a matter of debate. Telford may himself have had doubts—he was given to lengthy prayer on the days that the suspension chains were scheduled to take the weight of the bridge. Most of his bridges, however, still stand.

[2] Isambard Kingdom Brunel (1806–1859), perhaps the greatest engineer of the Industrial Revolution (*circa* 1760–1860) in terms of design ability, personality, power of execution and sheer willingness to take risks—the *Great Eastern*, for example, was five times larger than any previous ship ever built. He took the view that 'great things are not done by those who simply count the cost'. Brunel was a short man and self-conscious about his height; he favored tall top hats to make himself look taller.

Figure 6.1 Stress–strain curve for a metal.

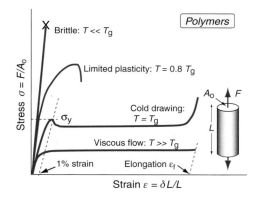

Figure 6.2 Stress–strain curve for a polymer.

Figure 6.3 Stress–strain curve for a ceramic.

beyond yield depends on the temperature relative to the glass temperature T_g. Well below T_g most polymers are brittle. As T_g is approached, plasticity becomes possible until, at about T_g, thermoplastics exhibit *cold drawing*: large plastic extension at almost constant stress during which the molecules are pulled into alignment with the direction of straining, followed by hardening and fracture when alignment is complete. At still higher temperatures, thermoplastics become viscous and can be molded; thermosets become rubbery and finally decompose.

The yield strength σ_y of a polymer–matrix composite is best defined by a set deviation from linear elastic behavior, typically 0.5%. Composites that contain fibers (and this includes natural composites like wood) are a little weaker (up to 30%) in compression than tension because the fibers buckle on a small scale.

Plastic strain, ε_{pl} is the permanent strain resulting from plasticity; thus it is the total strain ε_{tot} minus the recoverable, elastic, part:

$$\varepsilon_{pl} = \varepsilon_{tot} - \frac{\sigma}{E} \tag{6.1}$$

The *ductility* is a measure of how much plastic strain a material can tolerate. It is measured in standard tensile tests by the *elongation* ε_f (the tensile strain at break) expressed as a percentage (Figures 6.1 and 6.2). Strictly speaking, ε_f is not a material property because it depends on the sample dimensions—the values that are listed in handbooks and in the CES software are for a standard test geometry—but it remains useful as an indicator of the ability of a material to be deformed.

In Chapter 4, the area under the elastic part of the stress–strain curve was identified as the elastic energy stored per unit volume ($\sigma_y^2/2E$). Beyond the elastic limit *plastic work* is done in deforming a material permanently by yield or crushing. The increment of plastic work done for a small permanent extension or compression dL under a force F, per unit volume $V = AL_o$, is

$$dW_{pl} = \frac{F\,dL}{V} = \frac{F}{A_o}\frac{dL}{L_o} = \sigma\,d\varepsilon_{pl}$$

Thus, the plastic work per unit volume at fracture, important in energy-absorbing applications, is

$$W_{pl} = \int_0^{\varepsilon_f} \sigma\,d\varepsilon_{pl} \tag{6.2}$$

which is just the area under the stress–strain curve.

Ceramics and glasses are brittle at room temperature (Figure 6.3). They do have yield strengths, but these are so enormously high that, in tension, they are never reached: the materials fracture first. Even in compression ceramics and glasses

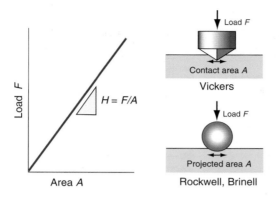

Figure 6.4 The hardness test. The Vickers test uses a diamond pyramid; the Rockwell and Brinell tests use a steel sphere.

crush before they yield. To measure their yield strengths, special tests that suppress fracture are needed. It is useful to have a practical measure of the strength of ceramics to allow their comparison with other materials. That used here is the *compressive crushing strength*, and since it is not true yield even though it is the end of the elastic part of the stress–strain curve, we call it the *elastic limit* and give it the symbol σ_{el}.

Tensile and compression tests are not always convenient: you need a large sample and the test destroys it. The hardness test (Figure 6.4) avoids these problems, although it has problems of its own. In it, a pyramidal diamond or a hardened steel ball is pressed into the surface of the material, leaving a tiny permanent indent, the size of which is measured with a microscope. The indent means that plasticity has occurred, and the resistance to it—a measure of strength—is the load F divided by the area A of the indent projected onto a plane perpendicular to the load:

$$H = \frac{F}{A} \tag{6.3}$$

The indented region is surrounded by material that has not deformed, and this constrains it so that H is larger than the yield strength σ_y; in practice it is about $3\sigma_y$. Strength, as we have seen, is measured in units of MPa, and since H is a strength it would be logical and proper to measure it in MPa too. But things are not always logical and proper, and hardness scales are among those that are not. A commonly used scale, that of *Vickers*, symbol H_v, uses units of kg/mm², with the result that

$$H_v \approx \frac{\sigma_y}{3} \tag{6.4}$$

Figure 6.5 shows conversions to other scales.

Figure 6.5 Common hardness scales compared with the yield strength.

The hardness test has the advantage of being non-destructive, so strength can be measured without destroying the component, and it requires only a tiny volume of material. But the information it provides is less accurate and less complete than the tensile test, so it is not used to provide critical design data.

6.3 The big picture: charts for yield strength

Strength can be displayed on material property charts. Two are particularly useful.

The strength–density chart

Figure 6.6 shows the yield strength σ_y or elastic limit σ_{el} plotted against density ρ. The range of strength for engineering materials, like that of the modulus, spans about six decades: from less than 0.01 MPa for foams, used in packaging and energy-absorbing systems, to 10^4 MPa for diamond, exploited in diamond tooling for machining and as the indenter of the Vickers hardness test. Members of each family again cluster together and can be enclosed in envelopes, each of which occupies a characteristic part of the chart.

Comparison with the modulus–density chart (Figure 4.7) reveals some marked differences. The modulus of a solid is a well-defined quantity with a narrow range of values. The strength is not. The strength range for a given class of metals, such as stainless steels, can span a factor of 10 or more, while the spread in stiffness

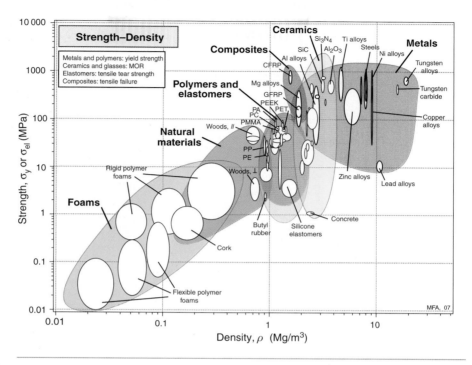

Figure 6.6   The strength–density chart.

is at most 10%. Since density varies very little (Chapter 4), the strength bubbles for metals are long and thin. The wide ranges for metals reflect the underlying physics of yielding and present designers with an opportunity for manipulation of the strength by varying composition and process history. Both are discussed later in this chapter.

Polymers cluster together with strengths between 10 and 100 MPa. The composites CFRP and GFRP have strengths that lie between those of polymers and ceramics, as one might expect since they are mixtures of the two. The analysis of the strength of composites is not as straightforward as for modulus in Chapter 4, though the same bounds (with strength replacing modulus) generally give realistic estimates.

The modulus–strength chart
Figure 6.7 shows Young's modulus, E, plotted against yield strength, σ_y or elastic limit σ_{el}. This chart allows us to examine a useful material characteristic, the *yield strain*, σ_y/E, meaning the strain at which the material ceases to be linearly elastic. On log axes, contours of constant yield strain appear as a family of straight parallel lines, as shown in Figure 6.7. Engineering polymers have large yield strains, between 0.01 and 0.1; the values for metals are at least a factor of 10 smaller. Composites and woods lie on the 0.01 contour, as good as the best metals.

Figure 6.7 The Young's modulus–strength chart. The contours show the strain at the elastic limit, σ_y/E.

Elastomers, because of their exceptionally low moduli, have values of σ_y/E in the range 1 to 10, much larger than any other class of material.

This chart has many other applications, notably in selecting materials for springs, elastic diaphragms, flexible couplings and snap-fit components. We explore these in Chapter 7.

6.4 Drilling down: the origins of strength and ductility

Perfection: the ideal strength

The bonds between atoms, like any other spring, have a breaking point. Figure 6.8 shows a stress–strain curve for a single bond. Here an atom is assumed to occupy a cube of side a_o (as was assumed in Chapter 4) so that a force F corresponds to a stress F/a_o^2. The force stretches the bond from its initial length a_o to a new length a, giving a strain $(a - a_o)/a_o$. When discussing the modulus in Chapter 4 we focused on the initial, linear part of this curve, with a slope equal to the modulus, E. Stretched further, the curve passes through a maximum and sinks to zero as

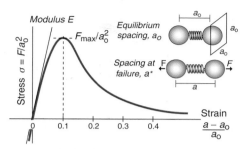

Figure 6.8 The stress–strain curve for a single atomic bond (it is assumed that each atom occupies a cube of side a_a).

the atoms lose communication. The peak is the bond strength—if you pull harder than this it will break. The same is true if you shear it rather than pull it.

The distance over which inter-atomic forces act is small—a bond is broken if it is stretched to more than about 10% of its original length. So the force needed to break a bond is roughly:

$$F \approx \frac{S\,a_o}{10} \tag{6.5}$$

where S, as before, is the bond stiffness. On this basis the *ideal strength* of a solid should therefore be roughly

$$\sigma_{ideal} \approx \frac{F_{max}}{a_o^2} = \frac{S}{10a_o} = \frac{E}{10}$$

(remembering that $E = S_o/a_o$, equation (4.17)).

or
$$\frac{\sigma_{ideal}}{E} \approx \frac{1}{10} \tag{6.6}$$

This doesn't allow for the curvature of the force–distance curve; more refined calculations give a ratio of 1/15.

Figure 6.9 shows σ_y/E for metals, polymers and ceramics. None achieve the ideal value of 1/10; most don't even come close. Why not? It's a familiar story: like most things in life, materials are imperfect.

Crystalline imperfection: defects in metals and ceramics
Crystals contain imperfections of several kinds. Figure 6.10 introduces the broad families, distinguished by their dimensionality. At the top left are *point*

Figure 6.9 The ideal strength is predicted to be about *E*/15, where *E* is Young's modulus. The figure shows σ_y/E with a shaded band at the ideal strength.

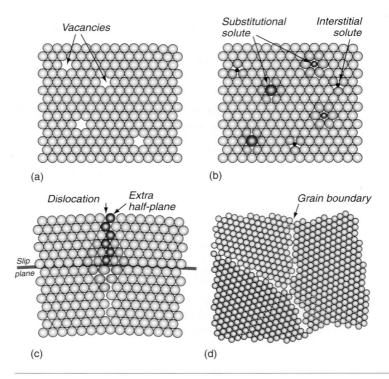

Figure 6.10 Defects in crystals. (a) Vacancies–missing atoms. (b) Foreign (solute) atom on interstitial and substitutional sites. (c) A dislocation–an extra half-plane of atoms. (d) Grain boundaries.

defects. All crystals contain *vacancies*, shown in (a): sites at which an atom is missing. They play a key role in diffusion, creep and sintering (Chapter 13), but we don't need them for the rest of this chapter because they do not influence strength. The others do.

No crystal is totally, 100%, pure and perfect. Some impurities are inherited from the process by which the material was made; more usually they are deliberately added, creating *alloys*: a material in which a second (or third or fourth) element is dissolved. 'Dissolved' sounds like salt in water, but these are solid solutions. Figure 6.10(b) shows both a *substitutional solid solution* (the dissolved atoms replace those of the host) and an *interstitial solid solution* (the dissolved atoms squeeze into the spaces or 'interstices' between the host atoms). The dissolved atoms or solute rarely have the same size as those of the host material, so they distort the surrounding lattice. The red atoms here are substitutional solute, some bigger and some smaller than those of the host; the cages of host atoms immediately surrounding them, shown green, are distorted. If the solute atoms are particularly small, they don't need to replace a host atom; instead, they dissolve *interstitially* like the black atoms in the figure, again distorting the surrounding lattice. So solute causes local distortion; this distortion is one of the reasons that alloys are stronger than pure materials, as we shall see in a moment.

Now to the key player, portrayed in Figure 6.10(c): the *dislocation*. 'Dislocated' means 'out of joint' and this is not a bad description of what is happening here. The upper part of the crystal has one more double-layer of atoms than the lower part (the double-layer is needed to get the top-to-bottom registry right). It is dislocations that make metals soft and ductile. Dislocations distort the lattice—here the green atoms are the most distorted—and because of this they have elastic energy associated with them. If they cost energy, why are they there? To grow a perfect crystal just one cubic centimeter in volume from a liquid or vapor, about 10^{23} atoms have to find their proper sites on the perfect lattice, and the chance of this happening is just too small. Even with the greatest care in assembling them, all crystals contain point defects, solute atoms and dislocations.

Most contain yet more drastic defects, among them grain boundaries. Figure 6.10(d) shows such boundaries. Here three perfect, but differently oriented, crystals meet; the individual crystals are called *grains*, the meeting surfaces are *grain boundaries*. In this sketch the atoms of the three crystals have been given different colors to distinguish them, but here they are the same atoms. In reality grain boundaries form in pure materials (when all the atoms are the same) and in alloys (when the mixture of atoms in one grain may differ in chemical composition from those of the next).

Now put all this together. The seeming perfection of the steel of a precision machine tool or of the polished case of a gold watch is an illusion: they are riddled with defects. Imagine all of the frames of Figure 6.10 superimposed and you begin to get the picture. Between them they explain diffusion, strength, ductility, electrical resistance, thermal conductivity and much more.

So defects in crystals are influential. For the rest of this section we focus on getting to know just one of them: the dislocation.

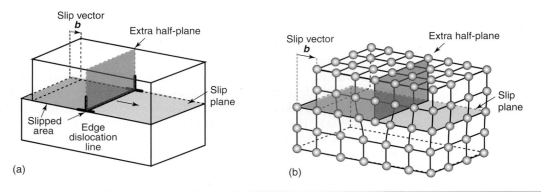

Figure 6.11   (a) Making a dislocation by cutting, slipping and rejoining bonds across a slip plane. (b) The atom configuration at an edge dislocation in a simple cubic crystal. The configurations in other crystal structures are more complex but the principle remains the same.

Dislocations and plastic flow

Recall that the strength of a perfect crystal computed from inter-atomic forces gives an 'ideal strength' around $E/15$ (where E is the modulus). In reality the strengths of engineering materials are nothing like this big; often they are barely 1% of it. This was a mystery until half way through the last century—a mere 60 years ago—when an Englishman, G.I. Taylor[3] and a Hungarian, Egon Orowan,[4] realized that a 'dislocated' crystal could deform at stresses far below the ideal. So what is a dislocation, and how does it enable deformation?

Figure 6.11(a) shows how to make a dislocation. The crystal is cut along an atomic plane up to the line shown as ⊥—⊥, the top part is slid across the bottom by one full atom spacing, and the atoms are reattached across the cut plane to give the atom configuration shown in Figure 6.11(b). There is now an extra half-plane of atoms with its lower edge along the ⊥—⊥ line, the *dislocation line*—the line separating the part of the plane that has slipped from the part that has not. This particular configuration is called an *edge* dislocation because it is formed by the edge of the extra half-plane, represented by the symbol ⊥.

[3] Geoffrey (G.I.) Taylor (1886–1975), known for his many fundamental contributions to aerodynamics, hydrodynamics and to the structure and plasticity of metals—it was he, with Egon Orowan, who realized that the ductility of metals implied the presence of dislocations. One of the greatest of contributors to theoretical mechanics and hydrodynamics of the 20th century, he was also a supremely practical man—a sailor himself, he invented (among other things) the anchor used by the Royal Navy.

[4] Egon Orowan (1901–1989), Hungarian/US physicist and metallurgist, who, with G.I. Taylor, realized that the plasticity of crystals could be understood as the motion of dislocations. In his later years he sought to apply these ideas to the movement of fault lines during earthquakes.

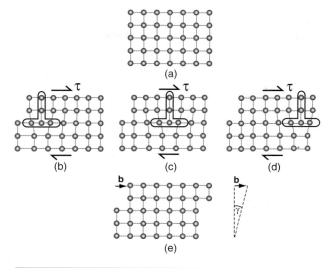

Figure 6.12   An initially perfect crystal is shown in (a). The passage of the dislocation across the slip plan, shown in the sequence (b), (c) and (d), shears the upper part of the crystal over the lower part by the slip vector **b**. When it leaves the crystal has suffered a shear strain γ.

When a dislocation moves it makes the material above the slip plane slide relative to that below, producing a shear strain. Figure 6.12 shows how this happens. At the top is a perfect crystal. In the central row a dislocation enters from the left, sweeps through the crystal and exits on the right. By the end of the process the upper part has slipped by **b**, the slip vector (or Burger's vector) relative to the part below. The result is the shear strain γ shown at the bottom.

There is another way to make a dislocation in a crystal. After making the cut in Figure 6.11(a), the upper part of the crystal can be displaced *parallel* to the edge of the cut rather than *normal* to it, as in Figure 6.13. That too creates a dislocation, but one with a different configuration of atoms along its line—one more like a corkscrew than like a squashed worm—and for this reason it is called a *screw dislocation*. We don't need the details of its structure; it is enough to know that its properties are like those of an edge dislocation except that when it sweeps through a crystal (moving normal to its line), the lattice is displaced parallel to the dislocation line, not normal to it. All dislocations are either edge or screw or *mixed*, meaning that they are made up of little steps of edge and screw. The line of a mixed dislocation can be curved but every part of it has the same slip vector **b** because the dislocation line is just the boundary of a plane on which a fixed displacement **b** has occurred.

It is far easier to move a dislocation through a crystal, breaking and remaking bonds only along its line as it moves, than it is to simultaneously break all the bonds in the plane before remaking them. It is like moving a heavy carpet by pushing a fold across it rather than sliding the whole thing at one go. In real

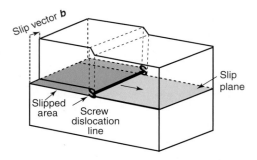

Figure 6.13   A screw dislocation. The slip vector **b** is parallel to the dislocation line **S**—**S**.

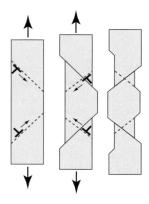

Figure 6.14   Dislocation motion causes extension

crystals it is easier to make and move dislocations on some planes than on others. The preferred planes are called *slip planes* and the preferred directions of slip in these planes are called *slip directions*. A slip plane is shown in gray and a slip direction as an arrow on the fcc and bcc unit cells of Figure 4.11.

Slip displacements are tiny—one dislocation produces a displacement of about 10^{-10} m. But if large numbers of dislocations traverse a crystal, moving on many different planes, the shape of a material changes at the macroscopic length scale. Figure 6.14 shows just two dislocations traversing a sample loaded in tension. The slip steps (here very exaggerated) cause the sample to get a bit thinner and longer. Repeating this millions of times on many planes gives the large plastic extensions observed in practice. Since none of this changes the average atomic spacing, the volume remains unchanged.

Why does a shear stress make a dislocation move?
Crystals resist the motion of dislocations with a friction-like resistance f per unit length—we will examine its origins in a moment. For yielding to take place the external stress must overcome the resistance f.

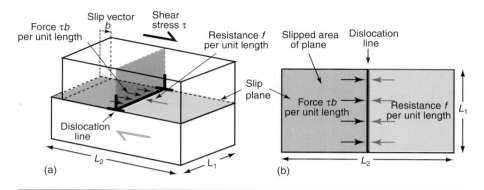

Figure 6.15 The force on a dislocation. (a) Perspective view. (b) Plan view of slip plane.

Imagine that one dislocation moves right across a slip plane, traveling the distance L_2, as in Figure 6.15. In doing so, it shifts the upper half of the crystal by a distance b relative to the lower half. The shear stress τ acts on an area $L_1 L_2$, giving a shear force $F_s = \tau L_1 L_2$ on the surface of the block. If the displacement parallel to the block is b, the force does work

$$W = \tau L_1 L_2\, b \qquad (6.7)$$

This work is done against the resistance f per unit length, or $f L_1$ on the length L_1, and it does so over a displacement L_2 (because the dislocation line moves this far against f), giving a total work against f of $f L_1 L_2$. Equating this to the work W done by the applied stress τ gives

$$\tau b = f \qquad (6.8)$$

This result holds for any dislocation—edge, screw or mixed. So, provided the shear stress τ exceeds the value f/b it will make dislocations move and cause the crystal to shear.

Line tension.
The atoms near the core of a dislocation are displaced from their proper positions, as shown by green atoms back in Figure 6.10(c), and thus they have higher potential energy. To keep the potential energy of the crystal as low as possible, the dislocation tries to be as short as possible—it behaves as if it had a *line tension*, T, like an elastic band. The tension can be calculated but it needs advanced elasticity theory to do it (the books listed under 'Further reading' give the analysis). We just need the answer. It is that the line tension, an energy per unit length (just as a surface tension is an energy per unit area), is

$$T \approx \frac{1}{2} E b^2 \qquad (6.9)$$

where E, as always, is Young's modulus. The line tension has an important bearing on the way in which dislocations interact with obstacles, as we shall see in a moment.

The lattice resistance

Where does the resistance to slip, f, come from? There are several contributions. Consider first the *lattice resistance*, f_i: the intrinsic resistance of the crystal structure to plastic shear. Plastic shear, as we have seen, involves the motion of dislocations. Pure metals are soft because the non-localized metallic bond does little to obstruct dislocation motion, whereas ceramics are hard because their more localized covalent and ionic bonds (which must be broken and reformed when the structure is sheared) lock the dislocations in place. When the lattice resistance is high, as in ceramics, further hardening is superfluous—the problem becomes that of suppressing fracture. On the other hand, when the lattice resistance f_i is low, as in metals, the material can be strengthened by introducing obstacles to slip. This is done by adding alloying elements to give *solid solution hardening* (f_{ss}), precipitates or dispersed particles giving *precipitation hardening* (f_{ppt}), other dislocations giving what is called *work hardening* (f_{wh}) or grain boundaries introducing *grain-size hardening* (f_{gb}). These techniques for manipulating strength are central to alloy design. We look at them more closely in the next section.

Plastic flow in polymers

At low temperatures, meaning below about $0.75T_g$, polymers are brittle. Above this temperature they become plastic. When pulled in tension, the chains slide over each other, unraveling, so that they become aligned with the direction of stretch, as in Figure 6.16(a), a process called *drawing*. It is harder to start drawing than to keep it going, so the zone where it starts draws down completely before propagating further along the sample, leading to profiles like that shown in the figure. The drawn material is stronger and stiffer than before, by a factor of about 8, giving drawn polymers exceptional properties, but because you can only draw fibers or sheet (by pulling in two directions at once) the geometries are limited.

Many polymers, among them PE, PP and nylon, draw at room temperature. Others with higher glass temperatures, such as PMMA, do not, although they draw well at higher temperatures. At room temperature they *craze*. Small crack-shaped regions within the polymer draw down. Because the crack has a larger volume than the polymer that was there to start with, the drawn material ends up as ligaments that link the craze surfaces, as in Figure 6.16(b). Crazes scatter light, so their presence causes whitening, easily visible when cheap plastic articles are bent. If stretching is continued, one or more crazes develop into proper cracks, and the sample fractures.

When crazing limits ductility in tension, large plastic strains may still be possible in compression by shear banding (Figure 6.16(c)). Within each band, shear takes place with much the same consequences for the shape of the sample as shear by dislocation motion. Continued compression causes the number of shear bands to increase, giving increased overall strain.

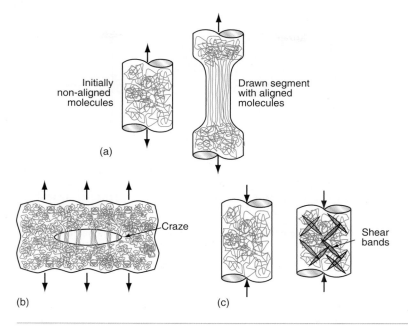

Initially non-aligned molecules

Drawn segment with aligned molecules

(a)

Craze

Shear bands

(b)

(c)

Figure 6.16 (a) Cold drawing—one of the mechanisms of deformation of thermoplastics. (b) Crazing—local drawing across a crack. (c) Shear banding.

6.5 Manipulating strength

Strengthening metals

The way to make crystalline materials stronger is to make it harder for dislocations to move. As we have seen, dislocations move in a pure crystal when the force τb per unit length exceeds the lattice resistance f_i. There is little we can do to change this—it is an intrinsic property like the modulus E. Other strengthening mechanisms add to it, and here there is scope for manipulation. Figure 6.17 introduces them. It shows the view of a slip plane from the perspective of an advancing dislocation: each strengthening mechanism presents a new obstacle course. In the perfect lattice shown in (a) the only resistance is the intrinsic strength of the crystal; solution hardening, shown in (b), introduces atom-size obstacles to motion; precipitation hardening, shown in (c), presents larger obstacles; and in work hardening, shown in (d), the slip plane becomes stepped and threaded with 'forest' dislocations.

Obstacles to dislocation motion increase the resistance f and thus the strength. To calculate their contribution to f, there are just two things we need to know about them: their spacing and their strength. Spacing means the distance L between them in the slip plane. The number of obstacles touching unit length of dislocation line is then

$$N_L = \frac{1}{L}$$

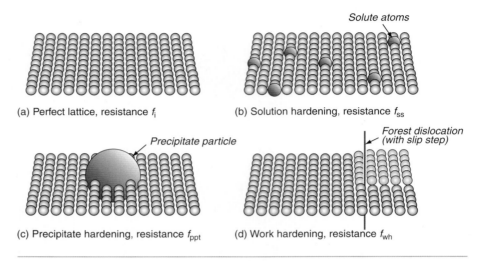

(a) Perfect lattice, resistance f_i

(b) Solution hardening, resistance f_{ss}

Solute atoms

Precipitate particle

Forest dislocation
(with slip step)

(c) Precipitate hardening, resistance f_{ppt}

(d) Work hardening, resistance f_{wh}

Figure 6.17   A 'dislocation-eye' view of the slip plane across which it must move.

Each individual obstacle exerts a *pinning force p* on the dislocation line—a resisting force per unit length of dislocation—so the contribution of the obstacles to the resistance f is

$$f = \frac{p}{L}$$

Thus, the added contribution to the shear stress τ needed to make the dislocation move is (from equation (6.8))

$$\Delta\tau = \frac{p}{b\,L} \tag{6.10}$$

The pinning is an elastic effect—it derives from the fact that both the dislocation and the obstacle distort the lattice elastically even though, when the dislocation moves, it produces plastic deformation. Because of this p, for any given obstacle in any given material, scales as $E\,b^2$, which has the units of force. The shear stress τ needed to force the dislocation through the field of obstacles then has the form

$$\tau = \alpha\,\frac{E\,b}{L} \tag{6.11}$$

where α is a dimensionless constant characterizing the obstacle strength.

Armed with this background we can explain strengthening mechanisms. We start with solid solutions.

Figure 6.18   (a) Solution hardening. (b) Precipitation or dispersion hardening. (c) Forest hardening (work hardening).

Solution hardening

Solid solution hardening is strengthening by deliberate additions of impurities or, more properly said, by *alloying* (Figure 6.18(a)). The addition of zinc to copper makes the alloy brass—copper dissolves up to 30% zinc. The zinc atoms replace copper atoms to form a *random substitutional solid solution*. The zinc atoms are bigger that those of copper and, in squeezing into the copper lattice, they distort it. This roughens the slip plane, so to speak, making it harder for dislocations to move, thereby adding an additional resistance f_{ss}, opposing dislocation motion. The figure illustrates that the concentration of solute, expressed as an atom fraction, is on average:

$$c = \frac{b^2}{L^2}$$

where L is the spacing of obstacles in the slip plane and b is the atom size. Thus,

$$L = \frac{b}{c^{1/2}}$$

Plugging this into equation (6.11) relates the contribution of solid solution to the shear stress required to move the dislocation:

$$\tau_{ss} = \alpha \, E \, c^{1/2} \tag{6.12}$$

τ_{ss} increases as the square root of the solute concentration. Brass, bronze and stainless steels, and many other metallic alloys, derive their strength in this way. They differ only in the extent to which the solute distorts the crystal, described by the constant α.

Dispersion and precipitate strengthening

A more effective way to impede dislocations is to *disperse* small, strong particles in their path. One way to make such a microstructure is to disperse small solid particles of a high melting point compound into a liquid metal, and to cast it to shape, trapping the particles in place—it is the way that metal–matrix composites such as Al–SiC are made. An alternative is to form the particles *in situ* by a *precipitation* process. If a solute (copper, say) is dissolved in a metal (aluminum, for instance) at high temperature when both are molten, and the alloy is solidified and cooled to room temperature, the solute precipitates as small particles, much as salt will crystallize from a saturated solution when it is cooled. An alloy of aluminum containing 4% copper, treated in this way, gives very small, closely spaced precipitates of the hard compound $CuAl_2$. Copper alloyed with a little beryllium, similarly treated, gives precipitates of the compound $CuBe$. Most steels are strengthened by precipitates of carbides, obtained in this way. The precipitates give a large contribution to f.

Figure 6.18(b) shows how particles obstruct dislocation motion. If the particles are too strong for the dislocation to slice through them, the force $\tau \, b$ pushes the dislocation between them, bending it to a tighter and tighter radius against its line tension (equation (6.9)). The radius is at a minimum when it reaches half the particle spacing, L; after that it can expand under lower stress. It is a bit like blowing up a bicycle inner tube when the outer tire has a hole in it: once you reach

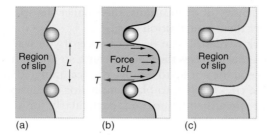

Figure 6.19 Successive positions of a dislocation as it bypasses particles that obstruct its motion. The critical configuration is that with the tightest curvature, shown in (b).

the pressure that balloons the inner tube through the hole, the balloon needs a smaller pressure to get bigger still. The *critical configuration* is the semicircular one: here the total force $\tau\, b\, L$ on one segment of length L is just balanced by the force $2T$ due to the line tension (equation (6.10)), acting on either side of the bulge, as in Figure 6.19. The dislocation escapes when

$$\tau_{\mathrm{ppt}} = \frac{2T}{b\, L} \approx \frac{E\, b}{L} \tag{6.13}$$

The obstacles thus exert a resistance of $f_{\mathrm{ppt}} = 2T/L$. Precipitation hardening is an effective way to increasing strength: precipitate-hardened aluminum alloys can be 15 times stronger than pure aluminum.

Work hardening
The rising part of the stress–strain curve of Figure 6.1 is caused by *work hardening*: it is caused by the accumulation of dislocations generated by plastic deformation. The *dislocation density*, ρ_{d}, is defined as the length of dislocation line per unit volume (m/m^3). Even in an annealed soft metal, the dislocation density is around 10^{10} m/m^3, meaning that a 1 cm cube (the size of a cube of sugar) contains about 10 km of dislocation line. When metals are deformed, dislocations multiply, causing their density to grow to as much as 10^{17} m/m^3 or more—100 million km per cubic centimeter. A moving dislocation now finds that its slip plane is penetrated by a forest of intersecting dislocations with an average spacing $L = \rho_{\mathrm{d}}^{-1/2}$ (since ρ_{d} is a number per unit area). Figure 6.18(c) suggests the picture. If a moving dislocation advances, it shears the material above the slip plane relative to that below, and that creates a little step called a *jog* in each forest dislocation. The jogs have potential energy—they are tiny segments of dislocation of length b—with the result that each exerts a pinning force $p = E\, b^2/2$ on the moving dislocation. Assembling these results into equation (6.10) gives

$$\tau_{\mathrm{wh}} \approx \frac{E\, b}{2} \sqrt{\rho_{\mathrm{d}}} \tag{6.14}$$

The greater the density of dislocations, the smaller the spacing between them, and so the greater their contribution to τ_{wh}.

All metals work harden. It can be a nuisance: if you want to roll thin sheet, work hardening quickly raises the yield strength so much that you have to stop and _anneal_ the metal (heat it up to remove the accumulated dislocations) before you can go on—a trick known to blacksmiths for centuries. But it is also useful: it is a potent strengthening method, particularly for alloys that cannot be heat-treated to give precipitation hardening.

Grain boundary hardening

Almost all metals are polycrystalline, made up of tiny, randomly oriented, crystals, or _grains_, meeting at grain boundaries like those of Figure 6.10(d). The grain size, D, is typically 10–100 μm. These boundaries obstruct dislocation motion. A dislocation in one grain—call it grain 1—can't just slide into the next (grain 2) because the slip planes don't line up. Instead, new dislocations have to nucleate in grain 2 with slip vectors that, if superimposed, match that of the dislocation in grain 1 so that the displacements match at the boundary. This gives another contribution to strength, τ_{gb}, that is found to scale as $D^{-1/2}$, giving

$$\tau_{gb} = \frac{k_p}{\sqrt{D}} \tag{6.15}$$

where k_p is called the Petch constant, after the man who first measured it. For normal grain sizes τ_{gb} is small and not a significant source of strength, but for materials that are microcrystalline ($D < 1$ μm) or nanocrystalline (D approaching 1 nm) it becomes significant.

Relationship between dislocation strength and yield strength

To a first approximation the strengthening mechanisms add up, giving a shear yield strength, τ_y, of

$$\tau_y = \tau_i + \tau_{ss} + \tau_{ppt} + \tau_{wh} + \tau_{gb} \tag{6.16}$$

Strong materials either have a high intrinsic strength, τ_i (like diamond), or they rely on the superposition of solid solution strengthening τ_{ss}, precipitates τ_{ppt} and work hardening τ_{wh} (like high-tensile steels). Nanocrystalline solids exploit, in addition, the contribution of τ_{gb}.

Before we can use this information, one problem remains: we have calculated the yield strength of one crystal, loaded in shear. We want the yield strength of a _polycrystalline material_ in _tension_. To link them there are two simple steps. First, a uniform tensile stress σ creates a shear stress on planes that lie at an angle to the tensile axis; dislocations will first move on the slip plane on which this shear stress is greatest. Figure 6.20 shows how this is calculated. A tensile force F acting on a rod of cross-section A, if resolved parallel to a plane with a normal

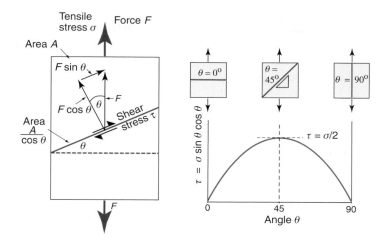

Figure 6.20   The resolution of stress. A tensile stress σ gives a maximum shear stress $\tau = \sigma/2$ on a plane at 45° to the tensile axis.

that lies at an angle θ to the axis of tension, gives a force $F \sin \theta$ in the plane. The area of this plane is $A/\cos \theta$, so the shear stress is

$$\tau = \frac{F \sin \theta}{A/\cos \theta} = \sigma \sin \theta \cos \theta$$

where $\sigma = F/A$ is the tensile stress. The value of τ is plotted against θ in the figure. The maximum lies at an angle of 45°, when $\tau = \sigma/2$.

Second, when this shear stress acts on an aggregate of crystals, some crystals will have their slip planes oriented favorably with respect to the shear stress, others will not. This randomness of orientation jacks up the strength by a further factor of 1.5 (called the Taylor factor — see the footnote on p. 122). Combining these results, the tensile stress to cause yielding of a sample that has many grains is approximately three times the shear strength of a single crystal:

$$\sigma_y \approx 3 \, \tau_y$$

Thus, the superposition of strengthening mechanisms in equation (6.16) applies equally to the yield strength, σ_y.

Strength and ductility of alloys

Of all the properties that materials scientists and engineers have sought to manipulate, the strength of metals and alloys is probably the most explored. It is easy to see why—Table 6.1 gives a small selection of the applications of metals and their alloys. Their importance in engineering design is enormous. The hardening mechanisms are often used in combination. This is illustrated graphically

Table 6.1   Metal alloys with typical applications, indicating the strengthening mechanisms used

Alloy	Typical uses	Solution hardening	Precipitation hardening	Work hardening
Pure Al	Kitchen foil			✓✓✓
Pure Cu	Wire			✓✓✓
Cast Al, Mg	Automotive parts	✓✓✓	✓	
Bronze (Cu–Sn), Brass (Cu–Zn)	Marine components	✓✓✓	✓	✓✓
Non-heat-treatable wrought Al	Ships, cans, structures	✓✓✓		✓✓✓
Heat-treatable wrought Al	Aircraft, structures	✓	✓✓✓	✓
Low-carbon steels	Car bodies, structures, ships, cans	✓✓✓		✓✓✓
Low alloy steels	Automotive parts, tools	✓	✓✓✓	✓
Stainless steels	Pressure vessels	✓✓✓	✓	✓✓✓
Cast Ni alloys	Jet engine turbines	✓✓✓	✓✓✓	

Symbols: ✓✓✓ = Routinely used. ✓ = Sometimes used.

for copper alloys in Figure 6.21. Good things, however, have to be paid for. Here the payment for increased strength is, almost always, loss of ductility so the elongation ε_f is reduced. The material is stronger but it cannot be deformed as much without fracture.

Strengthening polymers

In non-crystalline solids the dislocation is not a helpful concept. We think instead of some unit step of the flow process: the relative slippage of two segments of a polymer chain, or the shear of a small molecular cluster in a glass network. Their strength has the same origin as that underlying the lattice resistance: if the unit step involves breaking strong bonds (as in an inorganic glass), the materials will be strong and brittle, as ceramics are. If it only involves the rupture of weak bonds (the Van der Waals bonds in polymers, for example), it will be weak. Polymers too must therefore be strengthened by impeding the slippage of segments of their molecular chains. This is achieved by *blending*, by *drawing*, by *cross-linking* and by *reinforcement* with particles, fibers or fabrics.

A blend is a mixture of two polymers, stirred together in a sort of industrial food-mixer. The strength and modulus of a blend are just the average of those of the components, weighted by volume fraction (a rule of mixtures again). If one of these is a low molecular weight hydrocarbon, it acts as a plasticizer, reducing the modulus and giving the blend a leather-like flexibility.

Drawing is the deliberate use of the molecule-aligning effect of stretching, like that sketched in Figure 6.16(a), to greatly increase stiffness and strength in the direction of stretch. Fishing line is drawn nylon, Mylar film is a polyester with molecules aligned parallel to the film, and geotextiles, used to restrain earth banks, are made from drawn polyethylene.

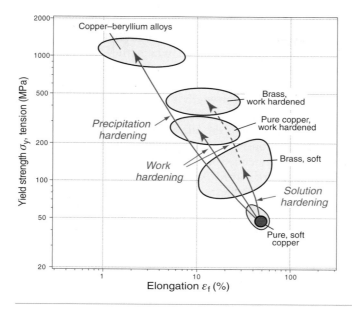

Figure 6.21 Strengthening mechanisms and the consequent drop in ductility, here shown for copper alloys. The mechanisms are frequently combined. The greater the strength, the lower the ductility (the elongation to fracture, ε_f).

Cross-linking, sketched in Figures 4.18 and 4.19, creates strong bonds between molecules that were previously linked by weak Van der Waals forces. Vulcanized rubber is rubber that has been cross-linked, and the superior strength of epoxies derives from cross-linking.

Reinforcement is possible with particles of cheap fillers—sand, talc or wood dust. Far more effective is reinforcement with fibers—usually glass or carbon—either continuous or chopped, as explained in Chapter 4.

6.6 Summary and conclusions

Load-bearing structures require materials with reliable, reproducible strength. There is more than one measure of strength. *Elastic design* requires that no part of the structure suffers plastic deformation, and this means that the stresses in it must nowhere exceed the *yield strength*, σ_y, of ductile materials or the *elastic limit* of those that are not ductile. *Plastic design*, by contrast, allows some parts of the structure to deform plastically so long as the structure as a whole does not collapse. Then two further properties become relevant: the *ductility*, ε_f, and the tensile strength, σ_{ts}, which are the maximum strain and the maximum stress the material can tolerate before fracture. The tensile strength is generally larger than the yield strength because of *work hardening*.

Charts plotting strength, like those plotting modulus, show that material families occupy different areas of material property space, depending on the strengthening mechanisms on which they rely. Crystal defects—particularly *dislocations*—are central to the understanding of these. It is the motion of dislocations that gives plastic flow in crystalline solids, giving them unexpectedly low strengths. When strength is needed it has to be provided by the strengthening mechanism that impedes dislocation motion.

First among these is the *lattice resistance*—the intrinsic resistance of the crystal to dislocation motion. Others can be deliberately introduced by alloying and heat treatment. *Solid solution hardening, dispersion* and *precipitation hardening, work hardening* and *grain boundary hardening* add to the lattice resistance. The strongest materials combine them all.

Non-crystalline solids—particularly polymers—deform in a less organized way by the pulling of the tangled polymer chains into alignment with the direction of deformation. This leads to *cold drawing* with substantial plastic strain and, at lower temperatures, to *crazing*. The stress required to do this is significant, giving polymers a considerable intrinsic strength. This can be enhanced by blending, cross-linking and reinforcement with particles or fibers to give the engineering polymers we use today.

6.7 Further reading

Ashby, M.F. and Jones, D.R.H. (2006) *Engineering materials*, Volumes I and II, Butterworth-Heinemann, Oxford UK. ISBN 7-7506-6380-4 and ISBN 0-7506-6381-2. *(An introduction to mechanical properties and processing of materials.)*

Cottrell, A.H. (1953) *Dislocations and Plastic Flow in Crystals*, Oxford University Press, Oxford, UK. *(Long out of print but worth a search: the book that first presented a coherent picture of the mechanisms of plastic flow and hardening.)*

Friedel, J. (1964) *Dislocations*, Addison-Wesley, Reading, MA, USA. Library of Congress No. 65-21133. *(A book that, with that of Cottrell, first established the theory of dislocations.)*

Hertzberg, R.W. (1989) *Deformation and Fracture of Engineering Materials*, 3rd edition, Wiley, New York, USA. ISBN 0-471-61722-9. *(A readable and detailed coverage of deformation, fracture and fatigue.)*

Hull, D. and Bacon, D.J. (2001) *Introduction to Dislocations*, 4th edition, Butterworth-Heinemann, Oxford, UK. ISBN 0-750-064681-0. *(An introduction to dislocation mechanics.)*

Hull, D. and Clyne, T.W. (1996) *An Introduction to Composite Materials*, 2nd edition, Cambridge University Press, Cambridge, UK. ISBN 0-521-38855-4. *(A concise and readable introduction to composites that takes an approach that minimizes the mathematics and maximizes the physical understanding.)*

Young, R.J. (1981) *Introduction to Polymers*, Chapman & Hall, London, UK. ISBN 0-412-22180-2. *(A good starting point for more information on the chemistry, structure and properties of polymers.)*

6.8 Exercises

Exercise E6.1	Sketch a stress–strain curve for a typical metal. Mark on it the yield strength σ_y, the tensile strength σ_{ts} and the ductility ε_f. Indicate on it the work done per unit volume in deforming the material up to a strain of $\varepsilon < \varepsilon_f$ (pick your own strain ε).
Exercise E6.2	What is meant by the ideal strength of a solid? Which material class most closely approaches it?
Exercise E6.3	Use the yield strength–density chart or the yield strength–modulus chart (Figures 6.6 and 6.7) to find:

- The metal with the lowest strength.
- The approximate range of strength of the composite GFRP.
- Whether there are any polymers that are stronger than wood measured parallel to the grain.
- How the strength of GFRP compares with that of wood.
- Whether elastomers, that have moduli that are far lower than polymers, are also far lower in strength.

Exercise E6.4	The lattice resistance of copper, like that of most FCC metals, is small. When 10% of nickel is dissolved in copper to make a solid solution, the strength of the alloy is 150 MPa. What would you expect the strength of an alloy with 20% nickel to be?
Exercise E6.5	A metal–matrix composite consists of aluminum containing hard particles of silicon carbide (SiC) with a mean spacing of $3\,\mu m$. The composite has a strength of 180 MPa. If a new grade of the composite with a particle spacing of $2\,\mu m$ were developed, what would you expect its strength to be?
Exercise E6.6	Nanocrystalline materials have grain sizes in the range 0.01–$0.1\,\mu m$. If the contribution of grain boundary strengthening in an alloy with grains of $0.1\,\mu m$ is 20 MPa, what would you expect it to be if the grain size were reduced to $0.01\,\mu m$?
Exercise E6.7	Polycarbonate, PC (yield strength 70 MPa) is blended with polyester (PET; yield strength 50 MPa) in the ratio 30%/70%. If the strength of blends follows a rule of mixtures, what would you expect the yield strength of this blend to be?

6.9 Exploring design with CES (use Level 2 Materials unless otherwise suggested)

Exercise E6.8	Find, by opening the records, the yield strengths of copper, brass (a solid solution of zinc in copper) and bronze (a solid solution of tin in copper). Report the mean values of the ranges that appear in the records. What explains the range within each record, since the composition is not a variable? What explains the differences in the mean values, when composition is a variable?
Exercise E6.9	Use a 'Limit' stage to find materials with a yield strength σ_y greater than 100 MPa and density ρ less than 2000 kg/m^3. List the results.
Exercise E6.10	Add two further constraints to the selection of the previous exercise. Require now that the material price be less than \$5/kg and the elongation be greater than 5%.
Exercise E6.11	Use the CES Level 3 database to select Polypropylene and its blended, filled and reinforced grades. To do so, open CES Edu Level 3, apply a 'Tree' stage selecting Polymers—Thermoplastics—Polypropylene (folder). Make a chart with Young's modulus E on the x-axis and yield strength σ_y on the y-axis. Label the records on the chart by clicking on them. Explain, as far as you can, the trends you see.
Exercise E6.12	Apply the same procedure as that of the last exercise to explore copper and its alloys. Again, use your current knowledge to comment on the origins of the trends.

6.10 Exploring the science with CES Elements

Exercise E6.13	The elastic (potential) energy per unit length of a dislocation is $0.5\, E\, b^2$ J/m. Make a bar chart of the energy stored in the form of dislocations, for a dislocation density of 10^{14} m/m^3. Assume that the magnitude of Burger's vector, b, is the same as the atomic diameter. (You will need to use the 'Advanced' facility in the axis-choice dialog box to make the function.) How do the energies compare with the cohesive energy, typically 5×10^4 MJ/m^3?
Exercise E6.14	Work hardening causes dislocations to be stored. Dislocations disrupt the crystal and have potential energy associated with them. It has been suggested that sufficient work hardening might disrupt the crystal so much that it

becomes amorphous. To do this, the energy associated with the dislocations would have to be about equal to the heat of fusion, since this is the difference in energy between the ordered crystal and the disordered liquid. The energy per unit length of a dislocation is $0.5\ E\ b^2$ J/m. Explore this in the following way:

(a) Calculate and plot the energy associated with a very high dislocation density of 10^{17} m/m^3 for the elements, i.e. plot a bar chart of $0.5 \times 10^{17}\ E\ b^2$ on the y-axis using twice atomic radius as equal to Burger's vector b. Remember that you must convert GPa into kPa and atomic radius from nm to m to get the energy in kJ/m^3.

(b) Now add, on the x-axis, the heat of fusion energy. Convert it from kJ/mol to kJ/m^3 by multiplying H_c by 1000/molar volume, with molar volume in m^3/kmol (as it is in the database). What, approximately, is the ratio of the dislocation energy to the energy of fusion? Would you expect this very high dislocation density to be enough to make the material turn amorphous?

Chapter 7
Bend and crush: strength-limited design

Elastic design, avoiding plasticity, ensures that the cabin of the car does not deform in a crash. Plasticity absorbs the energy of impact, and allows metals to be shaped and polymers to be molded. (Image of crash testing courtesy AutoNews; image of hot rolling courtesy of Tanis Inc., Delafield, WI.)

7.1 Introduction and synopsis

Stiffness-limited design, described in Chapter 5, is design to avoid excessive elastic deflection. Strength-limited design, our concern here, is design to avoid plastic collapse. That generally means design to avoid yield, arranging that the component remains elastic throughout, when it is called *elastic design*. Elastic design is not always possible or necessary: *local* yielding may be permissible provided *general* yield is avoided.

That is half the picture. The other half is design to permit controlled plastic collapse. The safety of modern cars relies on the front of the car absorbing the kinetic energy in a collision by plastic deformation. And the manufacturing processes of metal rolling, forging, extrusion and pressing use plastic flow. Here, strains are large, elastic deformation irrelevant, and the focus is on the forces and work necessary to achieve a prescribed change of shape (see cover picture).

Plasticity problems are solved in more than one way. When yield is to be avoided, we analyze the elastic state of stress in a component and make sure that this nowhere exceeds the yield strength. Full plasticity, by contrast, requires general yield. Then the mechanism of plastic collapse must be identified and the collapse load calculated by requiring that yield does occur where it needs to.

Chapter 5 introduced elastic solutions for common modes of loading: tension, compression, bending, torsion or internal pressure. Not surprisingly there are equivalent results for plastic design, and we start with these. As before, you don't need to know how to derive them, just where to find them and how to use them. We use them to develop material indices for strength-limited design and apply them via case studies.

7.2 Standard solutions to plastic problems

Yielding of ties and columns

A tie is a rod loaded in tension, a column is a rod loaded in compression. The state of stress within them is uniform, as was shown in Figure 5.1(a). If this stress, σ, is below the yield strength σ_y the component remains elastic; if it exceeds σ_y, it yields. Yield in compression is only an issue for short, squat columns. Slender columns and panels in compression are more likely to buckle elastically first (Chapter 5).

Yielding of beams and panels

The stress state in bending was introduced in Chapter 5. A bending moment M generates a linear variation of longitudinal stress σ across the section (Figure 7.1(a)) defined by

$$\frac{\sigma}{y} = \frac{M}{I} = E\kappa \tag{7.1}$$

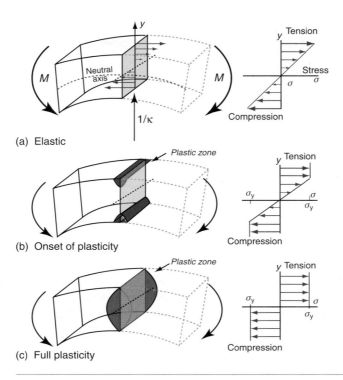

(a) Elastic

(b) Onset of plasticity

(c) Full plasticity

Figure 7.1   A beam loaded in bending. The stress state is shown on the right for purely elastic loading (a), the onset of plasticity (b), and full plasticity (c).

where y is the distance from the neutral axis, and the influence of the cross-section shape is captured by I, the second moment of area. For elastic deflection, we were interested in the last term in the equation—that containing the curvature κ. For yielding, it is the first term. The maximum longitudinal stress σ_{max} occurs at the surface (Figure 7.1(a)), at the greatest distance y_m from the neutral axis

$$\sigma_{max} = \frac{M\,y_m}{I} = \frac{M}{Z_e} \qquad (7.2)$$

The quantity $Z_e = I/y_m$ is called the *elastic section modulus* (not to be confused with the elastic modulus of the material, E). If σ_{max} exceeds the yield strength σ_y of the material of the beam, small zones of plasticity appear at the surface where the stress is highest, as in Figure 7.1(b). The beam is no longer elastic and, in this sense, is damaged even if it has not failed completely. If the moment is increased further, the linear profile is truncated—the stress near the surface

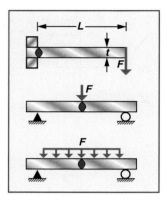

Figure 7.2   The plastic bending of beams.

remains equal to σ_y and plastic zones grow inwards from the surface. Although the plastic zone has yielded, it still carries load. As the moment increases further the plastic zones grow until they penetrate through the section of the beam, linking to form a *plastic hinge* (Figure 7.1(c)). This is the maximum moment that can be carried by the beam; further increase causes it to collapse by rotating about the plastic hinge.

Figure 7.2 shows simply supported beams loaded in bending. In the first, the maximum moment M is FL, in the second it is $FL/4$ and in the third $FL/8$. Plastic hinges form at the positions indicated in red when the maximum moment reaches the moment for collapse. This failure moment, M_f, is found by integrating the moment caused by the constant stress distribution over the section (as in Figure 7.1 (c), compression one side, tension the other)

$$M_f = \int_{section} b(y)\, y\, \sigma_y\, d_y = Z_p \sigma_y \tag{7.3}$$

where Z_p is the *plastic section modulus*. So two new functions of section shape have been defined for failure of beams: one for first yielding, Z_e, and one for full plasticity, Z_p. In both cases the moment required is simply $Z\sigma_y$. Values for both are listed in Figure 7.3. The ratio Z_p/Z_e is always greater than 1 and is a measure of the safety margin between initial yield and collapse. For a solid rectangle, it is 1.5, meaning that the collapse load is 50% higher than the load for initial yield. For efficient shapes, like tubes and I-beams, the ratio is much closer to 1 because yield spreads quickly from the surface to the neutral axis.

Yielding of shafts
We saw in Chapter 5 that a torque, T, applied to the ends of a shaft with a uniform circular section, and acting in the plane normal to the axis of the bar as in

Section shape	Area A m²	Elastic section modulus Z_e m³	Plastic section modulus Z_p m³
	$b\,h$	$\dfrac{bh^2}{6}$	$\dfrac{bh^2}{4}$
	$\pi\,r^2$	$\dfrac{\pi}{4}r^3$	$\dfrac{\pi}{3}r^3$
	$\pi(r_o^2-r_i^2)$ $\approx 2\pi\,rt$	$\dfrac{\pi}{4\,r_o}(r_o^4-r_i^4)$ $\approx \pi\,r^2 t$	$\dfrac{\pi}{3}(r_o^3-r_i^3)$ $\approx \pi\,r^2 t$

Figure 7.3   The area A, section modulus Z_e and fully plastic modulus Z_p for three simple sections.

Figure 7.4   Elastic torsion of shafts. The stress in the shaft depends on the torque T and the polar moment of area K. Helical springs are a special case of torsional loading.

Figure 7.4, produces a shear stress that increases linearly with distance r from the central axis:

$$\tau = \frac{T\,r}{K} = \frac{G\,\theta\,r}{K} \tag{7.4}$$

where K is the *polar second moment of area*. The resulting elastic deformation was described by the angle of twist per unit length θ/L. Failure occurs when the maximum surface stress exceeds the yield strength σ_y of the material. The maximum shear stress, τ_{max}, is at the surface and has the value

$$\tau_{max} = \frac{T\,R}{K} \qquad (7.5)$$

where R is the radius of the shaft. From Chapter 6, the yield stress in shear, k, is half the tensile yield stress, so first yield occurs when $\tau_{max} = \sigma_y/2$. When the torque is increased further, plasticity spreads inwards. The maximum torque that the shaft can carry occurs when $\tau = k$ over the whole section. Any greater torque than this causes the shaft to collapse in torsion by unrestrained rotation. For example, for a solid circular section, the collapse torque is

$$T = \frac{2}{3}\pi\,r^3\,k \qquad (7.6)$$

Helical springs are a special case of torsional loading (Figure 7.4): when the spring is loaded axially, the individual turns twist. It is useful to know the spring stiffness, S. If the spring has n turns of wire of shear modulus G, each of diameter d, wound to give a spring of radius R, the stiffness is

$$S = \frac{F}{u} = \frac{G\,d^4}{64\,n\,R^3}$$

where F is the axial force applied to the spring and u is its extension. The elastic extension is limited by the onset of plasticity. This occurs at the force

$$F_{crit} = \frac{\pi}{32}\frac{d^3\,\sigma_y}{R} \qquad (7.7)$$

Spinning disks (flywheels)

Spinning disks or rings store kinetic energy U (Figure 7.5). Centrifugal forces generate a radial tensile stress in the disk that reaches a maximum value σ_{max}. Analysis of a disk of density ρ, radius R and thickness t, rotating at an angular velocity ω radians/second, gives the kinetic energy and the maximum stress (when Poisson's ratio is taken as 1/3) as

$$U = \frac{\pi}{4}\rho t\,\omega^2\,R^4 \quad \text{and} \quad \sigma_{max} = 0.42\,\rho\,\omega^2\,R^2 \qquad (7.8)$$

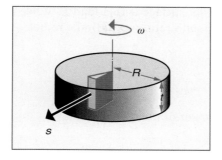

Figure 7.5   Spinning disks, as in flywheels and gyroscopes, carry radial tensile stress caused by centrifugal force.

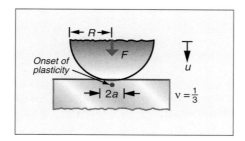

Figure 7.6   Contact stresses are another form of stress concentration. When elastic, the stresses and displacement of the surfaces towards each other can be calculated.

The disk yields when σ_{max} exceeds σ_y, and this defines the maximum allowable ω and limits the inertial energy storage.

Contact stresses

Contact stress analysis is important in design of rolling and sliding contacts as in bearings, gears and railway track. Yielding at contacts is closely linked to failure by wear and fatigue. When surfaces are placed in contact they touch at a few discrete points. If the surfaces are loaded, the contacts flatten elastically and the contact areas grow (Figure 7.6). The stress state beneath the contact is complex, first analyzed by the very same Hertz[1] for whom the unit of frequency is named. Consider a sphere of radius R made of material of Young's modulus

---

[1] Heinrich Rudolph Hertz (1821–1894), German physicist, discoverer of radio waves and how to generate them and inventor of the transmitter that started the radio age. In his spare time he dabbled in mechanics.

E and pressed against a flat surface with a load F. While the contact is elastic (and again assuming Poisson's ratio $\nu = 1/3$), the radius of the contact area is

$$a \approx 0.7 \left(\frac{FR}{E} \right)^{1/3} \tag{7.9}$$

and the relative displacement of the two bodies is

$$u \approx - \left(\frac{F^2}{E^2 R} \right)^{1/3} \tag{7.10}$$

For failure in contact, we need the maximum value of the shear stress, since it is this that causes first yield. It is beneath the contact at a depth of about $a/2$, and has the value

$$\tau_{max} = \frac{F}{2 \pi a^2} \tag{7.11}$$

If this exceeds the shear yield strength $k = \sigma_y/2$, a plastic zone appears beneath the center of the contact.

Stress concentrations

Holes, slots, threads and changes in section concentrate stress locally (Figure 7.7). Yielding will therefore start at these places, though as the bulk of the component is still elastic this initial yielding is not usually catastrophic. The same cannot be said for fatigue (Chapter 9), where stress concentrations are often implicated as the origins of failure.

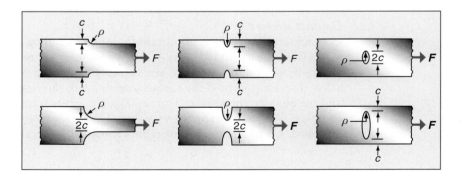

Figure 7.7   Stress concentrations. The change of section concentrates stress most strongly where the curvature of the surface is greatest.

We define the *nominal stress* in a component σ_{nom} as the load divided by the cross-section, ignoring features that cause the stress concentration. The *maximum local stress* σ_{max} is then found approximately by multiplying the nominal stress σ_{nom} by a *stress concentration factor* K_{sc}, where

$$K_{sc} = \frac{\sigma_{max}}{\sigma_{nom}} = 1 + \alpha \left(\frac{c}{\rho_{sc}}\right)^{1/2} \qquad (7.12)$$

Here ρ_{sc} is the minimum radius of curvature of the stress-concentrating feature and c is a characteristic dimension associated with it: either the half-thickness of the remaining ligament, the half-length of a contained notch, the length of an edge notch or the height of a shoulder, whichever is least (Figure 7.7). The factor α is roughly 2 for tension, but is nearer 1/2 for torsion and bending. Though inexact, the equation is an adequate working approximation for many design problems. More accurate stress concentration factors are tabulated in compilations such as Roark (see 'Further reading' at the end of this chapter).

As a simple example, consider a circular hole in a plate loaded in tension. The radius of curvature of the feature is the hole radius, $\rho_{sc} = R$, and the characteristic dimension is also the radius, $c = R$. From equation (7.12), the local stress next to the hole is thus three times the nominal tensile stress—however small the hole. Local yielding therefore occurs when the nominal stress is only $\sigma_y/3$.

7.3 Material indices for yield-limited design

Minimizing weight: a light, strong tie-rod
Many structures rely on tie members that must carry a prescribed tensile load without yielding—the cover picture of Chapter 4 showed two—often with the requirement that they be as light as possible. Consider a design that calls for a cylindrical tie-rod of given length L that must carry a tensile force F as in Figure 5.7(a), with the constraint that it must not yield, but remain elastic. The objective is to minimize its mass. The length L is specified but the cross-section area A is not (Table 7.1).

As before, we first seek an equation describing the quantity to be maximized or minimized. Here it is the mass m of the tie

$$m = AL\rho \qquad (7.13)$$

where A is the area of the cross-section and ρ is the density of the material of which it is made. We can reduce the mass by reducing the cross-section, but there is a constraint: the section area A must be sufficient to carry the tensile load F without yielding, requiring that

Table 7.1   Design requirements for the light tie

Function	• Tie-rod
Constraints	• Length L specified
	• Tie must support tensile load F without yielding
Objective	• Minimize the mass m of the tie
Free variables	• Choice of cross-section area, A
	• Choice of material

$$\frac{F}{A} \leq \sigma_y \qquad (7.14)$$

where σ_y is the yield strength. Eliminating A between these two equations gives

$$m \geq FL\left(\frac{\rho}{\sigma_y}\right) \qquad (7.15)$$

The lightest tie that will carry F safely is that made of the material with the smallest value of ρ/σ_y. We could define this as the material index of the problem, seeking a minimum, but as in Chapter 5 we will invert it, seeking materials with the largest values of

$$M_t = \frac{\sigma_y}{\rho} \qquad (7.16)$$

This index, the *specific strength*, is plotted as a line of slope 1 in the chart of Figure 7.8. A particular value of σ_y/ρ is identified, passing through the high-strength end of several major alloy systems—nickel alloys, high-strength steels, and aluminum and magnesium alloys. Titanium alloys are significantly better than the other metals; CFRP is better still. Ceramics and glasses have high values of M_t but are impractical as structural ties because of their brittleness.

Tension, then, is straightforward. The real problem in elastic design is seldom tension; it is *bending*. We therefore revisit the bending of panels and beams, applying a constraint on strength rather than stiffness.

Minimizing weight: light, strong panels

Figure 5.7(b) showed a panel supported at its edges, carrying a specified central load. The width b and span L are fixed, but we are free to choose the thickness h. The objective is to minimize the mass. Reducing the thickness reduces the mass, but it must be sufficient for the maximum stress to be below the elastic limit (Table 7.2).

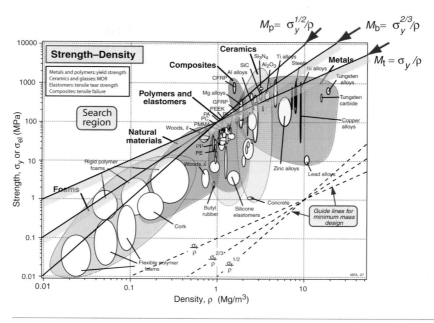

Figure 7.8 The strength–density chart with the indices σ_y/ρ, $\sigma_y^{2/3}/\rho$ and $\sigma_y^{1/2}/\rho$ plotted on it.

Table 7.2 Design requirements for the light panel

Function	• Panel in bending
Constraints	• Width b and span L specified
	• Panel must support bending load F without yielding
Objective	• Minimize the mass m of the panel
Free variables	• Choice of thickness h
	• Choice of material

The procedure is much as before—set up an equation for the mass; find an expression for the maximum stress (noting that for a rectangular section, $I = bh^3/12$); use this constraint to eliminate the free variable h and read off the material index. The analysis itself is left for the Exercises; the result is

$$M_{\mathrm{p}} = \frac{\sigma_y^{1/2}}{\rho} \tag{7.17}$$

This index is also shown by a shaded guideline in Figure 7.8. Now all the light alloys (Mg, Al and Ti) outperform steel, as do GFRP and wood. CFRP still leads the way.

Table 7.3   Design requirements for the light strong beam

Function	● Beam in bending
Constraints	● Span *L* specified, section shape square
	● Beam must support bending load *F* without yielding
Objective	● Minimize the mass *m* of the beam
Free variables	● Area *A* (or square section dimension *b*)
	● Choice of material

Light, strong beams: the effect of shape

In beam design we are free to choose the shape as well as the dimensions of the cross-section. First we consider beams of prescribed shape, with freedom to change their size in a self-similar way. Then we explore how much better we can make the design by using efficient shapes.

Start with the beam of square section $A = b \times b$, which may vary in size, loaded in bending over a span of fixed length L with a central load F (Figure 5.7(c) and Table 7.3).

The analysis is similar to the panel (see Exercises), with the modified second moment of area, $I = b^4/12 = A^2/12$. The resulting material index is

$$M_b = \frac{\sigma_y^{2/3}}{\rho} \tag{7.18}$$

The chart in Figure 7.8 shows that the slope of this index lies between the other two, so the competition between metals, wood and composites changes again, with CFRP still on top.

But, you will say, no one uses solid, square-section, beams for minimum mass design—and you are right. If you want to support bending loads it is better to choose a shape that uses less material to provide the same strength. Wooden floor-joists in houses are typically twice as deep as they are wide; standard steel and aluminum beams have an I-section or a box section; space frames, commonly, are made from tubes. Does this change the index? Well, yes and no. Start with the no.

Equation (7.18) was derived by analyzing a square beam, but the result holds for any *self-similar* shape, meaning one in which all dimensions remain in proportion as the size is varied. Shaping a given cross-sectional area from a solid square beam into a tube or an I-beam increases the second moment of area, without changing the cross-sectional area or the mass. In Chapter 5, it was shown that this provides greater stiffness with no mass penalty. It also makes the beam stronger. Equation (7.2) shows that a bending moment gives a maximum stress determined by the elastic section modulus, $Z_e = I/y_m$. Consequently the gain in strength by increasing I is not quite as great as the gain in stiffness—shaping the section to an I-beam increases I, but usually makes the beam deeper

Table 7.4   The effect of shaping on strength and mass of beams in different structural materials

Material	Maximum failure shape factor (failure moment relative to solid square beam) ϕ_B^y	Mass ratio by shaping (relative to solid square beam)
Steels	13	0.18
Al alloys	10	0.22
Composites (GFRP, CFRP)	9	0.23
Wood	3	0.48

too, increasing y_m. As before, we can define a 'shape factor' ϕ_B^y for strength—the ratio of Z_e for the shaped section to Z_e for the same area of material in a solid square section:

$$\phi_B^y = \frac{Z_e^{shaped}}{Z_e^{solid}}$$

Recall from Chapter 5 that materials are not all equally easy to shape; if they were, all could be given the same efficient shape and the index in equation (7.18) would be sufficient. Table 7.4 gives typical upper values of the shape factor for strength, for a range of competing materials.

The right-hand column of Table 7.4 shows the corresponding mass saving we might achieve by shaping and using less material to carry the same load in bending. Metals and composites can all be improved significantly (though again metals do a little better), but wood has more limited potential. So, when comparing materials for light, strong beams using the index in equation (7.18), the performance of wood should not be overestimated; other materials allow more efficient shapes.

Minimizing material cost or volume
When the objective is to minimize cost rather than weight, the indices change exactly as before. The objective function for the material cost C of the tie, panel or beam becomes

$$C = mC_m = ALC_m\rho \qquad (7.19)$$

where C_m is the cost per kg of the material. This leads to indices which are just those of equations (7.16)–(7.18) with ρ replaced by $C_m \rho$.

If instead the objective is to minimize volume, density is no longer relevant. The indices for ties, panels and beams are then the same as those for minimum mass with ρ deleted.

7.4 Case studies

Plasticity problems are of two types. In the first yield is avoided everywhere so that the entire component remains elastic, meeting the condition $\sigma < \sigma_y$ everywhere. Limited local plastic flow at stress concentrations may be allowed provided that, once it has happened, the condition $\sigma < \sigma_y$ is met everywhere (bedding down). In the second, full plasticity is the aim. Manufacturing processes such as metal forging and extrusion are well-managed plasticity; crash barriers and packaging, too, rely on full plasticity to absorb energy.

First a cautionary note. Carrying loads safely is not just a question of strength, but also of toughness—the resistance of the material to fracture. Strength and toughness are not the same thing—we explain why in the next chapter. For now, it is enough to know that some materials may appear to be good options for strength-limited design, but that they are impractical because they are too brittle in tension or in shock loading.

Corkscrew levers again: strength
The lever of the corkscrew of Figure 5.13 and described in Section 5.5 is loaded in bending. It needs some stiffness, but if it flexes slightly, no great harm is done. If, however, it yields, bending permanently before it extracts the cork, the user will not be happy. So it must also meet a strength constraint.

The cross-section is rectangular. As in Chapter 5 we make the assumption of self-similarity, meaning that we are free to change the scale of the section but not its shape. Then the criterion for selection is that of the index of equation (7.18), $M_b = \sigma_y^{2/3}/\rho$. The index is plotted in Figure 7.8, isolating materials for light, strong beams. The selection is almost the same as that for stiffness: CFRP, magnesium and aluminum alloys are the best choice.

But what about the stress-concentrating effect at the holes? A stress concentration factor K_{sc} means that yield starts when the nominal stress exceeds σ_y/K_{sc}. Changing the scale of the part does not change K_{sc} because it depends only on the shape of the defect—the ratio ρ/c (its value for a circular hole is 3, regardless of scale), with the result that the index remains unchanged and the selection remains valid.

Elastic hinges and couplings
Nature makes much use of elastic hinges: skin, muscle, cartilage all allow large, recoverable deflections. Man, too, designs with *flexural* and *torsional hinges*: ligaments that connect or transmit load between components while allowing limited relative movement between them by deflecting elastically (Figures 5.18 and 5.19). Which materials make good hinges?

Consider the hinge for the lid of a box. The box, lid and hinge are to be molded in one operation—there are no separate screws or pins. The hinge is a thin ligament of material that flexes elastically as the box is closed, as in Figure 5.18, but it carries no significant axial loads. Then the best material is the one that (for given ligament dimensions) bends to the smallest radius without yielding or failing.

Table 7.5   Design requirements for elastic hinges

Function	• Elastic hinge
Constraint	• No failure, meaning $\sigma < \sigma_y$ throughout the hinge
Objectives	• Maximize elastic flexure
Free variables	• Choice of material

When a ligament of thickness t is bent elastically to a radius R, the surface strain is

$$\varepsilon = \frac{t}{2R} \qquad (7.20)$$

and—since the hinge is elastic—the maximum stress is

$$\sigma = E\frac{t}{2R}$$

This must not exceed the yield or failure strength σ_y. Thus, the minimum radius to which the ligament can be bent without damage is

$$R \geq \frac{t}{2}\left[\frac{E}{\sigma_y}\right] \qquad (7.21)$$

The best material is the one that can be bent to the smallest radius—that is, the one with the greatest value of the index

$$M = \frac{\sigma_y}{E} \qquad (7.22)$$

Here we need the $\sigma_y - E$ chart (Figure 7.9). Candidates are identified by using the guideline of slope 1; a line is shown at the position $M = \sigma_y/E = 3 \times 10^{-2}$. The best choices for the hinge are all polymeric materials. The short-list includes polyethylene, polypropylene, nylon and, best of all, elastomers, though these may be too flexible for the body of the box itself. Cheap products with this sort of elastic hinge are generally molded from polyethylene, polypropylene or nylon. Spring steel and other metallic spring materials (like phosphor bronze) are possibilities: they combine usable σ_y/E with high E, giving flexibility with good positional stability (as in the suspensions of relays).

Figure 7.9 Materials for elastic hinges and springs. Polymers are the best choice for the former. High-strength steel, CFRP, and certain polymers and elastomers are the best choice for the latter.

Materials for springs

Springs come in many shapes (Figure 7.10) and have many purposes: axial springs (a rubber band, for example), leaf springs, helical springs, spiral springs, torsion bars. All depend on storing elastic energy when loaded, releasing it when unloaded again. The stored energy per unit volume in a material carrying a tensile stress σ was derived in Chapter 4—it is given by $\sigma^2/2E$. If the spring yields it deforms permanently and ceases to fulfill its function, so the maximum value of σ must not exceed σ_y, when the stored energy is $\sigma_y^2/2E$ per unit volume. The best material for a spring of minimum volume (Table 7.6) is therefore that with the greatest value of

$$M = \frac{\sigma_y^2}{E} \qquad (7.23)$$

While it is less obvious, the index is the same for leaf springs, torsional spring and coil springs—the best choice for one is the best choice for all.

Figure 7.10   Springs: leaf, helical, spiral and torsion bar. Springs store energy. The best material for a spring, regardless of its shape or the way it is loaded, is that of a material with a large value of σ_{el}^2/E.

Table 7.6   Design requirements for springs

Function	• Elastic spring
Constraint	• No failure, meaning stress below yield throughout the spring
Objectives	• Maximum stored elastic energy per unit volume
Free variables	• Choice of material

The choice of materials for springs of minimum volume is shown in Figure 7.9. A family of lines of slope 2 link materials with equal values of σ_y^2/E; those with the highest values of M lie towards the bottom right. The heavy line is one of the family; it is positioned so that a subset of materials is left exposed. The best choices are a *high-strength steel* lying near the top end of the line. Other materials are suggested too: *CFRP* (now used for truck springs), *titanium alloys* (good but expensive) and *nylon* (children's toys often have nylon springs), and, of course, *elastomers*. Note how the procedure has identified a candidate from almost every class of materials: metals, polymers, elastomers and composites.

Full plasticity: metal rolling

Plate, sheet, I-sections and sections like that of railroad track are shaped by rolling. Figure 7.11 shows the rolling of plate with an initial thickness t_0. The plate emerges from the rolls with a lesser thickness t_1, a reduction of $\Delta t = t_0 - t_1$. A lower bound for the torque T and power P required to do this is found from the plastic work, $\sigma_y \varepsilon_{pl}$ per unit volume that it takes to produce a plastic strain ε_{pl} of $\Delta t/t_0$. If the rolls rotate through $\Delta\theta$ a length $R\Delta\theta$ (and thus

Figure 7.11   Rolling.

of volume $V = R\Delta\theta t_0$ per unit width) is fed into the 'bite', where it is compressed to the emerging thickness t_1. Equating the work done by the pair of rolls, $2T\Delta\theta$, to the plastic work $V\sigma_y\varepsilon_{pl}$ gives the torque per roll:

$$T = \frac{1}{2}R\sigma_y\,\Delta t$$

(7.24)

The power P is just the torque times the angular velocity ω radians per second, giving

$$P = 2T\omega = R\omega\,\sigma_y\,\Delta t \qquad (7.25)$$

The torque and power increase with σ_y so hot rolling takes less power than cold rolling because σ_y is smaller and work hardening (the increase of yield strength with strain) is negligible.

The values of T and P calculated here are lower bounds. Most metal-working operations involve *redundant work* (plastic work that is not strictly necessary but happens anyway), friction as the compressed metal expands laterally and speeds up in the gap, sliding against the roll surfaces, and work hardening, all of which increase the forces and power that is needed.

7.5 Summary and conclusions

Elastic design is design to avoid yield. That means calculating the maximum stress, σ, in a loaded component and ensuring that it is less than σ_y, the yield strength. Standard solutions give this maximum stress for panels, beams, torsion bars, spinning disks and many other components in terms of their geometry and the loads applied to them. From these, indices are derived to guide material choice.

The indices depend on objectives such as minimizing mass or cost, much as they did for stiffness limited design in Chapter 5.

Sometimes, however, controlled plasticity is the aim. Then the requirement is that the stress must exceed the yield strength over the entire section of the component. Metal-forming operations such as forging, rolling and deep drawing rely on full plasticity. The forces and power they require scales with the yield strength of the material being shaped. Thus, soft metals such as lead, or pure aluminum or low-carbon steel are easy to shape by plastic deformation, but some of the hardest, like tool steels and the nickel based super-alloys used in gas turbines, have such high yield strengths that they have to be shaped in other ways.

7.6 Further reading

Ashby, M.F. (2005) *Materials Selection in Mechanical Design*, 3rd edition, Butterworth-Heinemann, Oxford, UK, Chapter 4. ISBN 0-7506-6168-2. (*Appendix A of this text is an expanded catalog of simple solutions to standard problems.*)

Gere, J.M. (2006) *Mechanics of Materials*, 6th edition, Thompson Publishing, Toronto, Canada. ISBN 0-534-41793-0. (*An intermediate level text on statics of structures by one of the fathers of the field; his books with Timoshenko introduced an entire generation to the subject.*)

Hosford, W.F. (2005) *Mechanical Behavior of Materials*, Cambridge University Press, Cambridge, UK. ISBN 0-521-84670-6. (*A text that nicely links stress–strain behavior to the micromechanics of materials.*)

Jenkins, C.H.M. and Khanna, S.K. (2006) *Mechanics of Materials*, Elsevier Academic, Boston, MA, USA. ISBN 0-12-383852-5. (*A simple introduction to mechanics, emphasizing design.*)

Riley, W.F., Sturges, L.D. and Morris, D.H. (2003) *Statics and Mechanics of Materials*, 2nd edition, McGraw-Hill, Hoboken, NJ, USA. ISBN 0-471-43446-9. (*An intermediate level text on the stress, strain and the relationships between them for many modes of loading. No discussion of micromechanics—response of materials to stress at the microscopic level.*)

Vable, M. (2002) *Mechanics of Materials*, Oxford University Press, Oxford, UK. ISBN 0-19-513337-4. (*An introduction to stress–strain relations, but without discussion of the micromechanics of materials.*)

Young W.C. (1989) *Roark's Formulas for Stress and Strain*, 6th edition, McGraw-Hill, New York, USA. ISBN 0-07-100373-8. (*This is the 'Yellow Pages' of formulae for elastic problems—if the solution is not here, it doesn't exist.*)

7.7 Exercises

Exercise E7.1 What is meant by the *elastic section modulus*, Z_e? A beam carries a bending moment M. In terms of Z_e, what is the maximum value M can take without initiating plasticity in the beam?

Exercise E7.2 Derive the index for selecting materials for a light, strong panel, equation (7.17), following the steps outlined above the equation in the text.

Exercise E7.3 Derive the index for selecting materials for a light, strong beam with a square cross-section, equation (7.18).

Exercise E7.4 Derive an index for selecting materials for a panel that meets a constraint on bending strength and is as thin as possible.

Exercise E7.5 Derive an index for selecting materials for a panel that meets a constraint on bending strength and is as cheap as possible.

Exercise E7.6 A centrifuge has a rotor that can be idealized as a uniform disk. It has a diameter of 200 mm and is made of a material of strength 450 MPa and density 7900 kg/m³. How fast can it be spun (in radians per second) before the stresses it carries exceed its yield strength?

Exercise E7.7 A material is sought for a high-speed centrifuge. The objective is to achieve as high an angular velocity ω of the centrifuge disk as possible. The constraint is that the stress created by the centrifugal force must not exceed the yield strength of the material of which it is made. Derive an index to guide the choice of material to allow the maximum ω.

Exercise E7.8 The engine of a car is mounted on four shear bolts designed to fail in shear in a front-end collision, detaching the engine from the car, if the deceleration exceeds 10 g. Assume that all four bolts carry the same load. The mass of the engine is 80 kg. Because of space limitations, the maximum diameter of each shear bolt cannot exceed 5 mm. (a) Calculate the constraint on the yield strength σ_y of the material of the shear bolt, assuming that the shear strength is $\sigma_y/2$ and that the mounting is done in such a way that the shear can take place without friction. (b) If, to save weight, the shear bolts have to be as light as possible, what metal (of those identified in the chart of Figure 7.8) would you choose to make them?

Exercise E7.9 A plate with a rectangular section 500 mm by 15 mm carries a tensile load of 50 kN. It is made of a ductile metal with a yield strength of 50 MPa. The plate contains an elliptical hole of length 100 mm and a minimum radius of 1 mm, oriented as shown in the diagram. What is:
(a) The nominal stress?
(b) The maximum stress in the plate?

Will the plate start to yield? Will it collapse completely?

Exercise E7.10 Valve springs for high-performance automobile engines must be light to minimize inertial loads since part of their mass moves with the valves; at high engine speeds the valves, if heavy, bounce out of contact with the valve itself ('valve bounce'), impeding the flow of gas into and out of the combustion chamber. Derive the index for light springs to guide choice for these, remembering that engine temperatures can reach 200°C.

7.8 Exploring design with CES (use Level 2 unless otherwise indicated)

Exercise E7.11 Use the 'Search' facility in CES to search for materials that are used for *springs*. Report what you find.

Exercise E7.12 Use the 'Search' facility in CES to search for materials that are used for *light springs*. Report what you find.

Exercise E7.13 Make a property chart with σ_y on one axis and E on the other. Use it to select materials for springs, using the index σ_y^2/E derived in the text. Which three metals emerge as the best metallic choices?

Exercise E7.14 A material is required for a spring that must be as light as possible. To be stiff enough it must also have a Young's modulus $E > 20\,\text{GPa}$. Make a bar chart with the index $\sigma_y^2/E\rho$ for selecting light springs derived in Exercise E7.10 (you will need to use the 'Advanced' facility in the axis-choice dialog box to do this). Add a 'Limit' stage to apply the constraint $E > 20\,\text{GPa}$. Hence find the two materials that are the best choices for this application.

Exercise E7.15 Exercise E7.10 describes the requirements for valve springs for high-performance engines. Apply the index derived there, $\sigma_y^2/\rho E$, by making an appropriate chart and plotting an appropriate selection line on it. Engines are hot: add a 'Limit' stage on maximum service temperature of 250°C. Hence select metals for this application.

Exercise E7.16 Abrasives have high hardness, H. Make a bar chart of hardness and identify the four materials with the highest values. They are prime choices for abrasive wheels and pastes.

Exercise E7.17 The text showed that the power required to roll a metal is proportional to its yield strength. Make a bar chart of yield strength, σ_y for metals. Open the record for low-carbon steel, find the range of its yield strength and take the average. Normalize all the yield strengths in the database by dividing them by this value, using the 'Advanced' option in the dialog box for selecting the axes, so that low-carbon steel now lies at the value 1 on the bar chart. Use the chart to read off how much less power is required to roll: (a) commercially pure zinc and (b) commercially pure lead at room temperature.

Exercise E7.18 Crash barriers, auto fenders and other protective structures rely on absorbing kinetic energy by plastic deformation. The energy W_{pl} absorbed in deforming a material to fracture (the area under the stress–strain curve) can be estimated approximately in CES as

$$W_{pl} = \frac{1}{2}(\sigma_y + \sigma_{ts})\varepsilon_f$$

Make a chart with density on the *x*-axis and W_{pl} on the *y*-axis by using the 'Advanced' facility in the axis-selection dialog box. Use a box selection to find the three materials that absorb the most energy. Rank them by price, using a 'Graph' stage to plot price.

Exercise E7.19   If the crash barrier of the last exercise is part of a vehicle, fuel is saved if it is light. We then want the materials with the largest value of W_{pl}/ρ, where ρ is the density. These are found by using a selection line of slope 1 on the chart made in the last exercise and selecting the materials above the line. In this application the materials must also have adequate stiffness so that they do not bend elastically too much, requiring a Young's modulus of at least 2.5 GPa. Apply this requirement using a 'Limit' stage, return to the chart and move the selection line until only three materials remain. Comment on the choice.

Exercise E7.20   Now explore what Level 3 of the database can do. Leave all the selection lines and criteria in example E7.19. Go to File > Change Database > CES Edu Level 3 > Open. A dialog box asks if you wish to apply this to your current project. Click on Yes. Then list the materials that now appear in the results window. Comment on the choice.

Chapter 8
Fracture and fracture toughness

8.1 Introduction and synopsis

It is easy to set a value on the engineering science that enables success, that makes things happen, much harder to value engineering science that prevents failure, that *stops* things happening. One of the great triumphs of recent engineering science has been the development from the 1960s onwards of a rigorous mechanics of material fracture. We have no numbers for the money and lives it has saved by preventing failures; all we know is that, by any measure, it is enormous. This chapter is about the ways in which materials fail when loaded progressively, and design methods to ensure that fracture won't happen unless you want it to.

Sometimes, of course, you do. Aircraft engines are attached to the wing by shear-bolts, designed to fail and shed the engine if it suddenly seizes. At a more familiar level, peel-top cans, seals on food containers and many other safety devices rely on controlled tearing or fracture. And processes like machining and cutting use a combination of plasticity and fracture.

We start by distinguishing *strength* from *toughness*. Toughness—resistance to fracture—requires a new material property, the *fracture toughness* developed in Section 8.3, to describe it. This new property is explored in Section 8.4 using charts like those we have already seen for modulus and strength. The underlying science mechanisms (Section 8.5) give insight into ways in which toughness can be manipulated (Section 8.6). The chapter ends in the usual way with a summary, Further reading and Exercises.

8.2 Strength and toughness

Strength and toughness? Why both? What's the difference?

Strength, when speaking of a material, is its resistance to plastic flow. Think of a sample loaded in tension. Increase the stress until dislocations sweep right across the section, meaning the sample just yields, and you measure the initial *yield strength*. Strength generally increases with plastic strain because of work hardening, reaching a maximum at the *tensile strength*. The area under the whole stress–strain curve up to fracture is the *work of fracture*. We've been here already—it was the subject of Chapter 5.

Toughness is the resistance of a material to the propagation of a crack. Suppose that the sample of material contained a small, sharp crack, as in Figure 8.1(a). The crack reduces the cross-section A and, since stress σ is F/A, it increases the stress. But suppose the crack is small, hardly reducing the section, and the sample is loaded as before. A tough material will yield, work harden and absorb energy as before—the crack makes no significant difference. But if the material is *not* tough (defined in a moment) then the unexpected happens; the crack suddenly propagates and the sample fractures at a stress that can be far below the yield strength. Design based on yield is common practice. The possibility of fracture

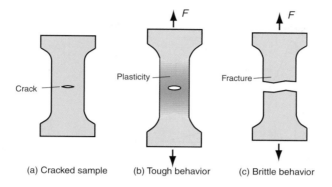

Figure 8.1   Tough and brittle behavior. The crack in the tough material, shown in (b), does not propagate when the sample is loaded; that in the brittle material propagates without general plasticity, and thus at a stress less than the yield strength.

at stresses below the yield strength is really bad news. And it has happened, on spectacular scales, causing boilers to burst, bridges to collapse, ships to break in half, pipelines to split and aircraft to crash. We get to that in Chapter 10.

So what is the material property that measures the resistance to the propagation of a crack? And just how concerned should you be if you read in the paper that cracks have been detected in the track of the railway on which you commute or in the pressure vessels of the nuclear reactor of the power station a few miles away? If the materials are tough enough you can sleep in peace. But what is 'tough enough'?

This difference in material behavior, once pointed out, is only too familiar. Buy a CD, a pack of transparent folders or even a toothbrush: all come in perfect transparent packaging. Try to get them out by pulling and you have a problem: the packaging is strong. But nick it with a knife or a key or your teeth and suddenly it tears easily. That's why the makers of shampoo sachets do the nick for you. What they forget is that the polymer of the sachet becomes tougher when wet, and that soapy fingers can't transmit much force. But they had the right idea.

Tests for toughness

If you were asked to devise a test to characterize toughness, you might dream up something like those of Figure 8.2: notch the material, then yank it or whack it till it breaks, measuring the energy to do so. Tests like these (there are many variants) are in fact used for ranking and as an acceptance procedure when taking delivery of a new batch of material. The problem is that they do not measure a true material property, meaning one that is independent of the size and shape of the test sample, so the energy measurements do not help with design. To get at the real, underlying, material properties we need the ideas of *stress intensity* and *fracture toughness*.

Figure 8.2 (a) The tear test. (b) The impact test. Both are used as acceptance tests and for quality control, but neither measures a true material property.

8.3 The mechanics of fracture

***Stress intensity* K_1 *and fracture toughness* K_{1c}**

Cracks concentrate stress. A remote stress σ applies a force that is transmitted through a cracked material in the way suggested by Figure 8.3. The local stress, σ_{local}, which is proportional to the number of lines of force crossing unit length of cross-section, is non-uniform, rising steeply as the crack tip is approached. The stress concentration factor, defined for notches and holes in Chapter 7, does not help here—it is relevant only for features with a finite radius of curvature. Cracks are sharp—the radius at the tip is essentially zero. Analysis of the stress field ahead of a sharp crack of length c shows that the local stress at a distance r from its tip caused by a remote uniform tensile stress σ is

$$\sigma_{local} = \sigma\left(1 + Y\sqrt{\frac{\pi c}{2\pi r}}\right) \tag{8.1}$$

where Y is a constant with a value near unity that depends weakly on the geometry of the cracked body. Far from the crack, where $r \gg c$, the local stress falls to the value σ, but near the tip, where $r \ll c$, it rises steeply, as

$$\sigma_{local} = Y\frac{\sigma\sqrt{\pi c}}{\sqrt{2\pi r}} \tag{8.2}$$

So for any given value of r the local stress scales as $\sigma\sqrt{\pi c}$ which therefore is a measure of the *intensity* of the local stress (the inclusion of the π is a convention

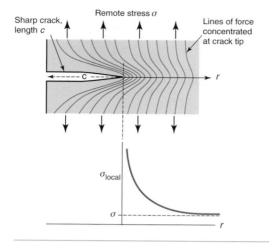

Remote stress σ

Sharp crack, length c

Lines of force concentrated at crack tip

σ_{local}

σ

r

Figure 8.3   Lines of force in a cracked body under load; the local stress is proportional to the number of lines per unit length, increasing steeply as the crack tip is approached.

used universally). This quantity is called the *mode 1 stress intensity factor* (the 'mode 1' means tensile fracture) and given the symbol K_1:

$$K_1 = Y\sigma\sqrt{\pi c} \qquad (8.3)$$

It has units of MPa.m$^{1/2}$.

For reasons explored below, cracks propagate when the stress intensity factor exceeds a critical value. This critical value is called the *fracture toughness*, K_{1c}. Figure 8.4 shows two sample geometries used to measure K_{1c} (there are others, described in Chapter 10). A sample containing a sharp crack of length c (if a surface crack) or $2c$ (if a contained crack) is loaded, recording the tensile stress σ^* at which it suddenly propagates. It is essential that the crack be sharp—not an easy thing to achieve—because if it is not, the part of the stress field with the highest stresses, where the lines of force in Figure 8.3 are closest together, is changed. There are ways of making sharp cracks for doing this test—books on fracture mechanics describe them—but let us simply assume we have one. The value of Y for this geometry is 1 provided $c \ll w$; when it is not, there is a small correction factor. Then the quantity K_{1c} is given by

$$K_{1c} = Y\sigma^*\sqrt{\pi c} \approx \sigma^*\sqrt{\pi c} \qquad (8.4)$$

Fracture toughness is a *material property*, and this means two things. The first is that its value is independent of the way it is measured—different test geometries,

Figure 8.4   Measuring fracture toughness, K_{1c}. Two test configurations are shown here; others are described in Chapter 10.

if properly conducted, give the same value of K_{1c} for any given material (Figure 8.4). The second is that it can be used for design, in ways described in Chapter 10.

Energy release rate G *and toughness* G_c

When a sample fractures, a new surface is created. Surfaces have energy, the *surface energy* γ, with units of joules[1] per square meter (typically $\gamma = 1 \text{ J/m}^2$). If you fracture a sample across a cross-section area A you make an area $2A$ of new surface, requiring an energy of at least $2A\gamma$ joules to do so. Consider first the question of the *necessary condition for fracture*. It is that sufficient external work be done, or elastic energy released, to at least supply the surface energy, γ per unit area, of the two new surfaces that are created. We write this as

$$G \geq 2\gamma \tag{8.5}$$

where G is called the *energy release rate*. In practice, it takes much more energy than 2γ because of plastic deformation round the crack tip. But the argument still holds: growing a crack costs energy G_c J/m^2 for the two surfaces—a sort of 'effective' surface energy, replacing 2γ. It is called, confusingly, the *toughness* (or the *critical strain energy release rate*). This toughness G_c is related to the *fracture toughness* K_{1c} in the following way.

---

[1] James Joule (1818–1889), English physicist, did not work on fracture or on surfaces, but his demonstration of the equivalence of heat and mechanical work linked his name to the unit of energy.

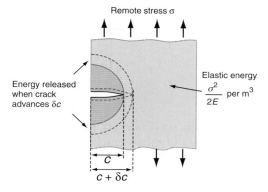

Remote stress σ

Energy released
when crack
advances δc

Elastic energy

$\dfrac{\sigma^2}{2E}$ per m³

c

$c + \delta c$

Figure 8.5 The release of elastic energy when a crack extends.

Think of a slab of material of unit thickness carrying a stress σ. The elastic energy stored in it (Chapter 4) is

$$U_{\mathrm{v}} = \frac{1}{2}\frac{\sigma^2}{E} \qquad (8.6)$$

per unit volume. Now put in a crack of length c, as in Figure 8.5. The crack relaxes the stress in a half-cylinder of radius about c—the reddish half-cylinder in the figure—releasing the energy it contained:

$$U(c) = \frac{1}{2}\frac{\sigma^2}{E} \cdot \frac{1}{2}\pi c^2 \qquad (8.7)$$

Suppose now that the crack extends by δc, releasing the elastic energy in the yellow segment. This energy must pay for the extra surface created, and the cost is $G_{\mathrm{c}}\,\delta c$. Thus, differentiating the last equation, the condition for fracture becomes

$$\delta U = \frac{\sigma^2 \pi c}{2E}\,\delta c = G_{\mathrm{c}}\delta c \qquad (8.8)$$

But $\sigma^2 \pi c$ is just $K_{1\mathrm{c}}^2$, so from equation (8.4), taking $Y = 1$,

$$\frac{K_{1\mathrm{c}}^2}{2E} = G_{\mathrm{c}} \qquad (8.9)$$

This derivation is an approximate one. A more rigorous (but much more complicated) one shows that the form of equation (8.9) is right, but that it is too small by exactly a factor of 2. Thus, correctly, the result we want (taking the square root) is:

$$K_{1c} = \sqrt{EG_c} \tag{8.10}$$

The crack tip plastic zone

The intense stress field at the tip of a crack generates a *process zone*: a plastic zone in ductile solids, a zone of micro-cracking in ceramics, a zone of delamination, debonding and fiber pull-out in composites. Within the process zone, work is done against plastic and frictional forces; it is this that accounts for the difference between the measured fracture energy G_c and the true surface energy 2γ. We can estimate the size of a plastic zone that forms at the crack tip as follows. The stress rises as $1/\sqrt{r}$ as the crack tip is approached (equation (8.2)). At the point where it reaches the yield strength σ_y the material yields (Figure 8.6) and—except for some work hardening—the stress cannot climb higher than this. The distance from the crack tip where $\sigma_{local} = \sigma_y$ is found by setting equation (8.2) equal to σ_y and solving for r. But the truncated part of the elastic stress field is redistributed, making the plastic zone larger. The analysis of this is complicated but the outcome is simple: the radius r_y of the plastic zone,

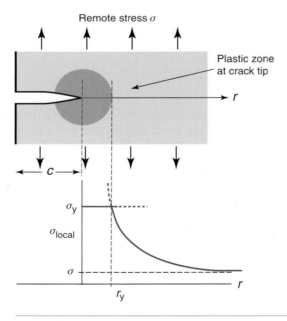

Figure 8.6   A plastic zone forms at the crack tip where the stress would otherwise exceed the yield strength σ_y.

allowing for stress redistribution, is twice the value found from equation (8.2), giving

$$r_y = 2\left(\frac{\sigma^2 \pi c}{2\pi\sigma_y^2}\right) = \frac{K_I^2}{\pi\sigma_y^2} \tag{8.11}$$

(taking $Y = 1$). Note that the size of the zone shrinks rapidly as σ_y increases: cracks in soft metals have large plastic zones; those in ceramics and glasses have small zones or none at all.

The property K_{1c} has well-defined values for brittle materials and for those in which the plastic zone is small compared to all dimensions of the test sample so that most of the sample is elastic. When this is not so, a more complex characterization is needed. In very ductile materials the plastic zone size exceeds the width of the sample; then the crack does not propagate at all—the sample simply yields.

When cracks are small, materials yield before they fracture; when they are large, the opposite is true. But what is 'small'? Figure 8.7 shows how the tensile failure stress varies with crack size. When the crack is small, this stress is equal to the yield stress; when large, it falls off according to equation (8.4), which we write (taking $Y = 1$ again) as

$$\sigma_f = \frac{K_{1c}}{\sqrt{\pi c}} \tag{8.12}$$

The transition from yield to fracture is smooth, as shown in the figure, but occurs around the intersection of the two curves, when $\sigma_f = \sigma_y$, giving the transition crack length

$$c_{crit} = \frac{K_{1c}^2}{\pi\sigma_y^2} \tag{8.13}$$

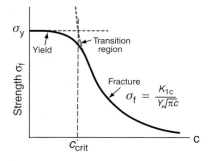

Figure 8.7   The transition from yield to fracture at the critical crack length c_{crit}.

Table 8.1   Approximate crack lengths for transition between yield and fracture

Material class	Transition crack length, c_{crit} (mm)
Metals	1–1000
Polymers	0.1–10
Ceramics	0.01–0.1
Composites	0.1–10

This is the same as the plastic zone size at fracture (equation (8.11)), when $K_1 = K_{1c}$.

Table 8.1 lists the range of values for the main material classes. These crack lengths are a measure of the *damage tolerance* of the material. Tough metals are able to contain large cracks but still yield in a predictable, ductile, manner. Ceramics (which always contain small cracks) fail in a brittle way at stresses far below their yield strengths. Glass can be used structurally, but requires careful treatment to prevent surface flaws developing. Polymers are perceived as tough, due to their resistance to impact when they are not cracked. But the table shows that defects less than 1 mm can be sufficient to cause some polymers to fail in a brittle manner.

8.4   Material property charts for toughness

The fracture toughness–modulus chart

The fracture toughness K_{1c} is plotted against modulus E in Figure 8.8. The range of K_{1c} is large: from less than 0.01 to over 100 MPa.m$^{1/2}$. At the lower end of this range are brittle materials, which, when loaded, remain elastic until they fracture. For these, linear elastic fracture mechanics works well, and the fracture toughness itself is a well-defined property. At the upper end lie the super-tough materials, all of which show substantial plasticity before they break. For these the values of K_{1c} are approximate but still helpful in providing a ranking of materials. The figure shows one reason for the dominance of metals in engineering; they almost all have values of K_{1c} above 15 MPa.m$^{1/2}$, a value often quoted as a minimum for conventional design.

The log scales of Figure 8.8 allow us to plot contours of *toughness*, G_c, the apparent fracture surface energy (since $G_c \approx K_{1c}^2/E$). The diagonal broken lines on the chart show that the values of the toughness start at 10^{-3} kJ/m^2 (about equal to the surface energy, γ) and range through almost five decades to over 100 kJ/m^2. On this scale, ceramics ($10^{-3} - 10^{-1}$ kJ/m^2) are much lower than polymers ($10^{-1} - 10$ kJ/m^2); this is part of the reason polymers are more widely used in engineering than ceramics, a point we return to in Chapter 10.

The fracture toughness–strength chart

Strength-limited design relies on the component yielding before it fractures. This involves a comparison between strength and toughness—Figure 8.9 shows

Figure 8.8 A chart of fracture toughness K_{Ic} and modulus E. The contours show the toughness, G_c.

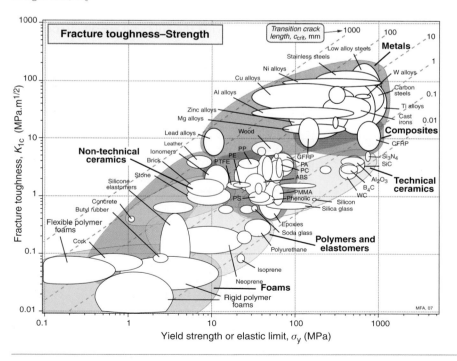

Figure 8.9 A chart of fracture toughness K_{1c} and yield strength σ_y. The contours show the transition crack size, c_{crit}.

them on a property chart. Metals are both strong and tough—that is why they have become the workhorse materials of mechanical and structural engineering.

The stress at which fracture occurs depends on both K_{1c} and the crack length c (equation (8.12)). The transition crack length c_{crit} at which ductile behavior is replaced by brittle is given by equation (8.13). It is plotted on the chart as broken lines labeled 'Transition crack length'. The values vary enormously, from near-atomic dimensions for brittle ceramics and glasses to almost a meter for the most ductile of metals like copper or lead. Materials towards the bottom right have high strength and low toughness; they *fracture before they yield.* Those towards the top left do the opposite: they *yield before they fracture.*

The diagram has application in selecting materials for the safe design of load-bearing structures (Chapter 10). The strength–fracture toughness chart is also useful for assessing the influence of composition and processing on properties.

8.5 Drilling down: the origins of toughness

Surface energy

The surface energy of a solid is the energy it costs to make it. It is an energy per unit area, units J/m^2. Think of taking a 1 m cube of material and cutting it in half to make two new surfaces, one above and one below, as in Figure 8.10. To do so we have to provide the cohesive energy associated with the bonds that previously connected across the cut. The atoms are bonded on all sides so the surface atoms lose one-sixth of their bonds when the cut is made. This means that we have to provide one-sixth of the cohesive energy H_c (an energy per unit

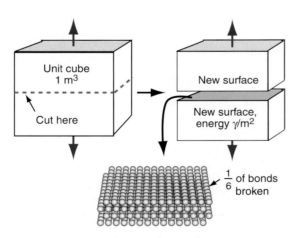

Figure 8.10   When new surface is created as here, atomic bonds are broken, requiring some fraction of the cohesive energy, H_c.

volume) to a slice $4r_o$ thick, were r_o is the atom radius, thus to a volume $4r_o \, m^3$. So the surface energy should be:

$$2\gamma \approx \frac{1}{6} H_c \cdot 4r_o \text{ or } \gamma \approx \frac{1}{3} H_c \cdot r_o$$

with H_c, typically, $3 \times 10^{10} \, J/m^3$ and r_o typically $10^{-10} \, m$, so surface energies are around $1 \, J/m^2$.

The toughness G_c cannot be less than 2γ. The chart of Figure 8.8 shows contours of G_c; for most materials its value is hundreds of times larger than 2γ. Where is the extra energy going? The answer is: into plastic work. We will examine that in more detail in a moment. First, cleavage fracture.

Brittle 'cleavage' fracture

Brittle fracture is characteristic of ceramics and glasses. These have very high yield strengths, giving them no way to relieve the crack tip stresses by plastic flow. This means that, near the tip, the stress reaches the ideal strength (about $E/15$, Chapter 6). That is enough to tear the atomic bonds apart, allowing the crack to grow as in Figure 8.11. And since $K_1 = \sigma \sqrt{\pi c}$ an increase in c means an increase in K_1, causing the crack to accelerate until it reaches the

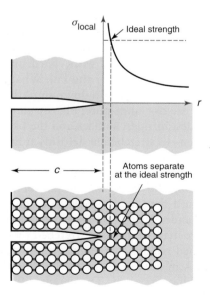

Figure 8.11   Cleavage fracture. The local stress rises as $1/\sqrt{r}$ towards the crack tip. If it exceeds that required to break inter-atomic bonds (the 'ideal strength') they separate, giving a cleavage fracture. Very little energy is absorbed.

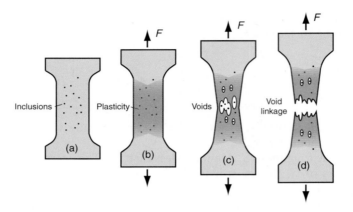

Figure 8.12   Ductile fracture. Plasticity, shown in red, concentrates stress on inclusions that fracture or separate from the matrix, nucleating voids that grow and link, ultimately causing fracture.

speed of sound—that is why brittle materials fail with a bang. Some polymers are brittle, particularly the amorphous ones. The crack tip stresses unzip the weak van der Waals bonds between the molecules.

Tough 'ductile' fracture

To understand how cracks propagate in ductile materials, think first of pulling a sample with no crack, as in Figure 8.12. Ductile metals deform plastically when loaded above their yield strength, work hardening until the tensile strength is reached. Thereafter, they weaken and fail. What causes the weakening? If ultra-pure, the metal may simply thin down until the cross-section goes to zero. Engineering alloys are not ultra-pure; almost all contain inclusions—small, hard particles of oxides, nitrides, sulfides and the like. As the material—here shown as a test specimen—is stretched, it deforms at first in a uniform way, building up stress at the inclusions, which act as stress concentrations. These either separate from the matrix or fracture, nucleating tiny holes. The holes grow as strain increases, linking and weakening the part of the specimen in which they are most numerous until they finally coalesce to give a *ductile fracture*. Many polymers, too, are ductile. They don't usually contain inclusions because of the way in which they are made. But when stretched they *craze*—tiny cracks open up in the most stretched regions, whitening them if the polymer is transparent, simply because the Van der Waals bonds that link their long chains to each other are weak and pull apart easily. The details differ but the results are the same: the crazes nucleate, grow and link to give a ductile fracture.

Return now to the cracked sample, shown in Figure 8.13. The stress still rises as $1/\sqrt{r}$ as the crack tip is approached, but at the point that it exceeds the yield strength σ_y the material yields and a plastic zone develops. Within the plastic

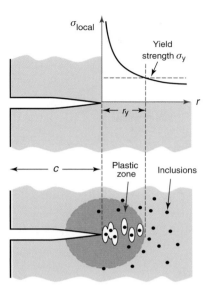

Figure 8.13 If the material is ductile a plastic zone forms at the crack tip. Within it voids nucleate, grow and link, advancing the crack in a ductile mode, absorbing energy in the process.

zone the same sequence as that of Figure 8.12 takes place: voids nucleate, grow and link to give a *ductile fracture*. The crack advances and the process repeats itself. The plasticity blunts the crack and the stress-concentrating effect of a blunt crack is less severe than that of a sharp one, so that at the crack tip itself the stress is just sufficient to keep plastically deforming the material there. This plastic deformation absorbs energy, increasing the toughness G_c.

The ductile-to-brittle transition

A cleavage fracture is much more dangerous than one that is ductile: it occurs without warning or any prior plastic deformation. At low temperatures some metals and all polymers become brittle and the fracture mode switches from one that is ductile to one of cleavage—in fact only those metals with an fcc structure (copper, aluminum, nickel and stainless steel, for example) remain ductile to the lowest temperatures. All others have yield strengths that increase as the temperature falls, with the result that the plastic zone at any crack they contain shrinks until it becomes so small that the fracture mode switches, giving a *ductile-to-brittle transition*. For some steels that transition temperature is as high as 0°C (though for most it is considerably lower), with the result that steel ships, bridges and oil rigs are more likely to fail in winter than in summer. Polymers, too, have a ductile-to-brittle transition, a consideration in selecting those that are to be used in freezers and fridges.

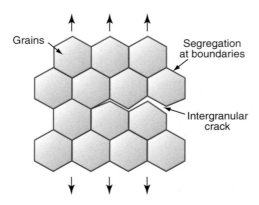

Figure 8.14   Chemical segregation can cause brittle intergranular cracking.

Embrittlement of other kinds

Change of temperature can lead to brittleness; so, too, can chemical segregation. When metals solidify, the grains start as tiny solid crystals suspended in the melt, and grow outwards until they impinge to form grain boundaries. The boundaries, being the last bit to solidify, end up as the repository for the impurities in the alloy. This grain boundary segregation can create a network of low-toughness paths through the material so that, although the bulk of the grains is tough, the material as a whole fails by brittle intergranular fracture (Figure 8.14). The locally different chemistry of grain boundaries causes other problems, such as corrosion (Chapter 17)—one way in which cracks can appear in initially defect-free components.

8.6 Manipulating properties: the strength–toughness trade-off

Metals

It is not easy to make materials that are both strong and tough. The energy absorbed by a crack when it advances, giving toughness, derives from the deformation that occurs in the plastic zone at its tip. Equation (8.11) showed that increasing the yield strength causes the zone to shrink and the smaller the zone, the smaller the toughness. Figure 8.15 shows strength and toughness for wrought and cast aluminum alloys, indicating the strengthening mechanisms. All the alloys have a higher strength and lower toughness than pure aluminum. The wrought alloys show a slight drop in toughness with increasing strength; for the cast alloys the drop is much larger because of intergranular fracture.

Separation within the plastic zone, allowing crack advance, results from the growth of voids that nucleate at inclusions (Figure 8.13). Toughness is increased with no loss of strength if the inclusions are removed, delaying the nucleation of the voids. 'Clean' steels, superalloys and aluminum alloys, made

Figure 8.15 The strength and toughness of wrought and cast aluminum alloys.

by filtering the molten metal before casting, have significantly higher toughness than those made by conventional casting methods.

Polymers and composites

The microstructures and properties of polymers are manipulated by cross-linking and by adjusting the molecular weight and degree of crystallinity. Figures 8.8 and 8.9 showed that the strengths of polymers span around a factor of 5, while fracture toughnesses spans a factor of 20. More dramatic changes are possible by *blending*, by adding *fillers* and by *reinforcement* with chopped or continuous fibers to form composites. Figure 8.16 shows how these influence the modulus E and fracture toughness K_{1c} of polypropylene (PP). Blending or co-polymerization with elastomers such as EPR or EDPM ('impact modifiers') reduces the modulus but increases the fracture toughness K_{1c} and toughness G_c. Filling with cheap powdered glass, talc or calcium carbonate more than doubles the modulus, but at the expense of some loss of toughness.

Reinforcement with glass or carbon fibers most effectively increases both modulus and fracture toughness K_{1c}, and in the case of glass, the toughness G_c as well. How is it that a relatively brittle polymer ($K_{1c} \approx 3\,\text{MPa.m}^{1/2}$) mixed with even more brittle fibers (glass $K_{1c} \approx 0.8\,\text{MPa.m}^{1/2}$) can give a composite K_{1c} as high as $10\,\text{MPa.m}^{1/2}$? The answer is illustrated in Figure 8.17. The fine fibers contain only tiny flaws and consequently have high strengths. When a crack grows in the matrix, the fibers remain intact and bridge the crack. This promotes

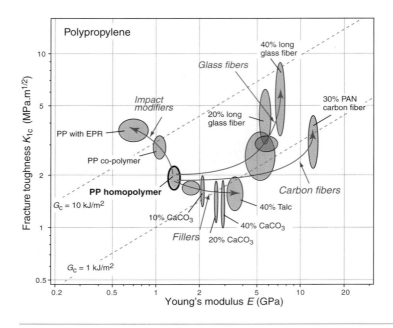

Figure 8.16 The strength and toughness of polypropylene, showing the effect of fillers, impact modifiers and fibers.

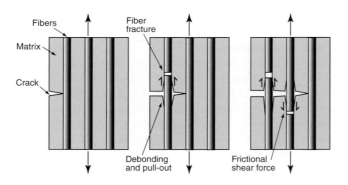

Figure 8.17 Toughening by fibers. The pull-out force opposes the opening of the crack.

multiple cracking—each contributing its own energy and thereby raising the overall dissipation. When the fibers do break, the breaks are statistically distributed, leaving ligaments of fiber buried in the matrix. Fiber pull-out, as the cracks open up, dissipates more energy by friction. Composites are processed to control adhesion between the fibers and the matrix to maximize the toughening by these mechanisms.

8.7 Summary and conclusions

Toughness is the resistance of a material to the propagation of a crack. Tough materials are forgiving: they tolerate the presence of cracks, they absorb impact without shattering and, if overloaded, they yield rather than fracture. The dominance of steel as a structural material derives from its unbeatable combination of low cost, high stiffness and strength, and high toughness.

Toughness is properly measured by loading a pre-cracked sample with one of a number of standard geometries, measuring the load at which the crack first propagates. From this is calculated the fracture toughness, K_{1c}, as the value of the stress intensity, K_1, at which the crack advances. Values of K_{1c} above about 15 MPa.m$^{1/2}$ are desirable for damage-tolerant design—design immune to the presence of small cracks. The charts of Figures 8.8 and 8.9 show that most metals and fiber-reinforced composites meet this criterion, but polymers and ceramics fall below it. We shall see in Chapter 10 that some designs require a high value not of K_{1c} but of the toughness $G_c = K_{1c}^2/E$, also shown in the chart of Figure 8.8. Polymers do well by this criterion and it is in applications that require it that they become a good choice. Ceramics are poor by this criterion too, making design with them much more difficult.

High toughness means that crack advance absorbs energy. It does so in the plastic zone that forms at the crack tip; the larger this becomes, the higher is the fracture toughness. As the material within the plastic zone deforms, voids nucleate at inclusions and link, advancing the crack; the cleaner the material, the fewer are the inclusions, contributing further to high toughness. The higher the yield strength, the smaller is the zone until, in ceramics, which have very high yield strengths, the zone becomes vanishingly small, the stresses approach the ideal strength and the material fails by brittle cleavage fracture.

This is the background we need for design. That comes in Chapter 10, after we have examined fracture under cyclic loads.

8.8 Further reading

Broek, D. (1981) *Elementary Engineering Fracture Mechanics*, 3rd edition, Martinus Nijhoff, Boston, USA. ISBN 90-247-2580-1. (*A standard, well-documented, introduction to the intricacies of fracture mechanics.*)

Ewalds, H.L. and Wanhill, R.J.H. (1984) *Fracture Mechanics*, Edward Arnold, London, UK. ISBN 0-7131-3515-8. (*An introduction to fracture mechanics and testing for both static and cyclic loading.*)

Hertzberg, R.W. (1989) *Deformation and Fracture of Engineering Materials*, 3rd edition, Wiley, New York, USA. ISBN 0-471-63589-8. (*A readable and detailed coverage of deformation, fracture and fatigue.*)

Kinloch, A.J. and Young, R.J. (1983) *Fracture Behavior of Polymers*, Elsevier Applied Science, London, UK. ISBN 0-85334-186-9. (*An introduction both to fracture mechanics as it is applied to polymeric systems and to the fracture behavior of different classes of polymers and composites.*)

Knott, J.F. (1973) *The Mechanics of Fracture*, Butterworths, London, UK. ISBN 0-408-70529-9. (*One of the first texts to present a systematic development of the mechanics of fracture; dated, but still a good introduction.*)

Tada, H., Paris, G. and Irwin, G.R. (2000) *The Stress Analysis of Cracks Handbook*, 3rd edition. ISBN 1-86058-304-0. (*Here we have another 'Yellow Pages', like Roark for stress analysis of uncracked bodies—this time of stress intensity factors for a great range of geometries and modes of loading.*)

8.9 Exercises

Exercise E8.1	What is meant by toughness? How does it differ from strength?
Exercise E8.2	Why does a plastic zone form at the tip of a crack when the cracked body is loaded in tension?
Exercise E8.3	Why is there a transition from ductile to brittle behavior at a transition crack length, c_{crit}?
Exercise E8.4	A tensile sample of width 10 mm contains an internal crack of length 0.3 mm. When loaded in tension the crack suddenly propagates when the stress reaches 450 MPa. What is the fracture toughness K_{1c} of the material of the sample? If the material has a modulus E of 200 GPa, what is its toughness G_c? (Assume $Y = 1$).
Exercise E8.5	Use the K_{1c}–E chart of Figure 8.8 to establish: • Whether CFRP has higher fracture toughness K_{1c} than aluminum alloys? • Whether polypropylene (PP) has a higher toughness G_c than aluminum alloys. • Whether polycarbonate (PC) has a higher fracture toughness K_{1c} than glass.
Exercise E8.6	Find epoxy, soda glass and GFRP (epoxy reinforced with glass fibers) on the chart of Figure 8.8 and read off an approximate mean value for the toughness G_c for each. Explain how it is that the toughness of the GFRP is so much larger than that of either of its components.
Exercise E8.7	Use the chart of Figure 8.8 to compare the fracture toughness, K_{1c}, of the two composites GFRP and CFRP. Do the same for their toughness, G_c. What do the values suggest about applications they might best fill?
Exercise E8.8	Use the K_{1c}–σ_y chart of Figure 8.9 to find: • The range of transition crack sizes for stainless steel. • The range of transition crack sizes for polycarbonate (PC). • The range of transition crack sizes for silicon nitride (Si_3N_4).

8.10 Exploring design with CES (use the Level 2 database unless otherwise stated)

Exercise E8.9	Make a bar chart with $\sigma_f = K_{1c}/\sqrt{\pi c}$ for an internal crack of length $2c = 1$ mm plotted on the y-axis (use the 'Advanced' facility to form the function Fracture toughness /((3.142 × 0.0005)^0.5). Which materials have the highest values? Add an axis of density, ρ. Use the new chart to find the two materials with highest values of σ_f/ρ.
Exercise E8.10	Suppose that the resolution limit of the non-destructive testing facility available to you is 1 mm, meaning that it can detect cracks of this length or larger. You are asked to explore which materials will tolerate cracks equal to or smaller than this without brittle fracture. Make a bar chart with $\sigma_f = K_{1c}/\sqrt{\pi c}$ for an internal crack of length $2c = 1$ mm plotted on the y-axis, as in the previous exercise. Add yield strength, σ_y, on the x-axis. The material will fracture in tension if $\sigma_f < \sigma_y$, and it will yield, despite being cracked, if $\sigma_f > \sigma_y$. Plot either on a print out of the chart or in CES (using the line selection tool) a line of slope 1 along which $\sigma_f = \sigma_y$. All the materials above the line will yield, all those below will fracture. Do age-hardened aluminum alloys lie above the line? Does CFRP?
Exercise E8.11	Find data for PVCs in the Level 3 database and make a plot like that of Figure 8.16 showing how fillers, blending and fibers influence modulus and toughness.

8.11 Exploring the science with CES Elements

Exercise E8.12	Exploring the origins of surface energy, γ J/m². The toughness, G_c, cannot be less than 2γ because two new surfaces are created when a material is fractured. What determines γ and how big is it? The text explained how bonds are broken and atoms separated when a new surface is created. It was shown in Section 8.4 that $$\gamma \approx \frac{1}{3}H_c \cdot r_o$$ where H_c is the cohesive energy in J/m³ and r_o is the atomic radius in m. To convert H_c in kJ/mol (as it is in the database) into these units, multiply it by

10^6/ molar volume and to convert r_o from nm into m multiply by 10^{-9}. All this can be done using the 'Advanced' facility in the axis-choice dialog box.

Make a chart with γ calculated in this way on the x-axis and the measured value, 'Surface energy, solid' on the y-axis, and see how good the agreement is.

Exercise E8.13 If you pull on an atomic bond, it breaks completely at a strain of about 0.1. If the atom spacing is a_o, then breaking it requires a displacement $\delta = 0.1a_o$ and doing so creates two new surfaces, each of area a_o^2. If the bond stiffness is S, then the work done to stretch the bonds to breaking point, and hence to create the new surfaces, per unit area, is

$$2\gamma \approx \frac{1}{a_o^2}\left(\frac{1}{2}S\delta\right) = \frac{Ea_o}{20} = \frac{1}{10}Er_o$$

(using equation (4.17) for S), where r_o is the atomic radius. Use the CES Elements database to explore whether real surface energies can be explained in this way. (Watch out for the units.)

Exercise E8.14 Observe the general magnitudes of surface energies γ: they are about 1.5 J/m^2. Thus, the minimum value for G_c should be about 2γ or 3 J/m^2. Return to the CES Edu Level 3 database and find the material with the lowest value of G_c, which you can calculate as K_{Ic}^2/E. Is it comparable with 2γ? Limit the selection to metals and alloys, polymers, technical ceramics and glasses only, using a 'Tree' stage (materials such as foam have artificially low values of G_c because they are mostly air).

Chapter 9
Shake, rattle and roll: cyclic loading, damage and failure

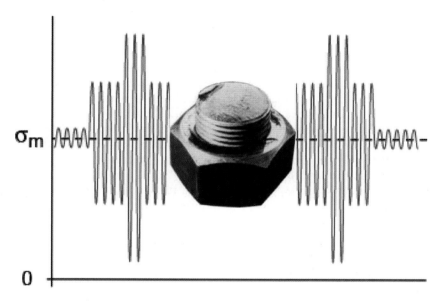

A bolt that has failed by fatigue. The fatigue crack initiated at the root of a thread, which acts as a notch, concentrating stress. (Image courtesy Bolt Science (www.Boltscience.com))

9.1 Introduction and synopsis

German engineering was not always what it is today. The rapidly expanding railway system of the mid-19th century was plagued, in Germany and elsewhere, by accidents, many catastrophic, caused by the failure of the axles of the coaches. The engineer August Wöhler[1] was drafted in to do something about it. It was his systematic tests that first established the characteristics of what we now call *fatigue*.

Repetition is tiring, the cause of many human mistakes and accidents. Materials, too, grow tired if repeatedly stressed, with failure as a consequence. This chapter is about the energy that is dissipated, and the damage and failure that can result, when materials are loaded in a cyclic, repetitive way. Even when the amplitude of the cycles is very small, some energy dissipation or *damping* occurs. Larger amplitudes cause the slow accumulation of damage, a little on each cycle, until a critical level is reached at which a crack forms. Continued cycling causes the crack to grow until the component suddenly fails. Fatigue failure is insidious—there is little sign that anything is happening until, bang, it does. So when the clip breaks off your pen or your office chair collapses, it is probably fatigue that is responsible (cover picture).

We start with cyclic loading at very small amplitudes and the energy loss or damping that goes with it. We then turn to the accumulation of damage and cracking that is associated with fatigue loading proper.

9.2 Vibration and resonance: the damping coefficient

Bells, traditionally, are made of bronze. They can be (and sometimes are) made of glass, and they could (if you could afford it) be made of silicon carbide. Metals, glasses and ceramics all, under the right circumstances, have low material damping or 'internal friction', an important material property when structures vibrate. By contrast, leather, wood (particularly when green or wet) and most foams, elastomers and polymers have high damping, useful when you want to kill vibration.

We are speaking here of elastic response. Until now we have thought of the elastic part of the stress–strain curve as linear and completely reversible so that the elastic energy stored on loading is completely recovered when the load is removed. No material is so perfect—some energy is always lost in a load–unload cycle. If you load just once, you might not notice the loss, but in vibration at, say, acoustic frequencies, the material is loaded between 20 and 20 000 times per second, as in stress cycle (a) of Figure 9.1. Then the energy loss becomes only too obvious.

---

[1] August Wöhler (1819–1914), German engineer, and from 1854 to 1889 Director of the Prussian Imperial Railways. It was Wöhler's systematic studies of metal fatigue that first gave insight into design methods to prevent it.

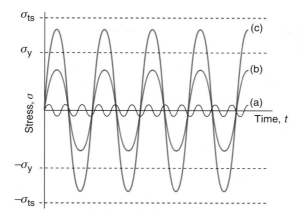

Figure 9.1   Cyclic loading. (a) Very low amplitude acoustic vibration. (b) High-cycle fatigue: cycling well below general yield, σ_y. (c) Low cycle fatigue: cycling above general yield (but below the tensile strength σ_{ts}).

The *mechanical loss coefficient* or *damping coefficient, η* (a dimensionless quantity), measures the degree to which a material dissipates vibrational energy. If an elastic material is loaded, energy is stored (Chapter 5). If it is unloaded, the energy is returned—it is how springs work. But materials have ways of cheating on this: the amount of energy returned is slightly less. The difference is called the *loss coefficient, η*. It is the fraction of the stored elastic energy that is not returned on unloading. If you seek materials for bells, you choose those with low η, but if you want to damp vibration, then you choose those with high.

9.3 Fatigue

The problem of fatigue

Low-amplitude vibration causes no permanent damage in materials. Increase the amplitude, however, and the material starts to suffer fatigue. You will, at some time or other, have used fatigue to flex the lid off a sardine can or to break an out-of-date credit card in two. The cyclic stress hardens the material and causes damage in the form of dislocation tangles to accumulate, from which a crack nucleates and grows until it reaches the critical size for fracture. Anything which is repeatedly loaded (like an oil rig loaded by the waves or a pressure vessel that undergoes pressure cycles), or rotates under load (like an axle), or reciprocates under load (like an automobile connecting rod), or vibrates (like the rotor of a helicopter) is a potential candidate for fatigue failure.

Figure 9.2 shows in a schematic way how the stress on the underside of an aircraft wing might vary during a typical flight, introducing some important

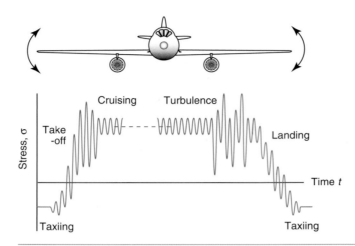

Figure 9.2   Schematic of stress cycling on the underside of a wing during flight.

practicalities of fatigue. First, the amplitude of the stress varies as the aircraft takes off, cruises at altitude, bumps its way through turbulence and finally lands. Second, on the ground the weight of the engines (and the wings themselves) bends them downwards, putting the underside in compression. Once in flight, aerodynamic lift bends them upwards, putting the underside in tension. The amplitude of the stress cycles and their mean values change during flight. Fatigue failure depends on both. And of course it depends on the total number of cycles. Aircraft wings bend to and fro at a frequency of a few hertz. In a trans-Atlantic flight, tens of thousands of loading cycles take place; in the lifetime of the aircraft, it is millions. For this reason, fatigue testing needs to apply tens of millions of fatigue cycles to provide meaningful design data.

The food can and credit card are examples of *low-cycle fatigue*, meaning that the component only survives for a small number of cycles. It is typical of cycling at stresses above the yield stress, σ_y, like that shown in Figure 9.1(c). More significant in engineering terms is *high-cycle fatigue*: here the stresses remain generally elastic and may be well below σ_y, as in cycle (b) of Figure 9.1; cracks nonetheless develop and cause failure, albeit taking many more cycles to do so, as in Wöhler's railway axles.

In both cases we are dealing with components that are initially undamaged, containing no cracks. In these cases, most of the fatigue life is spent generating the crack. Its growth to failure occurs only at the end. We call this *initiation-controlled* fatigue. Some structures contain cracks right from the word go, or are so safety-critical (like aircraft) that they are assumed to have small cracks. There is then no initiation stage—the crack is already there—and the fatigue life is *propagation controlled*, i.e. it depends on the rate at which the crack grows. A different approach to design is then called for—we return to this below.

Figure 9.3 Fatigue strength at 10^7 cycles, the endurance limit, σ_e.

High–cycle fatigue and the S–N curve

Figure 9.3 shows how fatigue characteristics are measured and plotted. A sample is cyclically stressed with amplitude $\Delta\sigma/2$ about a mean value σ_m, and the number of cycles to cause fracture is recorded. The data are presented as $\Delta\sigma$ – N_f ('S–N') curves, where $\Delta\sigma$ is the peak-to-peak range over which the stress varies and N_f is the number of cycles to failure. Most tests use a sinusoidally varying stress with an amplitude σ_a of

$$\sigma_a = \frac{\Delta\sigma}{2} = \frac{\sigma_{max} - \sigma_{min}}{2} \tag{9.1}$$

and a mean stress σ_m of

$$\sigma_m = \frac{\sigma_{max} + \sigma_{min}}{2} \tag{9.2}$$

all defined in the figure. Fatigue data are usually reported for a specified R-value:

$$R = \frac{\sigma_{min}}{\sigma_{max}} \tag{9.3}$$

An R-value of -1 means that the mean stress is zero; an R-value of 0 means the stress cycles from 0 to σ_{max}. For many materials there exists a *fatigue* or *endurance limit*, σ_e (units: MPa). It is the stress amplitude σ_a, about zero mean stress, below which fracture does not occur at all, or occurs only after a very large number ($N_f > 10^7$) cycles. Design against high-cycle fatigue is therefore very similar to strength-limited design, but with the maximum stresses limited by the endurance limit σ_e rather than the yield stress σ_y.

Experiments show that the high-cycle fatigue life is approximately related to the stress range by what is called Basquin's law:

$$\Delta\sigma \, N_f^b = C_1 \qquad (9.4)$$

where b and C_1 are constants; the value of b is small, typically 0.07 and 0.13. Dividing $\Delta\sigma$ by the modulus E gives the strain range $\Delta\varepsilon$ (since the sample is elastic):

$$\Delta\varepsilon = \frac{\Delta\sigma}{E} = \frac{C_1/E}{N_f^b} \qquad (9.5)$$

or, taking logs,

$$Log(\Delta\varepsilon) = -b \, Log(N_f) + Log(C_1/E)$$

This is plotted in Figure 9.4, giving the right-hand, high-cycle fatigue part of the curve with a slope of $-b$.

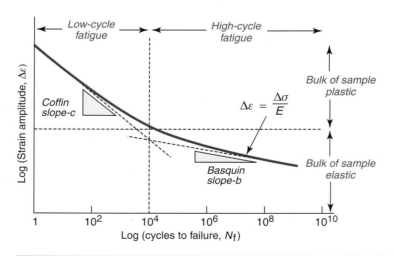

Figure 9.4   The low- and high-cycle regimes of fatigue and their empirical description of fatigue.

Low-cycle fatigue

In low-cycle fatigue the peak stress exceeds yield, so at least initially (before work hardening raises the strength), the entire sample is plastic. Basquin is no help to us here; we need another empirical law, this time that of Dr Lou Coffin:

$$\Delta\varepsilon^{pl} = \frac{C_2}{N_f^c} \tag{9.6}$$

where $\Delta\varepsilon^{pl}$ means the plastic strain range—the total strain minus the (usually small) elastic part. For our purposes we can neglect that distinction and plot it in Figure 9.4 as well, giving the left-hand branch. Coffin's exponent, c, is much larger than Basquin's; typically it is 0.5.

These laws adequately describe the fatigue failure of uncracked components cycled at constant amplitude about a mean stress of zero. But as we saw, real loading histories are often much more complicated (Figure 9.2). How do we make some allowance for variations in mean stress and stress range? Here we need yet more empirical laws, courtesy this time of Goodman and Miner. Goodman's rule relates the stress range $\Delta\sigma_{\sigma_m}$ for failure under a mean stress σ_m to that for failure at zero mean stress $\Delta\sigma_{\sigma_0}$:

$$\Delta\sigma_{\sigma_m} = \Delta\sigma_{\sigma_0}\left(1 - \frac{\sigma_m}{\sigma_{ts}}\right) \tag{9.7}$$

where σ_{ts} is the tensile stress, giving a correction to the stress range. Increasing σ_m causes a small stress range $\Delta\sigma_{\sigma_m}$ to be as damaging as a larger $\Delta\sigma_{\sigma_0}$ applied with zero mean (Figure 9.5(a)). The corrected stress range may then be plugged into Basquin's law.

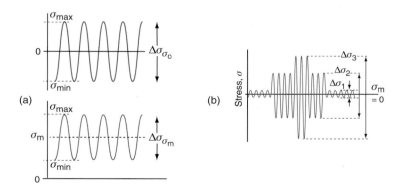

Figure 9.5   (a) The endurance limit refers to a zero mean stress. Goodman's law scales of the stress range to a mean stress σ_m. (b) When the cyclic stress amplitude changes, the life is calculated using Miner's cumulative damage rule.

The variable amplitude problem can be addressed approximately with Miner's *rule of cumulative damage*. Figure 9.5(b) shows an idealized loading history with three stress amplitudes (all about zero mean). Basquin's law gives the number of cycles to failure if each amplitude was maintained throughout the life of the component. So if N_1 cycles are spent at stress amplitude $\Delta\sigma_1$ a fraction $N_1/N_{f,1}$ of the available life is used up, where $N_{f,1}$ is the number of cycles to failure at that stress amplitude. Miner's rule assumes that damage accumulates in this way at each level of stress. Then failure will occur when the sum of the damage fractions reaches 1—that is, when

$$\sum_{i=1}^{n} \frac{N_i}{N_{f,i}} = 1 \qquad (9.8)$$

Goodman's law and Miner's rule are adequate for preliminary design, but they are approximate; in safety-critical applications, tests replicating service conditions are essential. It is for this reason that new models of cars and trucks are driven over rough 'durability tracks' until they fail—it is a test of fatigue performance.

The discussion so far has focused on initiation-controlled fatigue failure of uncracked components. Now it is time to look at those containing cracks.

Fatigue loading of cracked components
In fabricating large structures like bridges, ships, oilrigs, pressure vessels and steam turbines, cracks and other flaws cannot be avoided. Cracks in castings appear because of differential shrinkage during solidification and entrapment of oxide and other inclusions. Welding, a cheap, widely-used joining process, can introduce both cracks and internal stresses caused by the intense local heating. If the cracks are sufficiently large that they can be found it may be possible to repair them, but finding them is the problem. All non-destructive testing (NDT) methods for detecting cracks have a resolution limit; they cannot tell us that there are no cracks, only that there are none longer than the resolution limit, c_{lim}. Thus, it is necessary to assume an initial crack exists and design the structure to survive a given number of loadings. So how is the propagation of a fatigue crack characterized?

Fatigue crack growth is studied by cyclically loading specimens containing a sharp crack of length c like that shown in Figure 9.6. We define the cyclic stress intensity range, ΔK, using equation (8.4), as

$$\Delta K = K_{max} - K_{min} = \Delta\sigma\sqrt{\pi c} \qquad (9.9)$$

The range ΔK increases with time under constant cyclic stress because the crack grows in length: the growth per cycle, dc/dN, increases with ΔK in the way shown in Figure 9.7. The rate is zero below a threshold cyclic stress intensity

Figure 9.6   Cyclic loading of a cracked component. A constant stress amplitude $\Delta\sigma$ gives an increasing amplitude of stress intensity, $\Delta K = \Delta\sigma\sqrt{\pi c}$ as the crack grows in length.

Figure 9.7   Crack growth during cyclic loading.

ΔK_{th}, useful if you want to make sure it does not grow at all. Above it, there is a steady-state regime described by the Paris law:

$$\frac{dc}{dN} = A\,\Delta K^m \tag{9.10}$$

where K and m are constants. At high ΔK the growth rate accelerates as the maximum applied K approaches the fracture toughness K_{1c}. When it reaches K_{1c} the sample fails in a single load cycle.

Safe design against fatigue failure in potentially cracked components means calculating the number of loading cycles that can safely be applied without the crack growing to a dangerous length. We return to this in Chapter 10.

9.4 Charts for endurance limit

The most important single property characterizing fatigue strength is the endurance limit, σ_e, at 10^7 cycles and zero mean stress (an R-value of -1). Given this and the ability to scale it to correct for mean stress, and sum contributions when stress amplitude changes (equations (9.7) and (9.8)), enables design to cope with high-cycle fatigue.

Not surprisingly endurance limit and strength are related. The strongest correlation is with the tensile strength σ_{ts}, shown in the chart of Figure 9.8. The data for metals and polymers cluster around the line

$$\sigma_e \approx 0.33\,\sigma_{ts}$$

shown on the chart. For ceramics and glasses

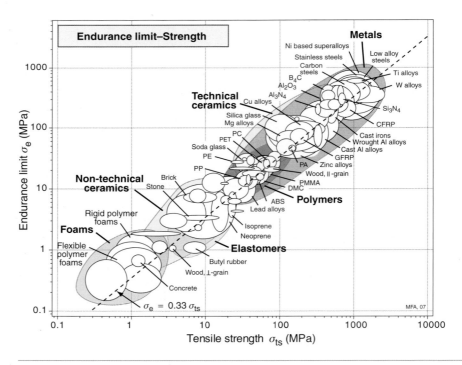

Figure 9.8   The endurance limit plotted against the tensile strength. Almost all materials fail in fatigue at stresses well below the tensile strength.

$$\sigma_e \approx 0.9\,\sigma_{ts}$$

In the next section we examine why.

9.5 Drilling down: the origins of damping and fatigue

Material damping: the mechanical loss coefficient
There are many mechanisms of material damping. Some are associated with a process that has a specific time constant; then the energy loss is centered about a characteristic frequency. Others are frequency independent; they absorb energy at all frequencies. In metals a large part of the loss is caused by small-scale dislocation movement: it is high in soft metals like lead and pure aluminum. Heavily alloyed metals like bronze and high-carbon steels have low loss because the solute pins the dislocations; these are the materials for bells. Exceptionally high loss is found in some cast irons, in manganese–copper alloys and in magnesium, making them useful as materials to dampen vibration in machine tools and test rigs. Engineering ceramics have low damping because the dislocations in them are immobilized by the high lattice resistance (which is why they are hard). Porous ceramics, on the other hand, are filled with cracks, the surfaces of which rub, dissipating energy, when the material is loaded. In polymers, chain segments slide against each other when loaded; the relative motion dissipates energy. The ease with which they slide depends on the ratio of the temperature T to the glass temperature, T_g, of the polymer. When $T/T_g < 1$, the secondary bonds are 'frozen', the modulus is high and the damping is relatively low. When $T/T_g > 1$, the secondary bonds have melted, allowing easy chain slippage; the modulus is low and the damping is high.

Fatigue damage and cracking
A perfectly smooth sample with no changes of section, and containing no inclusions, holes or cracks, would be immune to fatigue provided neither σ_{max} nor σ_{min} exceeds its yield strength. But that is a vision of perfection that is unachievable. Blemishes, small as they are, can be deadly. Rivet holes, sharp changes in section, threads, notches and even surface roughness concentrate stress in the way described in Chapter 7, Figure 7.7. Even though the general stress levels are below yield, the locally magnified stresses can lead to reversing plastic deformation. Dislocation motion is limited to a small volume near the stress concentration, but that is enough to cause damage that finally develops into a tiny crack.

In high-cycle fatigue, once a crack is present it propagates in the way shown in Figure 9.9(a). During the tensile part of a cycle a tiny plastic zone forms at the crack tip, stretching it open and thereby creating a new surface. On the compressive part of the cycle the crack closes again and the newly formed surface folds forwards, advancing the crack. Repeated cycles make it inch forward, leaving tiny ripples on the crack face marking its position on each cycle. These

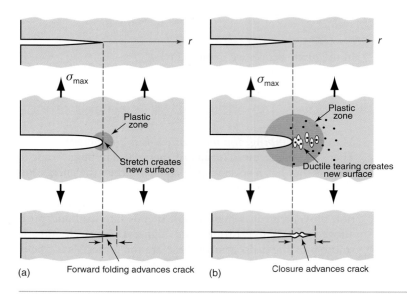

Figure 9.9   (a) In high-cycle fatigue a tiny zone of plasticity forms at the crack tip on each tension cycle; on compression the newly formed surface folds forwards. (b) In low-cycle fatigue the plastic zone is large enough for voids to nucleate and grow within it. Their coalescence further advances the crack.

'striations' are characteristic of a fatigue failure and are useful, in a forensic sense, for revealing where the crack started and how fast it propagated.

In low-cycle fatigue the stresses are higher and the plastic zone larger, as in Figure 9.9(b). It may be so large that the entire sample is plastic, as it is when you flex the lid of a tin to make it break off. The largest strains are at the crack tip, where plasticity now causes voids to nucleate, grow and link, just as in ductile fracture (Chapter 8).

9.6 Manipulating resistance to fatigue

Fatigue life is enhanced by choosing materials that are strong, making sure they contain as few defects as possible, and by giving them a surface layer in which the internal stresses are compressive.

Choosing materials that are strong
The chart of Figure 9.8 established the close connection between endurance limit and tensile strength. Most design is based not on tensile strength but on

Figure 9.10   The drop in fatigue ratio with increase in yield strength for an aluminum alloy and a steel, both of which can be treated to give a range of strengths.

yield strength, and here the correlation is less strong. We define the fatigue ratio, F_r, as

$$F_r = \frac{\text{Endurance limit } \sigma_e}{\text{Yield strength } \sigma_y}$$

Figure 9.10 shows data for F_r for an age-hardening aluminum alloy and a low alloy steel. Both can be heat-treated to increase their yield strength, and higher strength invites higher design stresses. The figure, however, shows that the fatigue ratio F_r falls as strength σ_y increases, meaning that the gain in endurance limit is considerably less than that in yield strength. In propagation-controlled fatigue, too, the combination of higher stress and lower fracture toughness implies that smaller crack sizes will propagate, causing failure (and the resolution of inspection techniques may not be up to the job).

Making sure they contain as few defects as possible
Mention has already been made of clean alloys—alloys with carefully controlled compositions that are filtered when liquid to remove unwanted particles that, in the solid, can nucleate fatigue cracks. Non-destructive testing using X-ray imaging or ultrasonic sensing detects dangerous defects, allowing the part to be

Figure 9.11   Shot peening, one of several ways of creating compressive surface stresses.

rejected. And certain secondary processes like hot isostatic pressing (HIPing), which heats the component under a large hydrostatic pressure of inert argon gas, can seal cracks and collapse porosity.

Providing a compressive surface stress

Cracks only propagate during the tensile part of a stress cycle; a compressive stress forces the crack faces together, clamping it shut. Fatigue cracks frequently start from the surface, so if a thin surface layer can be give an internal stress that is compressive, any crack starting there will remain closed even when the average stress across the entire section is tensile. This is achieved by treatments that plastically compress the surface by shot peening (Figure 9.11): leaf-springs for cars and trucks are 'stress peened', meaning that they are bent to a large deflection and then treated in the way shown in the figure; this increases their fatigue life by a factor of 5. A similar outcome is achieved by sandblasting or burnishing (local deformation with a smooth, polished tool), or diffusing atoms into the surface, expanding a thin layer, which, because it is bonded to the more massive interior, becomes compressed.

9.7 Summary and conclusions

Static structures like the Eiffel Tower or the Golden Gate Bridge stand for centuries; those that move have a much shorter life span. One reason for this is that materials are better at supporting static loads than loads that fluctuate. The long-term cyclic load a material can tolerate, σ_e, is barely one-third of its tensile strength, σ_{ts}. This sensitivity to cyclic loading—fatigue—was unknown until the mid-19th century, when a series of major industrial disasters made it the subject of intense study.

Empirical rules describing fatigue failure and studies of the underlying mechanisms have made fatigue failures less common, but they still happen, notably in rail track, rolling stock, engines and airframes. It is now known that cycling causes damage that slowly accumulates until a crack nucleates, grows slowly and suddenly runs unstably. This is pernicious behavior: the cracks are almost invisible until final failure occurs without warning.

The rules and data for their empirical constants, described in this chapter, give a basis for design to avoid fatigue failure. The scientific studies provide insight that guides the development of materials with greater resistance to fatigue failure. Surface treatments to inhibit crack formation are now standard practice. We return to these in the next chapter, in which the focus is on design issues.

9.8 Further reading

Hertzberg, R.W. (1989) *Deformation and Fracture of Engineering Materials*, 3rd edition, Wiley, New York, USA. ISBN 0-471-63589-8. *(A readable and detailed coverage of deformation, fracture and fatigue.)*

Suresh, S. (1998) *Fatigue of Materials*, 2nd edition, Cambridge University Press, Cambridge, UK. ISBN 0-521-57847-7. *(The place to start for an authoritative introduction to the materials science of fatigue in both ductile and brittle materials.)*

Tada, H., Paris, G. and Irwin, G.R. (2000) *The Stress Analysis of Cracks Handbook*, 3rd edition. ISBN 1-86058-304-0. *(Here we have another 'Yellow Pages', like Roark for stress analysis of uncracked bodies—this time of stress intensity factors for a great range of geometries and modes of loading.)*

9.9 Exercises

Exercise E9.1	What is meant by the mechanical loss coefficient, η, of a material? Give examples of designs in which it would play a role as a design-limiting property.
Exercise E9.2	What is meant by the endurance limit, σ_e, of a material?
Exercise E9.3	What is the fatigue ratio? If the tensile strength σ_{ts} of an alloy is 900 MPa, what, roughly, would you expect its endurance limit σ_e to be?
Exercise E9.4	The figure shows an S–N curve for AISI 4340 steel, hardened to a tensile strength of 1800 MPa.

- What is the endurance limit?
- If cycled for 100 cycles at an amplitude of 1200 MPa and a zero mean stress, will it fail?
- If cycled for 100 000 cycles at an amplitude of 900 MPa and zero mean stress, will it fail?
- If cycled for 100 000 cycles at an amplitude of 800 MPa and a mean stress of 300 MPa, will it fail?

Exercise E9.5   The high-cycle fatigue life, N_f, of an aluminum alloy is described by Basquin's law:

$$\frac{\Delta\sigma}{2} = \sigma_a = 480\left(N_f\right)^{-0.12}$$

(stress in MPa). How many cycles will the material tolerate at a stress amplitude σ_a of ± 70 MPa and zero mean stress? How will this change if the mean stress is 10 MPa? What if the mean stress is -10 MPa?

Exercise E9.6   The low-cycle fatigue of an aluminum alloy is described by Coffin's law:

$$\frac{\Delta\sigma}{2} = \sigma_a = 600\left(N_f\right)^{-0.33}$$

(stress in MPa). How many cycles will the material tolerate at a stress amplitude σ_a of ± 100 MPa?

Exercise E9.7   A material with a tensile stress $\sigma_{ts} = 350$ MPa is loaded cyclically about a mean stress of 70 MPa. If the stress range that will cause fatigue fracture in 10^5 cycles under zero mean stress is ± 60 MPa, what stress range about the mean of 70 MPa will give the same life?

Exercise E9.8   A component made of the AISI 4340 steel with a tensile strength of 1800 MPa and the S–N curve shown in Exercise E9.4 is loaded cyclically between 0 and

1200 MPa. What is the R-value and the mean stress, σ_m? Use Goodman's rule to find the equivalent stress amplitude for an R-value of -1, and read off the fatigue life from the S–N curve.

Exercise E9.9   A material has a threshold cyclic stress intensity ΔK_{th} of 2.5 MPa·m$^{1/2}$. If it contains an internal crack of length 1 mm will it be safe (meaning, no failure) if subjected to continuous cyclic range of tensile stress $\Delta\sigma$ of 50 MPa?

9.10 Exploring design with CES

Exercise E9.10   Make a bar chart of mechanical loss coefficient, η. Low loss materials are used for vibrating systems where damping is to be minimized—bells, high-frequency relays and resonant systems. High loss materials are used when damping is desired—sound deadening cladding for buildings, cars and machinery, for instance. Use the chart to find:

(a) The metal with the lowest loss coefficient.
(b) The metal with the highest loss coefficient.

Exercise E9.11   Use the 'Search' facility to search materials that are used for:

(a) Bells.
(b) Cladding.

Exercise E9.12   Use a 'Limit' stage, applied to the *Surface treatment* data table, to find surface treatment processes that enhance fatigue resistance. To do this:

(a) Change the selection table to Process Universe Level 2 Surface treatment, open a 'Limit' stage, locate Function of treatment and click on Fatigue resistance > Apply. Copy and report the results.
(b) Repeat, using the Level 3 Surface treatment data table.

Exercise E9.13   Explore the relationship between fatigue ratio and strength for a heat-treatable low alloy steel AISI 4340. The endurance limit σ_e is stored in the database under the heading 'Fatigue strength at 10^7 cycles'.

(a) Plot the fatigue ratio σ_e/σ_y against the yield strength σ_y.
(b) Plot the fatigue ratio σ_e/σ_{ts} against the tensile strength σ_{ts}.

Use Level 3 of the database, apply a 'Tree' stage to isolate the folder for the low alloy steel, AISI 4340, then make the two charts, hiding all the other materials. How do you explain the trends?

Chapter 10
Keeping it all together: fracture-limited design

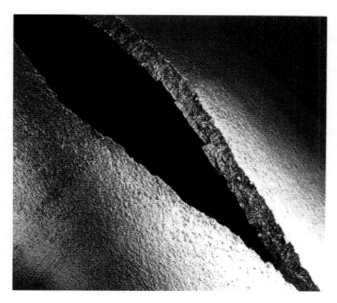

Fractured pipe (Image courtesy Prof. Robert Akid,
School of Engineering, Sheffield Hallam University,
Sheffield UK).

10.1 Introduction and synopsis

It is very hard to build a structure that is completely without cracks. As explained in Chapter 8, cracks caused by shrinkage in casting and welding, by the cracking of inclusions during rolling, or just caused by careless machining are common-place. And even if there are no cracks to start with, cyclic loading (Chapter 9) and corrosion (Chapter 17) can introduce them later.

This creates the need for design methods to deal with cracked structures. The idea of the *tensile stress intensity* K_1 caused by a crack was introduced in Chapter 8. It depends on crack length, component geometry and the way the component is loaded. We start with standard solutions for the stress intensity K_1 associated with generic configurations—there are others, but this is enough to get started. Cracks will not propagate if this K_1 is kept below the fracture toughness K_{1c} of the material of the structure.

This might suggest that the best material to resist fracture is the one with the highest K_{1c}, and in load-limited design it is. But sometimes the requirement is not to carry a given load without failure, but to store a given energy (springs) or allow a given deflection (elastic couplings) without failure. Then the best choice of material involves combinations of K_{1c} and Young's modulus E, as we will see in Section 10.3.

Cyclic loading, too, causes fracture. Chapter 9 explained how accumulating fatigue damage causes cracks to form and grow, a little on each cycle, until the length is such that the stress intensity exceeds K_{1c}, when fast fracture follows. Stress intensities in cyclic loading are given by the same standard solutions as for static loading. Cracks will not propagate if the cyclic stress intensity range, ΔK, is below the threshold range ΔK_{th}.

The chapter concludes with case studies illustrating how these ideas enable *no-fail* or *fail-safe* design.

10.2 Standard solutions to fracture problems

As we saw in Chapter 8, sharp cracks concentrate stress in an elastic body. To summarize, the local stress falls off as $1/r^{1/2}$ with radial distance r from the crack tip. A tensile stress σ, applied normal to the plane of a crack of length $2c$ contained in an infinite plate, gives rise to a local stress field σ_1 that is tensile in the plane containing the crack and given by

$$\sigma_1 = \frac{Y\sigma\sqrt{\pi c}}{\sqrt{2\pi r}} \tag{10.1}$$

where r is measured from the crack tip and Y is a constant. The mode 1 *stress intensity factor*, K_1, is defined as

$$K_1 = Y\,\sigma\sqrt{\pi c} \tag{10.2}$$

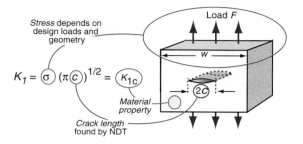

Figure 10.1 The key players in fracture-limited design: load and geometry, crack length and material.

The crack propagates when

$$K_1 = Y\,\sigma\sqrt{\pi c} = K_{1c} \tag{10.3}$$

where K_{1c} is the fracture toughness, a material property.

Equation (10.3), illustrated in Figure 10.1, summarizes the key factors. In the design of a component to avoid fracture, we can manipulate the geometry and loads, we can monitor it using non-destructive testing (NDT) to be sure it contains no crack larger than an acceptable value, and we can choose materials with adequate fracture toughness.

Expressions for K_1 for various geometries and modes of loading are given in Figure 10.2. When the stress is uniform and the crack length is small compared with all specimen dimensions, Y is equal to or close to 1. As the crack extends in a uniformly loaded component, it interacts with the free surfaces, requiring correction factors. If, in addition, the stress field is non-uniform (as it is in an elastically bent beam), Y differs from 1. More accurate approximations and other less common loading geometries can be found in the references listed in 'Further reading'.

10.3 Material indices for fracture-safe design

Among mechanical engineers there is a rule of thumb already mentioned: avoid materials with a fracture toughness K_{1c} less than 15 MPa.m$^{1/2}$. The K_{1c}–E chart of Figure 8.8 shows that almost all metals pass: most have values of K_{1c} in the range of 20–100 in these units. Just a few fail: white cast iron and some powder-metallurgy products have values as low as 10 MPa.m$^{1/2}$. Ordinary engineering ceramics have values in the range 1–6 MPa.m$^{1/2}$; mechanical engineers view them with deep suspicion. But engineering polymers have even smaller values of K_{1c} in the range 0.5–3 MPa.m$^{1/2}$ and yet engineers use them all the time. What is going on here?

Figure 10.2 Stress intensities K_1 associated with short cracks.

In a *load-limited* design—the structural member of a bridge or the wing-spar of an aircraft, for instance (Figure 10.3 (a, b))—the part will fail in a brittle way if the stress exceeds that given by equation (10.3). To maximize the load we want materials with highest values of

$$M_1 = K_{1c} \tag{10.4}$$

Figure 10.3   (a, b) Load-limited design. (c) Energy-limited design. (d) Displacement-limited design. Each must perform its function without fracturing.

But not all designs are load limited; some are *energy limited*, others are *deflection limited*. Then the criterion for selection changes. Consider, then, the other two scenarios sketched in Figure 10.3.

Energy-limited design

Not all design is load limited. Springs and containment systems for turbines and flywheels (to catch the bits if they disintegrate) are *energy* limited. Take the spring (Figure 10.3(c)) as an example. The elastic energy U_e stored in it is the integral over the volume of the energy per unit volume, U_e, where

$$U_e = \frac{1}{2} \sigma \varepsilon = \frac{1}{2} \frac{\sigma^2}{E}$$

The stress is limited by the fracture stress of equation (10.3) so that—if 'failure' means 'fracture'—the maximum energy per unit volume that the spring can store is

$$U_e^{max} = \frac{Y^2}{2\pi c} \left(\frac{K_{1c}^2}{E} \right)$$

For a given initial flaw size c, the energy is maximized by choosing materials with large values of

$$M_2 = \frac{K_{1c}^2}{E} \approx G_c \qquad (10.5)$$

where G_c is the toughness, defined in Chapter 8.

Displacement-limited design

There is a third scenario: that of *displacement*-limited design (Figure 10.3(d)). Snap-on bottle tops, elastic hinges and couplings are displacement limited: they must allow sufficient elastic displacement to permit the snap-action or flexure without failure, requiring a large failure strain ε_f. The strain is related to the stress by Hooke's law, $\varepsilon = \sigma/E$, and the stress is limited by the fracture equation (10.3). Thus, the failure strain is

$$\varepsilon_f = \frac{C}{\sqrt{\pi c_{max}}}\frac{K_{1c}}{E}$$

The best materials for displacement-limited design are those with large values of

$$M_3 = \frac{K_{1c}}{E} \tag{10.6}$$

Plotting indices on charts

Figure 10.4 shows the K_{1c}–E chart again. It allows materials to be compared by values of fracture toughness, M_1, by toughness, M_2, and by values of the deflection-limited index, M_3. As the engineer's rule of thumb demands, almost all metals have values of K_{1c} that lie above the 15 MPa.m$^{1/2}$ acceptance level for load-limited design, shown as a horizontal selection line in the figure. Polymers and ceramics do not.

Figure 10.4   The K_{1c}–E chart showing the three indices M_1, M_2 and M_3.

The line showing M_2 in Figure 10.4 is placed at the value $1\,kJ/m^2$. Materials with values of M_2 greater than this have a degree of shock resistance with which engineers feel comfortable (another rule of thumb). Metals, composites and many polymers qualify; ceramics do not. When we come to deflection-limited design, the picture changes again. The line shows the index $M_3 = K_{1c}/E$ at the value $10^{-3}m^{1/2}$. It illustrates why polymers find such wide application: when the design is deflection limited, polymers—particularly nylons (PA), polycarbonate (PC) and ABS—are better than the best metals.

The figure gives further insights. Mechanical engineers' love of metals (and, more recently, of composites) is inspired not merely by the appeal of their K_{1c} values. Metals are good by all three criteria (K_{1c}, K_{1c}^2/E and K_{1c}/E). Polymers have good values of K_{1c}/E and are acceptable by K_{1c}^2/E. Ceramics are poor by *all three* criteria. Herein lie the deeper roots of the engineers' distrust of ceramics.

10.4 Case studies

Forensic fracture mechanics: pressure vessels

An aerosol can is a pressure vessel. So, too, is a propane gas cylinder, the body of an airliner, the boiler of a power station and the containment of a nuclear reactor. Their function is to contain a gas under pressure—CO_2, propane, air, steam. Failure can be catastrophic. Think of the bang something as feeble as a party balloon can make, then multiply it by 10^{14}—yes, really—and you begin to get an idea of what happens when a large pressure vessel explodes. When talking pressure, safety is an issue. Here is an example.

A truck-mounted propane tank (Figure 10.5) was filled, then driven to the driver's home, where he parked it in the sun and went inside to have lunch,

Figure 10.5   A cylindrical pressure vessel with a cracked weld.

Table 10.1   Tank dimensions

Maximum design pressure, *p*	1.5 MPa
Wall thickness, *t*	14 mm (thicker at weld)
Outer diameter, 2*R*	1680 mm
Length, *L*	3710 mm

leaving the engine running. As he was drinking his coffee the tank exploded, causing considerable property damage and one death. The tank was made of rolled AISI 1030 steel plate, closed by a longitudinal weld and two domes joined to the rolled plate by girth welds. The failure occurred through the longitudinal weld, causing the tank to burst (see cover picture). Subsequent examination showed that the weld had contained a surface crack of depth 10 mm that had been there for a long time, as indicated by discoloration and by striations showing that it was growing slowly by fatigue each time the tank was emptied and refilled. At first sight it appears that the crack was the direct cause of the failure.

That's where fracture mechanics comes in. Measurements on a section of the tank wall gave a fracture toughness of 45 MPa.m$^{1/2}$. The dimensions of the tank are listed in Table 10.1. It was designed for a working pressure of 1.4 MPa, limited at 1.5 MPa by a pressure release valve.

The tensile stress in the wall of a thin-walled cylindrical pressure vessel of radius *R* and wall thickness *t* containing a pressure *p* (Figure 10.5) was given in Chapter 4. With the values listed above the stress in the tank wall is

$$\sigma = \frac{pR}{t} = 90 \text{ MPa} \tag{10.7}$$

A plate with a fracture toughness 45 MPa.m$^{1/2}$ containing a surface crack of depth 10 mm will fail at the stress

$$\sigma = \frac{K_{1c}}{1.1\sqrt{\pi c}} = 231 \text{ MPa}$$

This is 2.5 times higher than the stress in the tank wall. The pressure needed to generate this stress is 3.8 MPa—far higher than the safety limit of 1.5 MPa.

At first it appears that the calculations cannot be correct since the tank had a relief valve set at the working pressure, but the discrepancy prompted the investigators to look further. An inspection of the relief valve showed rust and corrosion, rendering it inoperative. Further tests confirmed that heat from the sun and from the exhaust system of the truck raised the temperature of the tank, vaporizing the liquefied gas and driving the pressure past the value of 3.8 MPa needed to make the crack propagate. The direct cause of the failure was the jammed pressure release valve. The crack would not have propagated at the normal operating pressure.

That story makes you wonder if there is a way to design such that there is warning before cracks propagate unstably. There is. It is called *fail-safe design*.

Fail-safe design

If structures are made by welding, it must be assumed that cracks are present. A number of techniques exist for detecting cracks and measuring their length without damaging the component or structure; they are called *non-destructive tests* (NDT). X-ray imaging is one—alloy wheels for cars are checked in this way. Ultrasonic testing is another—sound waves are reflected by a crack and can be analyzed to determine its size. Surface cracks can be revealed with fluorescent dyes. When none of these methods can be used, then proof testing—filling the pressure vessel with water and pressurizing it to a level above the planned working pressure—demonstrates that there are no cracks large enough to propagate during service. All these techniques have a resolution limit—a crack size c_{lim} below which detection is not possible. They do not demonstrate that the component or structure is crack free, only that there are no cracks larger than the resolution limit.

The first condition, obviously, is to design in such a way that the stresses are everywhere less than that required to make a crack of length c_{lim} propagate. Applying this condition gives the allowable pressure:

$$p \leq \frac{t}{R} \frac{K_{1c}}{\sqrt{\pi c_{lim}}} \tag{10.8}$$

The largest pressure (for a given R, t and c_{lim}) is carried by the material with the greatest value of

$$M_1 = K_{1c} \tag{10.9}$$

—our load-limited index of the last section.

But this design is not fail-safe. If the inspection is faulty or if, for some other reason, a crack of length greater than c_{lim} appears, catastrophe follows. Greater security is obtained by requiring that the crack will not propagate even if the stress is sufficient to cause general yield—for then the vessel will deform stably in a way that can be detected. This condition, called the *yield-before-break* criterion, is expressed requiring that the stress to cause fracture (from equation (10.3)) is less than the yield stress σ_y, giving

$$c_{max} \leq \frac{Y^2}{\pi} \left[\frac{K_{1c}}{\sigma_y} \right]^2 \tag{10.10}$$

Using this criterion, the tolerable crack size, and thus the integrity of the vessel, is maximized by choosing a material with the largest value of

$$M_5 = \frac{K_{1c}^2}{\sigma_y} \tag{10.11}$$

Large pressure vessels cannot always be X-rayed or ultrasonically tested, and proof testing them may be impractical. Further, cracks can grow slowly because of corrosion or cyclic loading, so that a single examination at the beginning of service life is not sufficient. Then safety can be ensured by arranging that a crack just large enough to penetrate both the inner and the outer surface of the vessel is still stable, because the leak caused by the crack can be detected. This condition, called the *leak-before-break* criterion, is achieved by setting $2c = t$ in equation (10.3), giving

$$\sigma = \frac{Y K_{1c}}{\sqrt{\pi t / 2}} \qquad (10.12)$$

The wall thickness t of the pressure vessel must also contain the pressure p without yielding. From equation (10.7), this means that

$$t \geq \frac{pR}{2\sigma_y}$$

Substituting this into the previous equation (with $\sigma = \sigma_f$) gives

$$p \leq \frac{4Y^2}{\pi R} \left(\frac{K_{1c}^2}{\sigma_y} \right)$$

The pressure is carried most safely by the material with the greatest value of

$$M_5 = \frac{K_{1c}^2}{\sigma_y} \qquad (10.13)$$

Both M_4 and M_5 could be made large by making the yield strength of the wall, σ_y, very small: lead, for instance, has high values of both, but you would not choose it for a pressure vessel. That is because the vessel wall must also be thin, both for economy of material and to keep it light. The thickness of the wall, from equation (10.7), is inversely proportional to the yield strength, σ_y. Thus, we wish also to maximize

$$M_6 = \sigma_y \qquad (10.14)$$

narrowing further the choice of material.

The way to explore these criteria is to use the $K_{1c} - \sigma_y$ chart introduced in Chapter 8. The indices M_1, M_4, M_5 and M_6 can be plotted onto it as lines of slope 0, 1, 1/2 and as lines that are vertical. Take 'leak-before-break' as an example. A diagonal line corresponding to a constant value of $M_5 = K_{1c}^2 / \sigma_y$ links materials with equal performance; those above the line are better. The line shown in the figure excludes everything but the toughest steels, copper, nickel and titanium alloys. A second selection line for M_6 at $\sigma_y = 200$ MPa narrows the selection further. Exercises at the end of this chapter develop further examples of the use of the others.

Boiler failures used to be commonplace—there are even songs about it. Now they are rare, though when safety margins are pared to a minimum (rockets) or

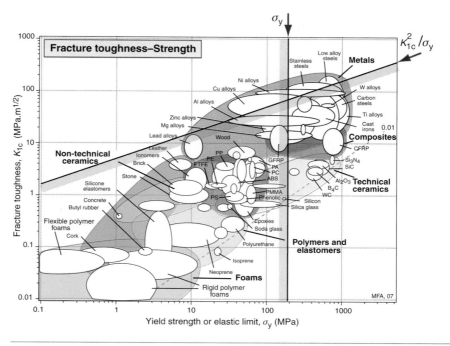

Figure 10.6   Selecting the best materials for leak-before-break design. The additional requirement of high yield strength is also shown. The best choices are the materials in the small area at the top right.

maintenance is neglected (the propane tank) pressure vessels still occasionally fail. This (relative) success is one of the major contributions of fracture mechanics to engineering practice.

Materials to resist fatigue: con-rods for high-performance engines

The engine of a family car is designed to tolerate speeds up to about 6000 rpm—100 revolutions per second. That of a Formula 1 racing car revs to about three times that. Comparing an F1 engine to that of a family car is like comparing a Rolex watch to an alarm clock, and their cost reflects this: about $200 000 per engine. Performance, here, is everything.

The connecting rods of a high-performance engine are critical components: if one fails the engine self-destructs. Yet to minimize inertial forces and bearing loads, each must weigh as little as possible. This implies the use of light, strong materials, stressed near their limits. When minimizing cost not maximizing performance is the objective, con-rods are made from cast iron. But here we want performance. What, then, are the best materials for such con-rods?

An F1 engine is not designed to last long—about 30 hours—but 30 hours at around 15 000 rpm means about 3×10^6 cycles of loading and that means high-cycle fatigue. For simplicity, assume that the shaft has a uniform section

Figure 10.7   A connecting rod.

of area A and length L (Figure 10.7) and that it carries a cyclic load $\pm F$. Its mass is

$$m = AL\,\rho \tag{10.15}$$

where ρ is the density. The fatigue constraint requires that

$$\frac{F}{A} \leq \sigma_e$$

where σ_e is the endurance limit of the material of which the con-rod is made. Using this to eliminate A in equation (10.15) gives an equation for the mass:

$$m \geq F\,\ell \left(\frac{\rho}{\sigma_e}\right)$$

containing the material index

$$M = \frac{\sigma_e}{\rho} \tag{10.16}$$

Materials with high values of this index are identified by creating a chart with σ_e and ρ as axes, applying an additional, standard, constraint that the fracture toughness exceeds $15\,\mathrm{MPa.m^{1/2}}$ (Figure 10.8). Materials near the top-left corner are attractive candidates: high-strength magnesium, aluminum and ultra-high-strength steels; best of all are titanium alloys and CFRP. This last material has been identified by others as attractive in this application. To go further we need S–N curves for the most attractive candidates. Figure 10.9 shows such a curve for the most widely used of titanium alloys: the alloy Ti–6Al–4V. The safe stress amplitude at $R = -1$ for a design life of 2.5×10^6 cycles can be read off: it is 620 MPa.

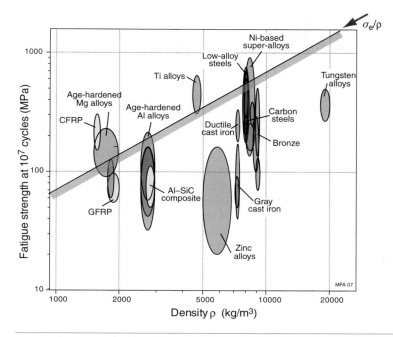

Figure 10.8 Endurance limit and density for high-strength metals and composites. (All materials with $K_{1c} < 15$ MPa.m$^{1/2}$ were screened out before making this chart.)

Figure 10.9 An S–N curve for the titanium alloy Ti–6Al–4V.

If the selection is repeated using the much larger CES Level 3 database, really exotic materials emerge: aluminum reinforced with SiC, boron or Al_2O_3 fibers, beryllium alloys and a number of high-performance carbon-reinforced composites. A con-rod made of CFRP sounds a difficult thing to make, but at least three prototypes have been made and tested. They use a compression strut (a CFRP tube) with an outer wrapping of filament-wound fibers to attach the bearing housings (made of titanium or aluminum) to the ends of the strut.

Rail cracking

At 12.23 p.m. on 7 October 2000, the 12.10 express from King's Cross (London) to Leeds (in the north of England) entered a curve just outside Hatfield at 115 mph—the maximum permitted for that stretch of line—and left the track. Four people died. More are killed on British roads every day, but the event brought much of the railway system, and through this, the country, to a near halt. It was clear almost immediately that poor maintenance and neglect had allowed cracks to develop in much of the country's track; it was an extreme case of this cracking that caused the crash.

Rail-head cracking is no surprise—rails are replaced regularly for that reason. The cracks develop because of the extreme contact stresses—up to 1 GPa—at the point where the wheels contact the rail (Figure 10.10(a)). This deforms the rail surface, and because the driving wheels also exert a shear traction, the steel surface is smeared in a direction opposite to that of the motion of the train, creating a very heavily sheared surface layer. Cracks nucleate in this layer after about 60 000 load cycles, and because of the shearing they lie nearly parallel to the surface, as in Figure 10.10(b). The deformation puts the top of the rail into compression, balanced by tension in the main web. Welded rails are also pre-tensioned to compensate for thermal expansion on a hot day, so this internal stress adds to the pre-tension that was already there.

The mystery is how these harmless surface cracks continue to grow when the stress in the rail head is compressive. One strange observation gives a clue: it is that the cracks propagate much faster when the rails are wet than when they are dry—as they remain in tunnels, for instance. Research following the Hatfield crash finally revealed the mechanism, illustrated in Figure 10.10(c). As a wheel approaches, a surface crack is forced open. If the rail is wet, water (shown in red) is drawn in. As the wheel passes the mouth of the crack, it is forced shut, trapping the water, which is compressed to a high pressure. Figure 10.2 lists the stress intensity of a crack containing an internal pressure p:

$$K_1 = p\sqrt{\pi c} \qquad (10.17)$$

So although the rail head is compressed, at the tip of the crack there is tension—enough to drive the crack forward. The crack inches forward and downward, driven by this hydraulic pressure, and in doing so it grows out of the compression field of the head of the rail and into the tensile field of the web. It is at this point that it suddenly propagates, fracturing the rail.

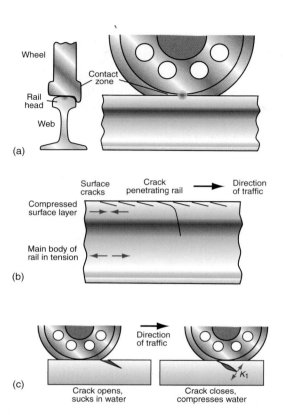

Figure 10.10   Rail cracking. (a) The rolling contact. (b) The surface cracks, one of which has penetrated the part of the rail that is in tension. (c) The mechanism by which the crack advances.

The answer—looking at equation (10.17)—is to make sure the surface cracks do not get too long; K_1 for a short pressurized crack is small, for a long one it is large. Proper rail maintenance involves regularly grinding the top surface of the track to remove the cracks, and when this has caused significant loss of section, replacing the rails altogether.

Fatigue crack growth: living with cracks
The crack in the LPG tank of the earlier case study showed evidence of striations: ripples marking the successive positions of the crack front as it slowly advanced, driven by the cyclic pressurizing of the tank. The cause of the explosion was over-pressure caused by poor maintenance, but suppose maintenance had been good and no over-pressure had occurred. How long would it have lasted before fatigue grew the crack to an unstable size? We calculate this size, c^*, by inverting equation (10.3):

$$c^* = \frac{K_{1c}^2}{Y^2 \pi \sigma^2} \qquad (10.18)$$

With $K_{lc} = 45$ MPa.m$^{1/2}$, $\sigma = 90$ MPa and $Y = 1$, the answer is $c^* = 80$ mm. How long will it take to grow that big? A week? A month? Twenty years?

This sort of question arises in assessing the safety of large plant that (because of welds) must be assumed to contain cracks: the casing of steam turbines, chemical engineering equipment, boilers and pipe-work. The cost of replacement of a large turbine casing is considerable and there is associated down-time. It does not make sense to take it out of service if, despite the crack, it is perfectly safe.

Fatigue crack growth was described in Chapter 9. Its rate is described by the equation

$$\frac{dc}{dN} = A \, \Delta K^m \tag{10.19}$$

in which the cyclic stress intensity range, ΔK, is

$$\Delta K = K_{max} - K_{min} = \Delta\sigma\sqrt{\pi c}$$

Inserting this into equation (10.19) gives

$$\frac{dc}{dN} = A \, \Delta\sigma^m (\pi c)^{m/2} \tag{10.20}$$

The residual life, N_R, is found by integrating this between $c = c_i$ (the initial crack length) and $c = c^*$, the value at which fast fracture will occur:

$$N_R = \int_0^N R \, dN = \frac{1}{A \, \Delta\sigma^m} \int_{c_i}^{c^*} \frac{dc}{(\pi c)^{m/2}} \tag{10.21}$$

We shall suppose the steel of which the tank is made has a crack-growth exponent $m = 4$ and a value of $A = 2.5 \times 10^{-6}$ when $\Delta\sigma$ is in MPa. Taking the initial crack length as 10 mm, the integral gives

$$N_R = \frac{1}{A \, \Delta\sigma^4}\left(\frac{1}{c_i} - \frac{1}{c^*}\right) = 7 \times 10^6 \text{ cycles}$$

Assuming that the tank is pressurized once per day, it will last forever.

The calculation illustrates how life-limited fatigue crack growth is calculated. It ignored the fact that the tank wall was only 12 mm thick. The crack will penetrate the wall, releasing propane and revealing its presence, long before it is 80 mm long. In this sense, the tank is fail-safe provided the stress amplitude does not exceed 1.5 MPa.

Designing for fracture
Manufacturers who distribute their products packaged in toughened envelopes, sheathed in plastic, or contained in aluminum or steel cans, need to design the containment so that the purchaser can get at the contents. As you will know, not all do—even getting the wrapper off a newly purchased CD is a pain. And

there is a more serious dimension. Producers of foods, drugs, chemicals, even washing powder, now package their products in 'break to access' packaging so that the purchaser knows that it has not been tampered with. How do you arrange protection and yet enable fracture? One part of the answer is to choose the right material; the other is to provide stress concentrations to focus stress on the break-lines or to use adhesives that have high shear strength but low peel resistance.

Start with material: materials with low ductility, in thin sheets, tear easily. If 'opening' means pulling in such a way as to tear (as it often does), then the first step is a to choose a material with adequate stiffness, strength and durability, but low ductility. The lids of top-opening drinks cans are made of a different alloy than the can itself for exactly this reason.

Next, stress concentration. This means reducing the section by grooving or serrating the package locally along the line where tearing is wanted. Toilet paper, as we all know, hardly ever tears along the perforations, but it was the right idea. The sardine can on the cover of this chapter is made of low-ductility aluminum alloy with a groove with a sharp radius of curvature along the tear line to provide a stress concentration factor (Chapter 7) of

$$K_{sc} \approx \left(1 + \frac{1}{2}\sqrt{\frac{c}{\rho}}\right) \tag{10.22}$$

where c is the groove depth and ρ its root radius and the factor ½ appears because the loading is shear rather than tension. A 0.2 mm groove with root radius of 0.02 mm gives a local stress that is 2.5 times higher than that elsewhere, localizing the tearing at the groove.

The peel-strip of a CD wrapper is not a groove; it is an additional *thicker* strip. How does this apparent reinforcement make it easier to tear open the package? Figure 7.7 provides the answer: it is because *any* sudden change of section concentrates stress. If the strip thickness is c and the radius where it joins the wrapping is ρ, the stress concentration is still given by equation (10.22).

The alternative to tearing is adhesive peeling. Figure 10.11 shows an adhesive joint before and during peeling. The adhesive has shear strength σ_s^* and toughness G_c^*. Adhesively bonded packaging must accept in-plane tension, since to protect the content it must support the mass of its contents and handling loads. The in-plane pull force F_t that the joint can carry without failing is

$$F_t = \sigma_s^* A = \sigma_s^* w L$$

where $A = wL$ is the area of the bonded surface. To open it, a peel force F_p is applied. The lower part of Figure 10.11 shows how F_p does work when the joint is peeled back by a distance δx, creating new surface of area $w\,\delta x$. This requires and energy $G_c^* w\,\delta x$ (since G_c^*, the toughness, is the energy to create unit area of new surface). The work done by F_p must provide this energy, giving

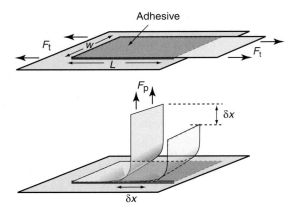

Figure 10.11   Peeling of an adhesive bond.

$$F_p \, \delta x = G_c^* \, w \, \delta x$$

Thus, $F_p = w \, G_c^*$ The ratio of the peel force to the tensile force is

$$\frac{F_p}{F_t} = \frac{G_c^*}{\sigma_s^* L}$$

Adhesive joints are designed to have a particular value for this ratio. The choice of adhesive sets the values of G_c^* and σ_s^*, allowing the length L to be chosen to give tensile strength with ease of peeling.

10.5 Summary and conclusions

Elastic deformation is recoverable. Plastic deformation is gradual and detectable before really bad things happen. Failure by fast fracture is none of these. It has been the cause of many great engineering disasters: collapsed bridges, burst boilers, rail accidents, aircraft crashes.

A component will fail by fast fracture if the stress intensity K_1 at any crack-like defect it contains exceeds the fracture toughness K_{1c} of the material of which it is made. The understanding of fast fracture and the development of design methods to deal with it are relatively new—before 1950 little was known. Both are now on a solid basis. This chapter introduced them, describing the ways in which K_1 is calculated, how K_{1c} is measured and the various scenarios in which it is relevant. It ended with examples of their application.

Much the same is true of failure by fatigue. A seemingly healthy component, one that has served its purpose well, fails without warning because an initially small and harmless crack has grown until it reached the size at which fast fracture takes over. Small components may, initially, be crack free, but cyclic loading can lead to crack nucleation and subsequent growth; design is then based on the S–N curve of the material. Larger structures, particularly those that are welded, cannot be assumed to be crack free; then the life is estimated from the predicted rate of crack growth, integrating it to find the number of cycles to grow the crack to a dangerous size.

Fracture, however, is not always bad. Enabling fracture allows you to get at the contents of shrink-wrapped packages, access food and drugs in tamper-proof containers and at drinks in pop-top cans. Here the trick is to design-in stress concentrations that create locally high stresses along the desired fracture path, and to choose a material with adequate strength but low ductility.

10.6 Further reading

Broek, D. (1981) *Elementary Engineering Fracture Mechanics*, 3rd edition, Martinus Nijhoff, Boston, USA. ISBN 90-247-2580-1. *(A standard, well-documented, introduction to the intricacies of fracture mechanics.)*

Ewalds, H.L. and Wanhill, R.J.H. (1984) *Fracture Mechanics*, Edward Arnold, London, UK. ISBN 0-7131-3515-8. *(An introduction to fracture mechanics and testing for both static and cyclic loading.)*

Hertzberg, R.W. (1989) *Deformation and Fracture of Engineering Materials*, 3rd edition, Wiley, New York, USA. ISBN 0-471-63589-8. *(A readable and detailed coverage of deformation, fracture and fatigue.)*

Hudson, C.M. and Rich, T.P. (eds) (1986) *Case Histories Involving Fatigue and Fracture Mechanics*, ASTM STP 918, American Society for Testing and Materials, USA. ISBN 0-8031-0485-5. *(A compilation of case studies of fatigue failures, many of them disastrous, analyzing what went wrong.)*

Suresh, S. (1998) *Fatigue of Materials*, 2nd edition, Cambridge University Press, Cambridge, UK. ISBN 0-521-57847-7. *(The place to start for an authoritative introduction to the materials science of fatigue in both ductile and brittle materials, with case studies.)*

Tada, H., Paris, G. and Irwin, G.R. (2000) *The Stress Analysis of Cracks Handbook*, 3rd edition. ISBN 1-86058-304-0. *(A comprehensive catalog of stress intensity factors for a great range of geometries and modes of loading.)*

10.7 Exercises

Exercise E10.1   Supersonic wind tunnels store air under high pressure in cylindrical pressure vessels—the pressure, when released, produces hypersonic rates of flow. The pressure vessels are routinely proof tested to ensure that they are safe. If such a cylinder, of diameter 400 mm and wall thickness 20 mm, made of a steel with a fracture toughness of 42 MPa.m$^{1/2}$, survives a proof test to 40 MPa (400 atmospheres), what is the length of the largest crack it might contain?

Exercise E10.2 You are asked to select a polymer to make a flexible coupling. The polymer must have a modulus greater than 2 GPa. The objective is to maximize the available flexure without fracture. Use the chart of Figure 10.4 to identify two good choices to meet these requirements. Are there any metals that are as good?

Exercise E10.3 Crash barriers like car fenders must absorb energy without total fracture. The most effective are those that deform plastically, absorbing energy in plastic work, but they are not reusable. Fenders that remain elastic spring back after impact. For practical reasons the material must have a modulus greater than 10 GPa. Use the chart of Figure 10.4 to find non-metallic materials for elastic fenders, assuming that the overriding consideration is that the displacement before fracture is as great as possible (the constraint on modulus ensures that it absorbs enough energy).

Exercise E10.4 Materials with high toughness G_c generally have high modulus. Sometimes, however, the need is for high toughness with low modulus, so that the component has some flexibility. Use the chart of Figure 10.4 to find the material (from among those on the chart) that has a modulus less than 0.5 GPa and the highest toughness G_c. List applications of this material that you can think of that exploit this combination of properties.

Exercise E10.5 If you want to support a precision optical system (laser metrology equipment, for instance) on a stable platform, you put it on a granite slab supported on end plinths to bring it to working height. (Granite can be ground to a flat surface and is thermally very stable and hard wearing.) The granite chosen for one such table has a fracture toughness of 0.9 MPa.m$^{1/2}$ and is known, from NDT procedures, to contain internal cracks up to 5 mm in length. If the table is 2 m long and 1 m deep, simply supported at its ends and must carry a uniformly distributed load of 2000 N on its upper surface (as in Figure 7.2 of the text), what is the minimum thickness the slab must have? Include the self-weight of the slab in the analysis. Assume that at least one of the cracks will lie in the part of the beam that carries the highest tensile stress—that is, at the lower surface (the density of granite is 2700 kg/m^3).

Exercise E10.6 The figure below shows, in (a), a cylindrical tie-rod with diameter 20 mm. The plan is to use it to carry a cyclic load with a stress range ± 200 kN. The figure also shows the S–N curve of the material of which it is to be made. Will it survive without failure for at least 10^5 cycles?

Exercise E10.7   The component of the previous exercise was made and tested. It failed in less than 10^5 cycles. A post-mortem revealed that it had fractured at the root of a threaded end, shown in (b) in the figure. The threads have a depth of 1.5 mm and a root radius of 0.2 mm. Given this additional information, how many cycles would you expect it to survive?

Exercise E10.8   An adhesive has a toughness $G_c = 100 \, \text{J/m}^2$ and a shear strength $\sigma_s = 0.1$ MPa. What must the dimensions of the bonded area of a lap-joint be if it is to carry an in-plane tensile F_t of 100 N but allow peeling at a force F_p of 5 N?

Exercise E10.9   The figure on next page shows a component to be made from the high-strength aerospace alloy Ti–6Al–4V. It will be loaded cyclically at a nominal stress range of ± 210 MPa. How long will it last?

10.8 Exploring design with CES (use Level 2, Materials, for all selections)

Exercise E10.10   Use the 'Search' facility in CES to find materials that are used for:

(a) Pressure vessels.
(b) Connecting rods.
(c) Rail track.

(Search using the singular—e.g. pressure vessel—since that will find the plural too.)

Exercise E10.11   You are asked, as in Exercise E10.2, to select a polymer to make a flexible coupling. The polymer must have a modulus E greater than 2 GPa. The objective is to maximize, the available flexure without fracture, and that means materials with high K_{1c}/E. Use a 'Limit' stage to impose the constraint on E, then use a 'Graph' stage to make a chart for K_{1c} and E, put on an appropriate selection line and move it until only three materials remain in the Results window. What are they? Rank them by price.

Exercise E10.12   You are asked to recommend materials that have yield strength above 500 MPa and perform best in a design based on the leak-before-break criterion. Construct an appropriate limit stage and chart, put on the necessary selection line and list the three materials you would recommend.

Exercise E10.13   Repeat the previous selection, applying instead a yield-before-break selection criterion.

Exercise E10.14   A material is sought for a high-performance con-rod, requiring that the index has a high value of the index σ_e/ρ, where σ_e is the endurance limit (the fatigue strength at 10^7 cycles). It must have enough toughness to tolerate stress concentrations, requiring that $K_{1c} > 15$ MPa.m$^{1/2}$. Make the appropriate selection stages and list the three materials that best meet the criteria.

Chapter 11

Rub, slither and seize: friction and wear

A disc brake (image courtesy Paul Turnbull, SEMTA, Watford, UK. www.gcseinenginering)

11.1 Introduction and synopsis

God, it is said, created materials, but their surfaces are the work of the devil. They are certainly the source of many problems. When surfaces touch and slide, there is friction; and where there is friction, there is wear. Tribologists—the collective noun for those who study friction and wear—are fond of citing the enormous cost, through lost energy and worn equipment, for which these two phenomena are responsible. It is certainly true that, if friction could be eliminated, the efficiency of engines, gearboxes, drive trains and the like would increase enormously; and if wear could be eradicated, they would also last much longer. But before accepting this negative image, one should remember that, without wear, pencils would not write on paper or chalk on blackboards; and without friction, one would slither off the slightest incline.

Tribological properties are not attributes of one material alone, but of one material sliding on another with—almost always—a third in between. The number of combinations is far too great to allow choice in a simple, systematic way. The selection of materials for bearings, drives and sliding seals relies heavily on experience. This experience is captured in reference sources (for which, see 'Further reading'); in the end it is these that must be consulted. But it does help to have a feel for the magnitude of friction coefficients and wear rates, and an idea of how these relate to material class. This chapter provides it.

11.2 Tribological properties

When two surfaces are placed in contact under a normal load F_n and one is made to slide over the other, a force F_s opposes the motion. This force F_s is proportional to F_n but does not depend on the area of the surface, facts discovered by none other than Leonardo da Vinci[1], then forgotten, then rediscovered 200 years later by Amontons[2], whose name they carry. The *coefficient of friction* μ is defined as in Figure 11.1:

$$\mu = \frac{F_s}{F_n} \tag{11.1}$$

When surfaces slide, they wear. Material is lost from both surfaces, even when one is much harder than the other. The *wear rate*, W, is conventionally defined as

---

[1] Leonardo da Vinci (1452–1519), Renaissance painter, architect, engineer, mathematician, philosopher and prolific inventor.
[2] Guillaume Amontons (1663–1705), French scientific instrument maker and physicist, perfector of the *clepsydra* or water clock—a device for measuring time by letting water flow from a container through a tiny hole.

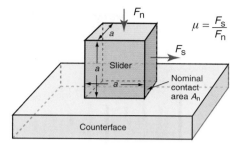

Figure 11.1 The definition of the coefficient of friction, μ. Here F_n is the force normal to the interface and F_s the force parallel to the interface required to maintain sliding.

$$W = \frac{\text{Volume of material removed from contact surface (m}^3)}{\text{Distance slid (m)}} \quad (11.2)$$

and thus has units of m^2. A more useful quantity, for our purposes, is the wear rate per unit area of surface, called the *specific wear rate*, Ω:

$$\Omega = \frac{W}{A_n} \quad (11.3)$$

which is dimensionless. It increases with bearing pressure P (the normal force F_n divided by A_n), such that

$$\Omega = k_a \frac{F_n}{A_n} = k_a P \quad (11.4)$$

where k_a, the *Archard wear constant*, has units of (MPa)$^{-1}$. It is a measure of the propensity of a sliding couple for wear: high k_a means rapid wear at a given bearing pressure.

11.3 Charting friction and wear

The coefficient of friction
Approximate values for the coefficient of friction for dry—that is, unlubricated—sliding of materials on a steel counterface are shown in Figure 11.2. The values depend on the counterface material, but the range is much the same: typically, $\mu \approx 0.5$. Certain combinations show much higher values, either because they seize when rubbed together (a soft metal rubbed on itself in a vacuum with

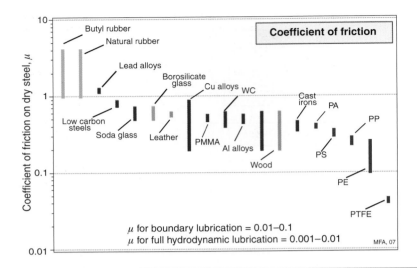

Figure 11.2   The coefficient of friction μ of materials sliding on an unlubricated steel surface.

no lubrication, for instance) or because one surface has a sufficiently low modulus that it conforms to the rough surface of the other (rubber on rough concrete). At the other extreme are sliding combinations with exceptionally low coefficients of friction, such as PTFE, or bronze bearings loaded with graphite, sliding on polished steel. Here the coefficient of friction falls as low as 0.04, though this is still high compared with friction for lubricated surfaces, as noted at the bottom of the diagram.

The wear rate – hardness chart

Figure 11.3 shows the Archard wear constant k_a plotted against hardness, here expressed in MPa. (Hardness in MPa $\approx$ 10 $\times$ Vickers hardness, H_V.) The bearing pressure P is the quantity specified by the design. The ability of a surface to resist a static contact pressure is measured by its hardness, so the *maximum* bearing pressure P_{max} is just the hardness H of the softer surface. Thus, the wear rate of a bearing surface can be written:

$$\Omega = k_a P = k_a \left(\frac{P}{P_{max}}\right) H$$

Two material properties appear in this equation: the wear constant k_a and the hardness, H—they are the axes of the chart. The diagonal contours show the product

$$K = k_a H \tag{11.5}$$

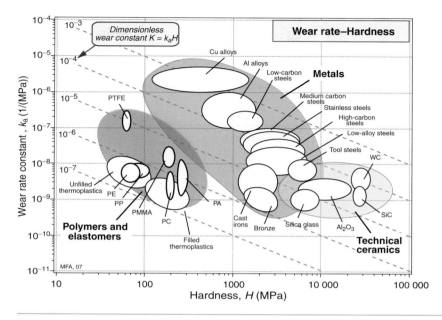

Figure 11.3 The normalized wear rate for lubricated sliding, k_a plotted against hardness H, here expressed in MPa rather than Vickers (H in MPa = 10 H_v). The chart gives an overview of the way in which common engineering materials behave.

The best materials for bearings for a given bearing pressure P are those with the lowest value of k_a—that is, those nearest the bottom of the diagram. On the other hand, an efficient bearing, in terms of size or weight, will be loaded to a safe fraction of its maximum bearing pressure—that is, to a constant value of P/P_{max}—and for these, materials with the lowest values of the product $k_a H$ are best. Materials of a given class (metals, for instance) tend to lie along a downward sloping diagonal across the figure, reflecting the fact that low wear rate is associated with high hardness.

11.4 The physics of friction and wear[3]

Friction

Surfaces, no matter how meticulously honed and polished, are never perfectly flat. The roughness, defined in Chapter 18, depends on how the surface was

---

[3] We owe our understanding of friction and wear to the work of many engineers and scientists, above all to the remarkably durable and productive collaboration between the Australian Philip Bowden (1903–1968) and David Tabor (1913–2005), English of Lithuanian descent, between 1936 and 1968 in Melbourne and Cambridge.

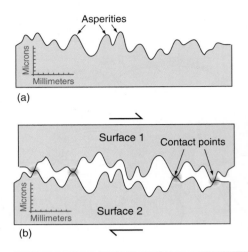

Figure 11.4   (a) The profile of a surface, much magnified vertically. (b) Two surfaces in contact touch only at asperities.

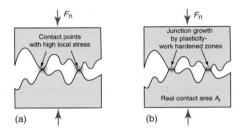

Figure 11.5   When surfaces first touch, high contact stresses appear, as in (a). These cause plasticity and junction growth until the junction area A_r can just support the load F_n without further plasticity, as in (b).

made—a sand-casting has a very rough surface, one that is precision-machined has a much smoother one. But none—*none*—are perfectly smooth. Magnified vertically, they look like Figure 11.4(a)—an endless range of asperity peaks and troughs. Thus, when two surfaces are placed in contact, no matter how carefully they have been crafted, they touch only at points where asperities meet as in Figure 11.4(b). The load F_n is supported solely by the contacting asperities. The real contact area A_r is only a tiny fraction of the apparent, nominal, area A_n.

What is this fraction? When they first touch the asperities deform elastically. But even small loads cause large contact stresses, enough to cause plastic deformation as in Figure 11.5(a). The contact points flatten, forming junctions with a total area A_r, as in Figure 11.5(b). The total load transmitted across the surface is

$$F_n = A_r \sigma_y \tag{11.6}$$

where σ_y is the yield strength. Thus, the real contact area is

$$A_r = \frac{F_n}{\sigma_y} \tag{11.7}$$

To see how small this is, return to the cube of Figure 11.1. Its mass is $m = \rho\, a^3$, so the force needed to support it is $F_n = \rho\, g\, a^3 = \rho\, g\, a\, A_n = A_r\, \sigma_y$, where $A_n = a^2$ is the nominal contact area and g is the acceleration due to gravity (9.81 m/s²). Thus, the ratio of real to nominal contact areas is

$$\frac{A_r}{A_n} = \frac{\rho g a}{\sigma_y}$$

Thus, a 100 mm cube of ordinary steel (density 7900 kg/m³, yield strength 200 MPa), weighing a hefty 7.9 kg, contacts any hard surface on which it rests over only 0.4 mm².

Now think of sliding one surface over the other. If the junctions weld together (as they do when surfaces are clean), it will need a shear stress F_s equal to the shear yield strength k of the material to shear them or, if the materials differ, one that is equal to the shear strength of the softer material. Thus, for unlubricated sliding

$$\frac{F_s}{A_r} \approx k$$

or, since $k \approx \sigma_y/2$,

$$F_s \approx \frac{1}{2} A_r \sigma_y \tag{11.8}$$

Dividing this by equation (11.6) gives

$$\mu = \frac{F_s}{F_n} \approx 0.5 \tag{11.9}$$

a value right in the middle of the spread in Figure 11.2.

So far we have spoken of *sliding* friction—F_s is the force to maintain a steady rate of sliding. If surfaces are left in static contact, the junctions tend to grow by creep, and the bonding between them becomes stronger so that the force to start sliding is larger than that to maintain it. This means that the coefficient of friction to start sliding, called the static coefficient, μ_s, is larger than that to sustain it, resulting in stick-slip behavior that can cause vibration in brakes, and is the way that the bow-hair causes a violin string to resonate.

Wear

We distinguish two sorts of sliding wear. In *adhesive wear*, characteristic of wear between the same or similar materials (copper on aluminum, for example), asperity tips, stuck together, shear off to give wear damage. In *abrasive wear*, characteristic of wear when one surface is much harder than the other (steel on plastic, say), the asperity tips of the harder material plough through the softer one, abrading it like sandpaper.

Figure 11.6 shows the way in which adhesion between surfaces causes wear fragments to be torn from the surface. To minimize the rate of wear we need to minimize the size of the fragments; that means minimizing the area of contact. Since $A_r = F_n/\sigma_y$, reducing the load or increasing σ_y (that is, the surface hardness) reduces wear.

Figure 11.7 shows abrasive wear. The asperities of the harder surface slice off segments of the softer one, like grating cheese. More commonly the problem is

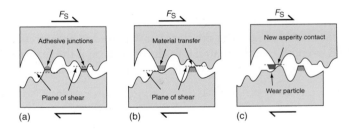

Figure 11.6   Adhesive wear: adhesion at the work-hardened junctions causes shear-off of material when the surfaces slide.

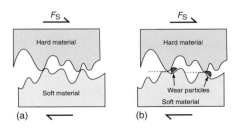

Figure 11.7   Abrasive wear: asperities of the harder material, or embedded hard particles, plough through the softer one, grinding off wear particles.

one of contamination: hard particles that sneak into the surface—dust, sand, oxidized wear particles—embed themselves in one surface and plough through the other. It is to prevent this that cars have air and oil filters.

11.5 Design and selection: materials to manage friction and wear

Bearings allow two components to move relative to each other in one or two dimensions while constraining them in the others. The key criteria are low friction coupled with the ability to support the bearing load F_n without yielding, overheating or wearing excessively. *Brakes* and *clutches*, by contrast, rely on high friction, but there is more to their design than that, as we shall see in a moment. But first, bearings.

Dry sliding, as we have seen, generally leads to friction with a coefficient μ of about 0.5. If the bearings of the drive-train of your car had a μ of this magnitude, negligible power would reach the wheels—almost all would be lost as heat in the bearing surfaces. In reality, about 15% of the power of the engine is lost in friction because of two innovations: *lubrication* and the replacement of sliding by *rolling bearings*.

Lubrication

Most lubricants are viscous oils containing additives with polar groups (fatty acids with $-OH^-$ groups not unlike ordinary soap) that bind to metal surfaces to create a chemisorbed boundary layer (Figure 11.8). A layer of the lubricant between the surfaces allows easy shear, impeded only by the viscosity of the oil, giving a coefficient of fraction in the range 0.001–0.01. When the sliding surfaces are static, the bearing load F_n tends to squeeze the oil out of the bearing, but the bonding of the additive is strong enough to resist, leaving the two surfaces separated by a double-layer of lubricating molecules. This *boundary-layer lubrication* is sufficient to protect the surfaces during start-up and sliding at low loads, but it is not something to rely on for continuous use or high performance. For this you need *hydrodynamic lubrication*.

Figure 11.8   The formation of a boundary layer of chemisorbed, polar molecules; the red dot represents the polar group, the tail the hydrocarbon chain attached to it.

(a)

(b)

Figure 11.9   The distribution of pressure under a sliding and a rotating bearing with hydrodynamic lubrication.

The viscosity of the oil is both a curse and a blessing—a curse because it exerts a drag (and thus friction) on the sliding components; a blessing because, without it, hydrodynamic lubrication would not be possible. Figure 11.9 explains (an explanation first provided by Reynolds[4]). Sliding bearings are designed so that the surfaces are slightly inclined. Oil is dragged along the wedge-shaped region between them from the wide to the narrow end, as in Figure 11.9(a). The rise in pressure as the oil is compressed forces the surfaces apart against the force of the normal load carried by the bearing. The maximum pressure is reached just behind the center of the slider, where the color is most intense in the figure. The same principle governs the self-alignment of a rotating journal bearing, as in Figure 11.9(b). Again, the lubricant is dragged by viscosity into the wide end of the wedge-shaped region and swept towards the narrow end, where the rise of pressure pushes the journal back onto the axis of the bearing. The result is that the surfaces never touch but remain separated by a thin film of oil or other lubricating fluid.

---

[4] Osborne Reynolds (1842–1912), Northern Irishman, brilliant mathematician and engineer but awful lecturer, known for his contributions to lubrication, turbulence and for his celebrated Number—the dimensionless group describing the transition from laminar to turbulent flow.

At high temperatures, hydrocarbon lubricants decompose. Solid lubricants— PTFE, molybdenum disulfide (MoS_2) and graphite (one form of carbon)— extend the operating temperature to 500°C. PTFE, good to about 250°C, has an intrinsically low coefficient of friction. Moly-disulfide (up to 300°C) and graphite (to 500°C) rely on their lamellar (layer-like) crystal structure, able to shear easily on certain crystal planes, to provide coefficients of friction well below 0.1. All these can be applied as lubricants or incorporated into the pores of porous bearings (made by sintering powdered metal like bronze and cast iron) to give self-lubricating bearings.

That's the background. So what materials are good for bearings? That depends—plane bearings or roller bearings?

Plane bearings

The surfaces of sliding bearings, if properly lubricated, do not touch. Then almost any material that meets the other constraints of the design will do—steel, cast iron, brass, bronze. The key material properties are strength (to carry the bearing loads), corrosion resistance and conformability—the ability to adapt to slight misalignment. Full hydrodynamic lubrication breaks down under static loads or when sliding velocities are small, and long-term reliance on boundary lubrication is risky. Bearing materials are designed to cope with this.

When soft metals slide over each other (lead on lead, for instance), the junctions are weak but their area of contact is large so μ is large. When hard metals slide (e.g. steel on steel), the junctions are small but they are strong, and again μ is large. The way to get low friction with high strength is to support a soft metal with a hard one, giving weak junctions of small area. White-metal bearings, for example, consist of soft lead or tin supported in a matrix of a stronger material; leaded-bronze bearings consist of a matrix of strong, corrosion-resistant bronze containing particles of soft lead that smear out to give an easily sheared surface film when asperities touch; polymer-impregnated porous bearings are made by partly sintering bronze and then forcing a polymer (usually PTFE) into its pores. Bearings like these are not designed to run dry, but if they do, the soft component saves them, at least for a while.

Rolling bearings

In rolling element bearings the load is carried by a set of balls or rollers trapped between an inner and an outer *race* by a *cage*, as in Figure 11.10. When the loading is radial the load is carried at any instant by half the balls or rollers, but not equally: the most heavily loaded one carries about $5/Z$ times the applied load, where Z is the total number of balls or rollers in the bearing. Its contact area with the race is small, so the contact stresses are high.

The main requirement is that the deformation of the components of the bearing remain at all times elastic, so the criterion for material choice is that of high hardness. That usually means a high-carbon steel or a tool steel. If the environment is corrosive or the temperature high, stainless steel or ceramic (silicon carbide or silicon nitride) are used. When the loads are very light the races can be

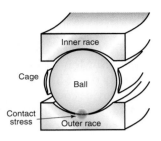

Figure 11.10   A rolling-contact bearing. The contact stresses are high, requiring materials with high hardness and resistance to fatigue failure.

made from thermoplastic—polypropylene or acetal—with the advantage that no lubrication is needed (cheap bike pedals sometimes have such bearings).

When rolling element bearings fail, it is usually because of fatigue—the repeated loading and unloading of ball and race as the bearing rotates causes cracks to nucleate, splitting the balls or pitting the races. Inclusions in the material are sources of fatigue cracks, so steels for rolling bearings are processed to be as clean and inclusion free as possible by processing under vacuum.

High friction: materials for brakes and clutches

Applying friction sounds easy—just rub. Brakes for horse-drawn carriages did just that—spoon-shaped blocks of wood, sometimes faced with leather (bigger μ), were wound down with a screw to rub on the rim of the carriage wheel. But brake materials today do much more.

The friction coefficient of materials for brake linings is—surprisingly—not special: μ is about 0.4. The key thing is not μ but the ability to apply it without vibration, wear or fade. That is not easy when you reckon that, to stop a one-tonne car from 100 kph (60 mph) requires that you find somewhere to put about 0.4 MJ. That's enough (if you don't lose any) to heat 1 kg of steel to 800°C. Brake materials must apply friction at temperatures that can approach such values, and they must do so consistently and without the stick-slip that causes 'brake squeal'—not acceptable in even the cheapest car. That means materials that can tolerate heat, conduct it away to limit temperature rise and—curiously—lubricate, since it is this that quenches squeal.

For most of the history of the automobile, brake linings were made of compacted asbestos (able to tolerate 2000°C) mixed with brass or copper particles or wires for heat conduction and strength. When it became apparent that asbestos dust—the product of brake wear—was perhaps not so good for you, new materials appeared. They are an example of hybrid material design methods: no one material can do what you want, so we synthesize one that does.

Today's materials are a synthesis: a matrix, an abrasive additive to increase friction, a reinforcement—one to help conduct heat if possible—and a lubricant

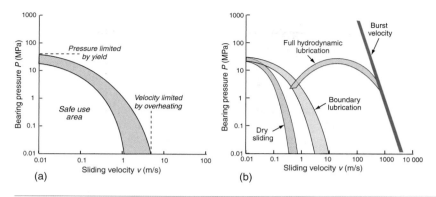

Figure 11.11   (a) A P–v curve for a bearing material. (b) P–v envelopes for various lubricated and unlubricated sliding, ultimately limited by the disintegration of the bearing under centrifugal force at very high velocities.

to suppress vibration. Intimate details are trade secrets, but here is the idea. The brake pads on your bike are made up of a synthetic rubber with particles of a cheap silicate to reduce pad wear and give wet friction. Those on your car have a phenolic matrix with particles of silicates, silicon carbide (sandpaper grit) to control friction and graphite or MoS_2 as a lubricant. Those on a military jet, a 747 or an F1 car are carbon or ceramic—here high temperature is the problem; replacement every 10 days is acceptable if they stand the heat (try calculating how much aluminum you could melt with the energy dissipated in bringing a 200-tonne 747 to rest from 200 kph).

Waging war on wear

As explained earlier, the rate of wear increases with the bearing pressure P (equation (11.4)). It also increases with sliding velocity v because the faster the sliding, the more work is done against friction and it all turns into heat. A given bearing material has an acceptable P–v envelope within which it is usable; venture outside it and seizure or catastrophic wear awaits you (Figure 11.11(a)). To extend the envelope upwards, we select materials with higher strength; to extend it laterally, we choose materials with higher thermal conductivity and melting point—or lower friction.

Lubrication, of course, reduces friction. Figure 11.11(b) shows that this does not increase the admissible bearing pressure—that is a static property of the material—but it does increase the permissible sliding velocity. Boundary lubrication expands the envelope a little; hydrodynamic lubrication expands it much more—but note its drop-off at low velocities because there is not enough sliding speed to build up the pressure profiles of Figure 11.9.

At very high rotational velocities, approached in ultra-centrifuges, gyroscopic navigation systems and super-flywheels for energy storage, there is an ultimate cut-off. It is the speed at which centrifugal forces exceed the strength of the material and the whole system disintegrates.

Hard materials have low wear rates but they tend to be brittle. Soft ones are tougher but they wear rapidly. Wear is a surface property, toughness a property of the bulk. So the way to fight wear is to take a tough material and give it a hard surface coating, using techniques of surface treatment. High-carbon steels are surface hardened by rapidly heating with a flame, an electron beam or a laser beam and then quenching the surface. *Carburizing*, *nitriding* and *boriding* give wear resistance to components like crankshafts or ball races by diffusing carbon, nitrogen or boron atoms into a thin surface layer, where they react to give particles of carbides, nitrides or borides. Hard, corrosion-resistant layers of alloys rich in tungsten, cobalt or chromium can be sprayed onto surfaces, but a refinishing process is then necessary to restore dimensional precision. Hard ceramic coatings of alumina (Al_2O_3) or of tungsten or titanium carbide (WC, TiC) can be applied by plasma spraying to give both wear resistance and resistance to chemical attack, and chemical and physical vapor deposition methods allow surface coatings of other metals and ceramics, including diamond-like carbon.

Records for all these treatments (and more) are contained in the CES software under the heading 'Surface treatment'. The software allows selection of coatings to meet given design requirements, of which wear resistance is one of the more important. We return to these in Chapter 18 on selecting processes.

11.6 Summary and conclusions

When surfaces slide, there is friction and there is wear. Eliminate all friction and a spinning wheel would spin forever. Eliminate all wear and it would last forever too. Both are about as probable as living forever while eating nothing. So we have to learn to live with them: to use them when they are useful, to minimize the problems they cause when they are a nuisance.

The remarkable thing about friction is that, if you slide one surface over another, the force to do so depends only on the normal force pushing the surfaces together, not on the area of the surface. That is because surfaces are not perfectly smooth; they touch only at points (flattened a little by the pressure)—just enough of them to support the normal force. Changing the area does not change the number of contact points. The sliding force is the force to shear the contact points, so it, too, is independent of area.

The points of contact are where wear happens. When surfaces slide the points shear off, becoming wear particles. This process can be useful—it helps surfaces bed down and conform by what is called 'run-in'. But if the wear particles or particles from elsewhere (dust, grit) get between the surfaces, the grinding action continues and the surfaces wear.

Soft materials—lead, polyethylene, PTFE, graphite—shear easily. Impregnating a sliding material with one of these reduces friction—skis and snow boards have polyethylene surfaces for just this reason. Better still is hydrodynamic lubrication: floating the sliding surfaces apart by dragging fluid between them. Flotation requires that the surfaces slide; if they stop the fluid is squeezed out. Clever lubricants overcome this by bonding to the surfaces, leaving a thin boundary layer that is enough to separate the surfaces until hydrodynamic lubrication gets going.

Friction, wear and corrosion cause more damage to mechanisms than anything else. They are the ultimate determinants of the life of products—or they would be, if we kept them until they wore out. But in today's society, many products are binned before they are worn out; many when they still work perfectly well. More on that subject in Chapter 20.

11.7 Further reading

Bowden, F.P. and Tabor, D. (1950 and 1964) *The Friction and Lubrication of Solids*, Parts 1 and 2, Oxford, University Press, Oxford, UK. ISBN 0-19-851204-X. *(Two volumes that establish the cornerstones of tribology: contact between solids; friction of metals and non-metals; frictional temperature rise; boundary lubrication and other chemical effects; adhesion; sliding wear; and hardness.)*

Hutchings, I.M. (1992) *Tribology: Friction and Wear of Engineering Materials*, Arnold, London, UK. ISBN 0-340-56184-X. *(An introduction to the mechanics and materials science of friction, wear and lubrication.)*

Neale, M.J. (1995) *The Tribology Handbook*, 2nd edition, Butterworth-Heinemann, Oxford, UK. ISBN 0-7506-1198-7. *(A tribological bible: materials, bearings, drives, brakes, seals, lubrication failure and repair.)*

11.8 Exercises

Exercise E11.1	Define the coefficient of friction. Explain why it is independent of the area of contact of the sliding surfaces.
Exercise E11.2	Woods—both resinous wood from the pine family and hardwoods like *lignum vitae* (it is so dense it sinks in water)—were, historically, used for bearings; even clocks had hardwood bearings. Why? Think of your own ideas, then research the Web.
Exercise E11.3	Give examples, based on your experience in sport (tennis, golf, swimming, skiing, rock climbing, hang-gliding, etc.) of instances in which friction is wanted and when it is not.
Exercise E11.4	Now a more challenging one. Do the same, based again on your experience in sport, of instances in which wear is desirable and in which it is not.
Exercise E11.5	What are the characteristics of materials that are a good choice for use as brake pads?

Exercise E11.6 A 30 mm diameter plane bearing of length 20 mm is to be made of a material with the $P - v$ characteristics shown in Figure 11.11 (b). If the bearing load is 300 N and the maximum rotation rate is 500 rpm, is it safe to run it dry? If not, is boundary lubrication sufficient? (Remember that the bearing pressure is the load divided by the projected area normal to the load.)

Exercise E11.7 A bronze statue weighing 4 tonnes with a base of area 0.8 m² is placed on a granite museum floor. The yield strength of the bronze is 240 MPa. What is the true area of contact, A_r, between the base and the floor?

Exercise E11.8 The statue of the previous example was moved to a roof courtyard with a lead floor. The yield strength of the lead is 6 MPa. What now is the true area of contact?

Exercise E11.9 How would you measure the true area of contact A_r between two surfaces pressed into contact? Use your imagination to devise ways.

11.9 Exploring design with CES (use Level 2 unless otherwise stated)

Exercise E11.10 Use the 'Search' facility to find materials that are used as ball bearings. (Search on the singular—ball bearing—since that picks up the plural as well.)

Exercise E11.11 Use the 'Search' facility to find materials that are used as for brake pads. Do the same for brake discs.

Exercise E11.12 Use a 'Limit' selection stage applied to the Surface Treatment data table to find processes that enhance wear resistance. (To do this, click on 'Select' in the main toolbar, then on 'Selection Data', just below it. Select Level 2 Surface treatment in the dialog box. Open a 'Limit' stage, open Function of treatment, and click on Wear resistance.) Explore the record for laser-based methods. What are its typical uses?

Exercise E11.13 Follow the same procedure as that of the previous example to search for processes used to control friction. Explore the record for grinding and mechanical polishing. What are grinding wheels made of?

Chapter 12
Agitated atoms: materials and heat

A heat exchanger. (Image courtesy of the International Copper Development Association)

Chapter contents

12.1 Introduction and synopsis

Thermal properties quantify the response of materials to heat. Heat, until about 1800, was thought to be a physical substance called 'caloric' that somehow seeped into things when they were exposed to flame. It took the American Benjamin Thompson[1], backed up by none other than the formidable Carnot[2], to suggest what we now know to be true: that heat is atoms or molecules in motion. In gases, they are flying between occasional collisions with each other. In solids, by contrast, they vibrate about their mean positions; the higher the temperature, the greater the amplitude of vibrations. From this perception emerges all of our understanding of thermal properties of solids: their heat capacity, expansion coefficient, conductivity, even melting.

Heat affects mechanical and physical properties too. As temperature rises, materials expand, the elastic modulus decreases, the strength falls and the material starts to creep, deforming slowly with time at a rate that increases as the melting point is approached until, on melting, the solid loses all stiffness and strength. This we leave for Chapter 13.

Thermal design, the ultimate topic of this chapter, is design to cope properly with the effects of heat or, where possible, to exploit them. The chapter opening page shows an example: a copper heat exchanger, designed to transfer heat efficiently between two circulating fluids.

12.2 Thermal properties: definition and measurement

Two temperatures, the *melting temperature*, T_m, and the *glass temperature*, T_g (units for both: kelvin[3], K, or centigrade, °C) are fundamental points of reference because they relate directly to the strength of the bonds in the solid. Crystalline solids have a sharp melting point, T_m. Non-crystalline solids do not; the glass temperature T_g characterizes the transition from true solid to very viscous liquid. It is helpful, in engineering design, to define two further temperatures: the *maximum* and *minimum service temperatures*, T_{max} and T_{min} (units for both: K or °C). The first tells us the highest temperature at which the material can be used continuously without oxidation, chemical change or excessive

---

[1] Benjamin Thompson (1753–1814), later Lord Rumford, while in charge of the boring of cannons at the Watertown Arsenal near Boston, Mass., noted how hot they became and formulated the 'mechanical equivalent of heat'—what we now call the heat capacity.

[2] Sadi Nicolas Léonhard Carnot (1796–1832), physicist and engineer, formulator of the Carnot cycle and the concept of entropy, the basis of the optimization of heat engines. He died of cholera, a fate most physicists today are, mercifully, spared.

[3] William Thompson, Lord Kelvin (1824–1907), Scottish mathematician who contributed to many branches of physics; he was known for a self-confidence that led him to claim (in 1900) that there was nothing more to be discovered in physics—he already knew it all.

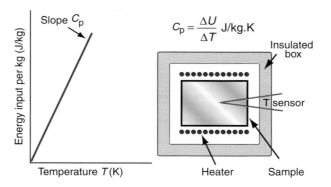

Figure 12.1 Measuring heat capacity, C_p. Its units are J/kg.K.

distortion becoming a problem. The second is the temperature below which the material becomes brittle or otherwise unsafe to use.

It costs energy to heat a material up. The energy to heat 1 kg of a material by 1 K is called the *heat capacity* or *specific heat*, and since the measurement is usually made at constant pressure (atmospheric pressure) it is given the symbol C_p. Heat is measured in joules[4], symbol J, so the units of specific heat are J/kg · K. When dealing with gases, it is more usual to measure the heat capacity at constant volume (symbol C_v) and for gases this differs from C_p. For solids the difference is so slight that it can be ignored, and we shall do so here. C_p is measured by the technique of calorimetry, which is also the standard way of measuring the glass temperature T_g. Figure 12.1 shows how, in principle, this is done. A measured quantity of energy (here, electrical energy) is pumped into a sample of material of known mass. The temperature rise is measured, allowing the energy/kg.K to be calculated. Real calorimeters are more elaborate than this, but the principle is the same.

Most materials expand when they are heated (Figure 12.2). The thermal strain per degree of temperature change is measured by the *linear thermal expansion coefficient*, α. It is defined by

$$\alpha = \frac{1}{L}\frac{dL}{dT} \tag{12.1}$$

where L is a linear dimension of the body. If the material is anisotropic it expands differently in different directions, and two or more coefficients are required. Since strain is dimensionless, the units of α are K^{-1} or, more conveniently, 'microstrain/K', i.e. $10^{-6}\ K^{-1}$.

[4] James Joule (1818–1889), English physicist, who, motivated by theological belief, sought to establish the unity of the forces of nature. His demonstration of the equivalence of heat and mechanical work did much to discredit the caloric theory.

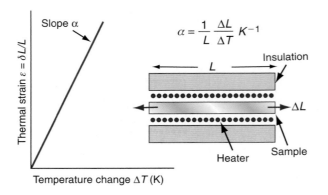

Figure 12.2   Measuring the thermal expansion coefficient, α. Its units are 1/K or, more usually, 10^{-6}/K (microstrain/K).

Figure 12.3   Measuring the thermal conductivity, λ. Its units are W/m · K.

Power is measured in watts[5]; 1 watt (W) is 1 joule/second. The rate at which heat is conducted through a solid at steady-state (meaning that the temperature profile does not change with time) is measured by the *thermal conductivity, λ* (units: W/m.K). Figure 12.3 shows how it is measured: by recording the heat flux q (W/m^2) flowing through the material from a surface at higher temperature

---

[5] James Watt (1736–1819), instrument maker and inventor of the condenser steam engine (the idea came to him whilst 'walking on a fine Sabbath afternoon'), which he doggedly developed. Unlike so many of the characters footnoted in this book, Watt, in his final years, was healthy, happy and famous.

T_1 to a lower one at T_2 separated by a distance x. The conductivity is calculated from Fourier's[6] law:

$$q = -\lambda \frac{\mathrm{d}T}{\mathrm{d}x} = \lambda \frac{(T_1 - T_2)}{x} \qquad (12.2)$$

(where q is the heat flux per unit area (units: W/m^2))

Thermal conductivity, as we have said, governs the flow of heat through a material at steady-state. The property governing transient heat flow (when temperature varies with time) is the *thermal diffusivity, a* (units: m^2/s). The two are related by

$$a = \frac{\lambda}{\rho C_\mathrm{p}} \qquad (12.3)$$

where ρ is the density and C_p is, as before, the heat capacity. The thermal diffusivity can be measured directly by measuring the time it takes for a temperature pulse to traverse a specimen of known thickness when a heat source is applied briefly to the one side; or it can be calculated from λ (and ρC_p) via equation (12.3).

12.3 The big picture: thermal property charts

Three charts give an overview of thermal properties and their relationship to strength.

Thermal expansion, α, and thermal conductivity, λ

Figure 12.4 maps thermal expansion, α, and thermal conductivity, λ. Metals and technical ceramics lie towards the lower right: they have high conductivities and modest expansion coefficients. Polymers and elastomers lie at the upper left: their conductivities are 100 times less and their expansion coefficients 10 times greater than those of metals. The chart shows contours of λ/α, a quantity important in designing against thermal distortion—a subject we return to in Section 12.6. An extra material, Invar (a nickel alloy), has been added to the chart because of its uniquely low expansion coefficient at and near room temperature, a consequence of a trade-off between normal expansion and a contraction associated with a magnetic transformation.

Thermal conductivity, λ, and thermal diffusivity, a

Figure 12.5 shows the room temperature values of thermal conductivity, λ, and thermal diffusivity, a, with contours of volumetric heat capacity ρC_p, equal to the ratio of the two, λ/α (from equation (12.3)). The data span almost five decades in λ and a. Solid materials are strung out along the line

---

[6] Baron Jean Baptiste Joseph Fourier (1768–1830), mathematician and physicist; he nearly came to grief during the French Revolution but survived to become one of the savants who accompanied Napoleon Bonaparte in his conquest of Egypt.

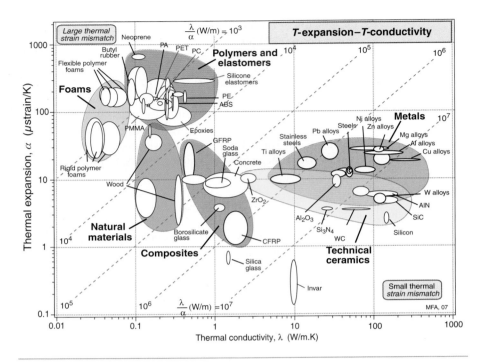

Figure 12.4 The linear expansion coefficient, α, plotted against the thermal conductivity, λ. The contours show the thermal distortion parameter λ/α.

$$\rho C_p \approx 3 \times 10^6 \, \text{J/m}^3 \cdot \text{K} \qquad (12.4)$$

meaning that the heat capacity per unit volume, ρC_p, is almost constant for all solids, something to remember for later. As a general rule, then,

$$\lambda = 3 \times 10^6 \, a \qquad (12.5)$$

(λ in W/m.K and a in m^2/s). Some materials deviate from this rule: they have lower-than-average volumetric heat capacity. The largest deviations are shown by porous solids: foams, low-density firebrick, woods and the like. Because of their low density they contain fewer atoms per unit volume and, averaged over the volume of the structure, ρC_p is low. The result is that, although foams have low *conductivities* (and are widely used for insulation because of this), their *thermal diffusivities* are not necessarily low. This means that they don't transmit much heat, but they do change temperature quickly.

Thermal conductivity λ and yield strength σ_y
The final chart, Figure 12.6, shows thermal conductivity, λ and strength, σ_y. Metals, particularly the alloys of copper, aluminum and nickel, are both strong

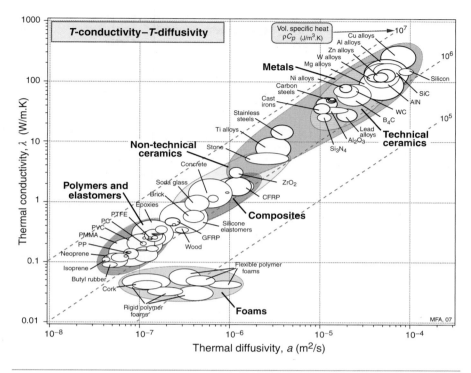

Figure 12.5 The thermal conductivity, λ, plotted against the thermal diffusivity, $a = \lambda/\rho C_p$. The contours show the specific heat per unit volume ρC_p.

and good conductors, a combination of properties we seek for applications such as heat exchangers—one we return to in Section 12.6.

12.4 Drilling down: the physics of thermal properties

Heat capacity

We owe our understanding of heat capacity to Albert Einstein[7] and Peter Debye[8]. Heat, as already mentioned, is atoms in motion. Atoms in solids vibrate about their mean positions with an amplitude that increases with temperature. Atoms in solids can't vibrate independently of each others because they are coupled by their inter-atomic bonds; the vibrations are like standing elastic waves. Figure 12.7

[7] Albert Einstein (1879–1955), Patent Officer, physicist and campaigner for peace, one of the greatest scientific minds of the 20th century; he was forced to leave Germany in 1933, moving to Princeton, NJ, where his influence on US defence policy was profound.

[8] Peter Debye (1884–1966), Dutch physicist and Nobel Prize winner, did much of his early work in Germany until in 1938, harassed by the Nazis, he moved to Cornell, NY, where he remained until his death.

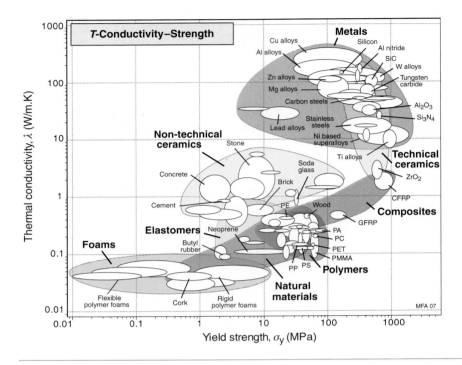

Figure 12.6   Thermal conductivity λ and strength σ_y.

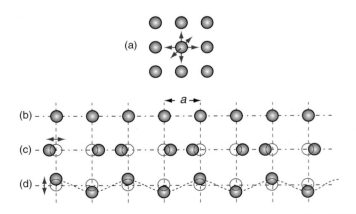

Figure 12.7   (a) An atom vibrating in the 'cage' of atoms that surround it with three degrees of freedom. (b) A row of atoms at rest. (c) A longitudinal wave. (d) One of two transverse waves (in the other the atoms oscillate normal to the page).

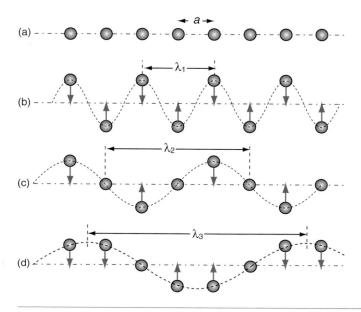

Figure 12.8   Thermal energy involves atom vibrations. There is one longitudinal mode of vibration and two transverse modes, one of which is shown here. The shortest meaningful wavelength, $\lambda_1 = 2a$, is shown in (b).

shows how, along any row of atoms, there is the possibility of one longitudinal mode and two transverse modes, one in the plane of the page and one normal to it. Some of these have short wavelengths and high energy, others long wavelengths and lower energy (Figure 12.8). The shortest possible wavelength, λ_1, is just twice the atomic spacing; the other vibrations have wavelengths that are longer. In a solid with N atoms there are N discrete wavelengths, and each has a longitudinal mode and two transverse modes, $3N$ modes in all. Their amplitudes are such that, on average, each has energy k_BT, where k_B is Boltzmann's constant[9], 1.38×10^{-23} J/K. If the volume occupied by an atom is Ω, then the number of atoms per unit volume is $N = 1/\Omega$ and the total thermal energy per unit volume in the material is $3k_BT/\Omega$. The heat capacity per unit volume, ρC_p, is the *change* in this energy per kelvin change in temperature, giving:

$$\rho C_p = \frac{3k_B}{\Omega} \text{ J/m}^3\text{K} \tag{12.6}$$

[9] Ludwig Boltzmann (1844–1906), born and worked in Vienna at a time when that city was the intellectual center of Europe; his childhood interest in butterflies evolved into a wider interest in science, culminating in his seminal contributions to statistical mechanics.

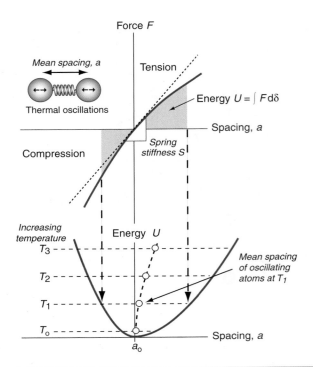

Figure 12.9   Thermal expansion results from the oscillation of atoms in an unsymmetrical energy well.

Atomic volumes do not vary much; all lie within a factor of 3 of the value $2 \times 10^{-29}/m^3$, giving a volumetric heat capacity

$$\rho C_p \approx 2 \times 10^6 \ J/m^3K \qquad (12.7)$$

The chart of Figure 12.5 showed that this is indeed the case.

Thermal expansion
If a solid expands when heated (and almost all do) it must be because the atoms are moving further apart. Figure 12.9 shows how this happens. It looks very like an earlier one, Figure 4.22, but there is a subtle difference: the force–spacing curve, a straight line in the earlier figure, is not in fact quite straight; the bonds become stiffer when the atoms are pushed together and less stiff when they are pulled apart. Atoms vibrating in the way described earlier oscillate about a mean spacing that increases with the amplitude of oscillation, and thus with increasing temperatures. So thermal expansion is a nonlinear effect; if the bonds between atoms were linear springs, there would be no expansion.

The stiffer the springs, the steeper is the force–displacement curve and the narrower is the energy well in which the atom sits, giving less scope for expansion. Thus, materials with high modulus (stiff springs) have low expansion coefficient, those with low modulus (soft springs) have high expansion—indeed, to a good approximation

$$\alpha = \frac{1.6 \times 10^{-3}}{E} \qquad (12.8)$$

(E in GPa, α in K^{-1}). It is an empirical fact that all crystalline solids expand by about the same amount on heating from absolute zero to their melting point: it is about 2%. The expansion coefficient is the expansion per degree kelvin, meaning that

$$\alpha \approx \frac{0.02}{T_m} \qquad (12.9)$$

For example, tungsten, with a melting point of around 3330°C (3600 K) has $\alpha = 5 \times 10^{-6}/°C$, while lead, with a melting point of about 330°C (600 K, six times lower) expands six times more ($\alpha = 30 \times 10^{-6}/°C$). Equation (12.9) (with T_m in kelvin, of course) is a remarkably good approximation for the expansion coefficient. Note in passing that equations (12.8) and (12.9) indicate that E is expected to scale with T_m—try exploring how well these approximations hold using the CES Elements database.

Thermal conductivity

Heat is transmitted through solids in three ways: by thermal vibrations; by the movement of free electrons in metals; and, if they are transparent, by radiation. Transmission by thermal vibrations involves the propagation of *elastic waves*. When a solid is heated the heat enters as elastic wave packets or *phonons*. The phonons travel through the material and, like any elastic wave, they move with the speed of sound, c_o ($c_o = \sqrt{E/\rho}$). If this is so, why does heat not diffuse at the same speed? It is because phonons travel only a short distance before they are scattered by the slightest irregularity in the lattice of atoms through which they move, even by other phonons. On average they travel a distance called the *mean free path*, ℓ_m, before bouncing off something, and this path is short: typically less than 0.01 μm (10^{-8} m).

We calculate the conductivity by using a *net flux model*, much as you would calculate the rate at which cars accumulate in a car park by counting the rate at which they enter and subtracting the rate at which they leave. Phonon conduction can be understood in a similar way, as suggested by Figure 12.10. Here a rod with unit cross-section carries a uniform temperature gradient dT/dx between its ends. Phonons within it have three degrees of freedom of motion (they can travel in the $\pm x$, the $\pm y$ and the $\pm z$ directions). Focus on the midplane M–M. On average, one-sixth of the phonons are moving in the $+x$ direction;

Figure 12.10   The transmission of heat by the motion of phonons.

those within a distance ℓ_m of the plane will cross it from left to right before they are scattered, carrying with them an energy $\rho C_p (T + \Delta T)$ where T is the temperature at the plane M–M and $\Delta T = (dT/dx)\ell_m$. Another one-sixth of the phonons move in the $-x$ direction and cross M–M from right to left, carrying an energy $\rho C_p (T - \Delta T)$. Thus, the energy flux q J/m².s across unit area of M–M per second is

$$q = -\frac{1}{6}\rho C_p\, c_o\left(T + \frac{dT}{dx}\,\ell_m\right) + \frac{1}{6}\rho C_p\, c_o\left(T - \frac{dT}{dx}\,\ell_m\right)$$

$$= -\frac{1}{3}\rho C_p\, \ell_m\, c_o\, \frac{dT}{dx}$$

Comparing this with the definition of thermal conductivity (equation (12.2)), we find the conductivity to be

$$\lambda = \frac{1}{3}\rho C_p\, \ell_m\, c_o \tag{12.10}$$

Sound waves, which, like phonons, are elastic waves, travel through the same bar without much scattering. Why do phonons scatter so readily? It's because waves are scattered most by obstacles with a size comparable with their wavelengths—that is how diffraction gratings work. Audible sound waves have wavelengths of meters, not microns; they are scattered by buildings (which is why you can't always tell where a police siren is coming from) but not by the atom-scale obstacles that scatter phonons, with wavelengths starting at two atomic spacings (Figure 12.8).

Elastic waves contribute little to the conductivity of pure metals such as copper or aluminum because the heat is carried more rapidly by the free electrons. Equation (12.10) still applies but now C_p, c_o and ℓ_m become the thermal capacity, the velocity and the mean free path of the electrons. Free electrons also conduct electricity, with the result that metals with high electrical conductivity also have high thermal conductivity (the Wiedemann–Franz law), as will be illustrated in Chapter 14.

12.5 Manipulating thermal properties

Thermal expansion

Expansion, as already mentioned, can generate thermal stress and cause distortion, undesirable in precise mechanisms and instruments. The expansion coefficient α, like modulus or melting point, depends on the stiffness and strength of atomic bonds. There's not much you can do about these, so the expansion coefficient, normally, is beyond our control. Some few materials have exceptionally low expansion—borosilicate glass (Pyrex), silica and carbon fibers, for example—but they are hard to use in engineering structures. There is one exception: the family of alloys called Invars with very low expansion. They achieve this by the trick of canceling the thermal expansion (which is, in fact, happening) with a contraction caused by the gradual loss of magnetism as the material is heated.

Thermal conductivity and heat capacity

Equation (12.10) describes thermal conductivity. Which of the terms it contains can be manipulated and which cannot? The volumetric heat capacity ρC_p, as we have seen, is almost the same for all solid materials. The sound velocity $c_o \approx \sqrt{E/\rho}$) depends on two properties that are not easily changed. That leaves the mean free path, ℓ_m, of the phonons or, in metals, of the electrons. Introducing *scattering centers* such as impurity atoms or finely dispersed particles achieves this. Thus, the conductivity of pure iron is 80 W/m.K, while that of stainless steel—iron with up to 30% of nickel and chromium—is only 18 W/m.K. Certain materials can exist in a glassy state, and here the disordered nature of the glass makes every molecule a scattering center and the mean free path is reduced to a couple of atom spacings; the conductivity is correspondingly low.

Chapter 6 showed how strength is increased by defects such as solute atoms, precipitates and work hardening. Some of these reduce the thermal

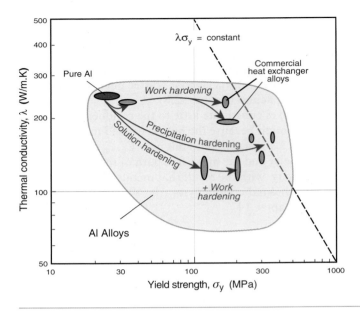

Figure 12.11   Thermal conductivity and strength for aluminum alloys. Good materials for heat exchangers have high values of $\lambda\sigma_y$, indicated by the dashed line.

conductivity. Figure 12.11 shows a range of aluminum alloys hardened by each of the three main mechanisms: solid solution, work hardening and precipitation hardening. Work hardening strengthens significantly without changing the conductivity much, while solid solution and precipitation hardening introduce more scattering centers, giving a drop in conductivity.

We are not, of course, restricted to fully dense solids, making density ρ a possible variable after all. In fact, the best thermal insulators are porous materials—low-density foams, insulating fabric, cork, insulating wool, etc.—that take advantage of the low conductivity of still air trapped in the pores (λ for air is 0.02 W/m.K).

12.6 Design to exploit thermal properties

All materials have a thermal expansion coefficient α, a specific heat C_p and a thermal conductivity λ. Material property charts are particularly helpful in choosing materials for applications that use these. Certain thermal properties, however, are shown only by a few special materials: shape-memory behavior is an example. There is no point in making charts for these because so few materials have them, but they are still of engineering interest. Examples that use them are described nearer the end of the section.

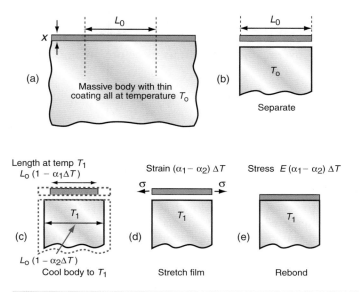

Figure 12.12 Thermal stresses in thin films arise on cooling or heating when their expansion coefficients differ. Here that of the film is α_1 and that of the substrate, a massive body, is α_2.

Managing thermal stress

Most structures, small or large, are made of two or more materials that are clamped, welded or otherwise bonded together. This causes problems when temperatures change. Railway track will bend and buckle in exceptionally hot weather (steel, high α, clamped to mother earth with a much lower α). Overhead transmission lines sag for the same reason—a problem for high-speed electric trains. Bearings seize, doors jam. Thermal distortion is a particular problem in equipment designed for precise measurement or registration like that used to make masks for high-performance computer chips, causing loss of accuracy when temperatures change.

All of these derive from *differential thermal expansion*, which, if constrained (clamped in a way that stops it happening) generates *thermal stress*. As an example, many technologies involve coating materials with a thin surface layer of a different material to impart wear resistance, or resistance to corrosion or oxidation. The deposition process often operates at a high temperature. On cooling, the substrate and the surface layer contract by different amounts because their expansion coefficients differ and this puts the layer under stress. This residual stress calculated as follows.

Think of a thin film bonded onto a component that is much thicker than the film, as in Figure 12.12(a). First imagine that the layer is detached, as in Figure 12.12(b). A temperature drop of ΔT causes the layer to change in length by $\delta L_1 = \alpha_1 L_0 \Delta T$. Meanwhile, the substrate to which it was previously bonded contracts by $\delta L_2 = \alpha_2 L_0 \Delta T$. If $\alpha_1 > \alpha_2$ the surface layer shrinks more than the substrate. If we want to stick the film back on the much-more-massive substrate, covering the same surface as before, we must stretch it by the strain

$$\varepsilon = \frac{\delta L_1 - \delta L_2}{L_0} = \Delta T(\alpha_1 - \alpha_2)$$

This requires a stress[10] in the film of

$$\alpha_1 = E_1(\alpha_1 - \alpha_2)\Delta T \tag{12.11}$$

The stress can be large enough to crack the surface film. The pattern of cracks seen on glazed tiles arises in this way.

So if you join dissimilar materials you must expect thermal stress when they are heated or cooled. The way to avoid it is to avoid material combinations with very different expansion coefficients. The α–λ chart of Figure 12.4 has expansion α as one of its axes: benign choices are those that lie close together on this axis, dangerous ones are those that lie far apart.

But avoiding materials with α mismatch is not always possible. Think of joining glass to metal—Pyrex to stainless steel, say—a common combination in high vacuum equipment. Their expansion coefficients can be read from the α–λ chart of Figure 12.4: for Pyrex (borosilicate glass), $\alpha = 4 \times 10^{-6}$/K; for stainless steel, $\alpha = 20 \times 10^{-6}$/K: a big mismatch. Vacuum equipment has to be 'baked out' to desorb gases and moisture, requiring that it be heated to about 150°C, enough for the mismatch to crack the Pyrex. The answer is to grade the joint with one or more materials with expansion that lies between the two: first join the Pyrex to a glass of slightly higher expansion and the stainless to a metal of lower expansion—the chart suggests soda glass for the first and a nickel alloy for the second—and then diffusion-bond these to each other, as in Figure 12.13. The graded joint spreads the mismatch, lowering the stress and the risk of damage.

There is an alternative, although it is not always practical. It is to put a compliant layer—rubber, for instance—in the joint; it is the way windows are mounted in cars. The difference in expansion is taken up by distortion in the rubber, transmitting very little of the mismatch to the glass or surrounding steel.

Thermal sensing and actuation
Thermal expansion can be used to *sense* (to measure temperature) and to *actuate* (to open or close valves or electrical circuits, for instance). The direct thermal displacement $\delta = \alpha L_0 \Delta T$ is small; it would help to have a way to magnify it. Figure 12.14 shows one way to do this. Two materials, deliberately chosen to have different expansion coefficients α_1 and α_2, are bonded together in the form

---

[10] Strictly, E should be replaced by $E/(1 + \nu)$, where ν is Poisson's ratio, because the loading is bi-axial.

Figure 12.13   A graded joint. The high expansion stainless steel is bonded to a nickel alloy of lower expansion, the low expansion Pyrex (a borosilicate glass) is bonded to a glass with higher expansion, and the set are assembled to give a stepwise graded expansion joint.

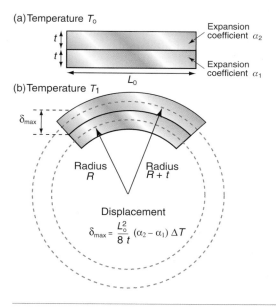

Figure 12.14   The thermal response of a bi-material strip when the expansion coefficients of the two materials differ. The configuration magnifies the thermal strain.

of a bi-material strip, of thickness $2t$, as in Figure 12.14(a). When the temperature changes by $\Delta T = (T_1 - T_o)$, one expands more than the other, causing the strip to bend. The mid-thickness planes of each strip (dotted in Figure 12.14(b)) differ in length by $L_o (\alpha_2 - \alpha_1)\Delta T$. Simple geometry then shows that

$$\frac{R + t}{R} = 1 + (\alpha_2 - \alpha_1)\Delta T$$

from which

$$R = \frac{t}{(\alpha_2 - \alpha_1)\Delta T} \qquad (12.12)$$

The resulting upward displacement of the center of the bi-material strip (assumed to be thin) is

$$\delta_{max} = \frac{L_o^2}{8t}(\alpha_2 - \alpha_1)\Delta T \qquad (12.13)$$

A large aspect ratio L_o/t produces a large displacement, and one that is linear in temperature.

Thermal gradients

There is a subtler effect of expansion that causes problems even in a single material. Suppose a chunk of material is suddenly cooled by ΔT (by dropping it hot into a bath of cold water, for example) or suddenly heated by ΔT (by dropping it cold into boiling water). The surface, almost immediately, is forced to the temperature of the bath, as in Figure 12.15. It contracts (if cooled) or expands (if heated) by δL such that $\delta L/L_o = \alpha \Delta T$. The inside of the material has not

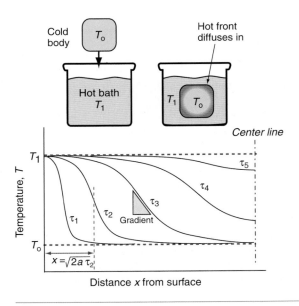

Figure 12.15   The diffusion of heat into an initially cold body.

expanded yet—there has not been time for the heat to diffuse into it. That means there is a *temperature gradient,* and even though the material has just one expansion coefficient, there are different strains in different places. Since the surface is stuck to the interior, it is constrained, so thermal stresses appear in it just as they did in the thin film of Figure 12.12. In brittle materials, these stresses can cause fracture. The ability of a material to resist this, its *thermal shock resistance,* ΔT_s (units: K or °C), is the maximum sudden change of temperature to which such a material can be subjected without damage.

Even if the stresses do not cause failure, temperature gradients lead to distortion. Which properties minimize this? Figure 12.15 shows the temperature profile in the quenched body at successive times $\tau_1, \tau_2 \ldots \tau_5$. A cooling or heating front propagates inwards from the surface by the process of heat diffusion discussed earlier—and while it is happening there is a temperature gradient and consequent thermal stress. If the material remains elastic, then the stresses fade away when the temperature is finally uniform—though the component may by then have cracked. If the material partially yields, the stresses never go away, even when the temperature profile has smoothed out, because the surface has yielded but the center has not. This *residual stress* is a major problem in metal processing, particularly welding, causing distortion and failure in service.

The important new variable here is *time*—it is this that distinguishes the contours in Figure 12.15. The material property that determines this time is the thermal diffusivity a that we encountered earlier:

$$a = \frac{\lambda}{\rho C_p} \tag{12.14}$$

with units of m²/s. The bigger the conductivity, λ, the faster the diffusivity, a, but why is ρC_p on the bottom? If this quantity, the heat capacity per unit volume, is large, a lot of heat has to diffuse in or out of unit volume to change the temperature much. The more that has to diffuse, the longer it takes to do so, so that the diffusivity a is reduced. If you solve *transient heat flow* problems (of which this is an example), a general result emerges: it is that the characteristic distance x (in meters) that heat diffuses in a time τ (in seconds) is

$$x \approx \sqrt{2a\tau} \tag{12.15}$$

This is a particularly useful little formula, and although approximate (because the constant of proportionality, here $\sqrt{2}$, actually depends somewhat on the geometry of the body into which the heat is diffusing), it is perfectly adequate for first estimates of heating and cooling times.

Thermal distortion due to thermal gradients is a problem in precision equipment in which heat is generated by the electronics, motors, actuators or sensors that are necessary for its operation: different parts of the equipment are at different temperatures and so expand by different amounts. The answer is to make the equipment from materials with low expansion α (because this minimizes the expansion difference in a given temperature gradient) and with high conductivity λ (because this spreads the heat further, reducing the steepness of the gradient). It can be shown that the best choice is that of materials with high values of the ratio λ/α, just as our argument suggests.

The α–λ chart of Figure 12.4 guides selection for this too. The diagonal contours show the quantity λ/α. Materials with the most attractive values lie towards the bottom right. Copper, aluminum, silicon, aluminum nitride and tungsten are good—they distort the least; stainless steel and titanium are much less good.

Conduction with strength: heat exchangers

A heat exchanger transfers heat from one fluid to another while keeping the fluids physically separated. You find them in central heating systems, power-generating plants, and cooling systems in car and marine engines. Heat is transferred most efficiently if the material separating the fluids is kept thin and if the surface area for heat conduction is large—a good solution is a tubular coil. At least one of the fluids is under pressure, so the tubes must be strong enough to withstand it. Here there is a conflict in the design: thinner walls conduct heat faster, thicker walls are stronger. How is this trade-off optimized by choosing the tube material?

Consider the idealized heat exchanger made of thin-walled tubes shown in Figure 12.16. The tubes have a given radius r, a wall thickness t (which may be varied) and must support an internal pressure of p, with a temperature difference of ΔT between the fluids. The objective is to maximize the heat transferred per unit surface area of tube wall. The heat flow rate per unit area is given by

$$q = -\lambda \frac{dT}{dx} \approx \lambda \frac{\Delta T}{t} \tag{12.16}$$

Figure 12.16   A heat exchanger. It must transmit heat yet withstand the pressure p.

The stress due to the internal pressure in a cylindrical thin-walled tube (Chapter 10, equation (10.7)) is

$$\sigma = p\frac{R}{t}$$

This stress must not exceed the yield strength σ_y of the material of which it is made, so the minimum thickness t needed to support the pressure p is

$$t = p\frac{R}{\sigma_y} \qquad (12.17)$$

Combining equations (12.16) and (12.17) to eliminate the free variable t gives

$$q = \lambda\sigma_y\frac{\Delta T}{pR} \qquad (12.18)$$

The best materials are therefore those with the highest values of the index $\lambda\sigma_y$. They are the ones at the top right of the λ–σ_y chart of Figure 12.6—metals and certain ceramics. Ceramics are brittle and difficult to shape into thin-walled tubes, leaving metals as the only viable choice. The chart shows that the best of these are the alloys of copper and of aluminum. Both offer good corrosion resistance, but aluminum wins where cost and weight are to be kept low—in automotive radiators, for example. Commercial aluminum alloys used for this application are identified in Figure 12.11. They do not offer quite the highest values of $\lambda\,\sigma_y$ amongst the aluminum alloys—other factors are at work here. One is the continuous exposure to temperature during service, which can soften the precipitation-hardened alloys; another is that an efficient manufacturing route for thin-walled multi-channel tubing is extrusion, for which high ductility is needed.

Many aluminum alloys corrode in salt water. Marine heat exchangers use copper alloys even though they are heavier and more expensive because of their resistance to corrosion in seawater.

Insulation: thermal walls

Before 1940 few people had central heating or air-conditioning. Now most have both, and saunas and freezers and lots more. All consume power, and consuming power is both expensive and hard on the environment (Chapter 20). To save on both, it pays to insulate. Consider, as an example, the oven of Figure 12.17. The power lost per m^2 by conduction through its wall is

$$q = \lambda\frac{(T_i - T_o)}{t}\ \text{W/m}^2 \qquad (12.19)$$

where t is the thickness of the insulation, T_0 is the temperature of the outside and T_i that of the inside. For a given wall thickness t the power consumption is

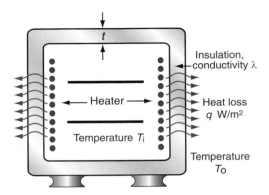

Figure 12.17 A heated chamber with heat loss by conduction through the insulation.

minimized by choosing materials with the lowest possible λ. The chart of Figure 12.6 shows that these are foams, largely because they are about 95% gas and only 5% solid. Polymer foams are usable up to about 170°C; above this temperature even the best of them lose strength and deteriorate. When insulating against heat, then, there is an additional constraint that must be met: that the material has a maximum service temperature T_{max} that lies above the planned temperature of operation. Metal foams are usable to higher temperatures but they are expensive and—being metal—they insulate less well than polymers. For high temperatures ceramic foams are the answer. Low-density firebrick is relatively cheap, can operate up to 1000°C and is a good thermal insulator. It is used for kiln and furnace linings.

Using heat capacity: storage heaters

The demand for electricity is greater during the day than during the night. It is not economic for electricity companies to reduce output, so they seek instead to smooth demand by charging less for off-peak electricity. Storage heaters exploit this, using cheap electricity to heat a large block of material from which heat is later extracted by blowing air over it.

Storage heaters also fill other more technical roles. When designing supersonic aircraft or re-entry vehicles, it is necessary to simulate the conditions they encounter—hypersonic air flow at temperatures up to 1000°C. The simulation is done in wind tunnels in which the air stream is heated rapidly by passing it over a previously heated thermal mass before it hits the test vehicle, as in Figure 12.18. This requires a large mass of heat-storage material, and to keep the cost down the material must be cheap. The objective, both for the home heater and for the wind tunnel, is to store as much heat per unit cost as possible.

The energy stored in a mass m of material with a heat capacity C_p when heated through a temperature interval ΔT is

$$Q = mC_p\Delta T \tag{12.20}$$

Figure 12.18   A wind tunnel with storage heating. The thermal mass is pre-heated, giving up its heat to the air blast when the fan is activated.

The cost of mass m of material with a cost per kg of C_m is

$$C = mC_m$$

Thus, the heat stored per unit material cost is

$$\frac{Q}{C} = \frac{C_p}{C_m} \tag{12.21}$$

The best choice of material is simply that with the largest heat capacity, C_p, and lowest price, C_m. There is one other obvious constraint: the material must be able to tolerate the temperature to which it will be heated—say, $T_{max} > 150°C$ for the home heater, $T_{max} > 1000°C$ for the air tunnel. It is hard to do this selection with the charts provided here, but easy to do it with the CES software: apply the limit on T_{max} then use the index

$$M = \frac{C_p}{C_m}$$

to isolate the cheapest candidates. You will find that the best choice for the home heater is concrete (cast to a shape that provides some air channels). For the wind tunnel, in which temperatures are much higher, the best choice is a refractory brick with air channels for rapid heat extraction.

Using phase change: thermal buffers
When materials vaporize, or melt, or change their crystal structure while remaining solid (all of which are known as 'phase changes'), many things happen, and

they can be useful. They change in volume. If solid, they may change in shape. And they absorb or release heat, the *latent heat L* (J/kg) of the phase change, without changing temperature.

Suppose you want to keep something at a fixed temperature without external power. Perfect insulation is not just impractical, it is impossible. But—if keeping things warm is the problem—the answer is to surround them by a liquid that solidifies at the temperature that you want to maintain. As it solidifies it releases its latent heat of fusion, holding the temperature precisely at its solidification temperature (i.e. its melting point) until all the liquid is solid—it is the principle on which some food warmers and 'back-woods' hand warmers work. The reverse is also practical: using latent heat of melting to keep things cool. For this, choose a solid that melts, absorbing latent heat, at the temperature you want to maintain—'cool bag' inserts for holiday travel work in this way. In both cases, change of phase is used as a *thermal buffer*.

Shape-memory and super-elastic materials

Almost all materials expand on heating and shrink on cooling. Some few have a different response, one that comes from a solid-state phase change. *Shape-memory materials* are alloys based on titanium or nickel that can, while still solid, exist in two different crystal configurations with different shapes, called *allotropes*. If you bend the material it progressively changes structure from the one allotrope to the other, allowing enormous distortion. Shape-memory alloys have a critical temperature. If below this critical temperature, the distorted material just stays in its new shape. Warm it up and, at its critical temperature, the structure reverts to the original allotrope and the sample springs back to its original shape. This has obvious application in fire alarms, sprinkler systems and other temperature-controlled safety systems, and they are used for all of these.

But what if room temperature is already above this critical temperature? Then the material does not sit waiting for heat, but springs back to its original shape as soon as released, just as if it were elastic. The difference is that the phase change allowed distortions which are larger, by a factor of 100 or more, than straightforward elasticity without the phase change. The material is *super-elastic*. It is just what is needed for eyeglass frames that get sat on and otherwise mistreated, or springs that must allow very large displacements at constant restoring force—and these, indeed, are the main applications for these materials.

12.7 Summary and conclusions

All materials have:

- A specific heat, C_p: the amount of energy it takes to heat unit mass by unit temperature.
- A thermal expansion, α: the change in its dimensions with change of temperature.

- A thermal conductivity λ and thermal diffusivity a: the former characterizing how fast heat is transmitted through the solid, the latter how long it takes for the temperature, once perturbed, to settle back to a steady pattern.
- Characteristic temperatures of its changes of phase or behavior: for a crystalline solid, its melting point T_m, and for non-crystalline solids, a glass transition temperature T_g; for design in all materials, an empirical temperature may be defined, the maximum service temperature T_{max}, above which, for reasons of creep or degradation, continuous use is impractical.
- Latent heats of melting and of vaporization: the energy absorbed or released at constant temperature when the solid melts or evaporates.

These properties can be displayed as material property charts—examples of which are utilized in this chapter. The charts illustrate the great range of thermal properties and the ways they are related. To explain these we need an understanding of the underlying physics. The key concepts are that thermal energy is atomic-level vibration, that increasing the amplitude of this vibration causes expansion, and that the vibrations propagate as phonons, giving conduction. In metals the free electrons—the ones responsible for electrical conduction—also transport heat, and they do so more effectively than phonons, giving metals their high thermal conductivities. Thermal properties are manipulated by interfering with these processes: alloying to scatter phonons and electrons; foaming to dilute the solid with low-conductivity, low-specific-heat gases.

The thermal response can be a problem in design—thermal stress, for instance, causes cracking. It can also be useful—thermal distortion can be used for actuation or sensing. It is the specific heat that makes an oven take 15 minutes to warm up, but it also stores heat in a way that can be recovered on demand. Good thermal conduction is not what you want in a coffee cup, but when it cools the engine of your car or the chip in your computer it is a big help.

This chapter has been about the thermal properties of materials. Heat, as we have said, also changes other properties, mechanical and electrical. We look more deeply into that in Chapter 13.

12.8 Further reading

Cottrell, A.H. (1964) *The Mechanical Properties of Matter*, Wiley, London, UK. Library of Congress number 64-14262. *(The best introduction to the physical understanding of the properties of materials, now out of print but worth searching for.)*

Hollman, J.P. (1981) *Heat Transfer*, 5th edition, McGraw-Hill, New York, USA. ISBN 0-07-029618-9. *(A good starting point: an elementary introduction to the principles of heat transfer.)*

Tabor, D. (1969) *Gases, Liquids and Solids*, Penguin Books, Harmondsworth, UK. ISBN 14-080052-2. *(A concise introduction to the links between inter-atomic forces and the bulk properties of materials.)*

12.9 Exercises

Exercise E12.1	Define specific heat. What are its units? How would you calculate the specific heat per unit volume from the specific heat per unit mass? If you wanted to select a material for a compact heat storing device, which of the two would you use as a criterion of choice?
Exercise E12.2	What two metals would you choose from the λ–a chart of Figure 12.5 to maximize the thermal displacement of a bi-metallic strip actuator? If the bimetallic strip has thickness 2 mm and average thermal diffusivity a of $5 \times 10^{-5}/°C$, how long will it take to respond when the temperature suddenly changes?
Exercise E12.3	A new alloy has a density of 7380 kg/m³. Its specific heat has not yet been measured. How would you make an approximate estimate of its value in the normal units of J/kg.K? What value would you then report?
Exercise E12.4	The same new alloy of the last exercise has a melting point of 1540 K. Its thermal expansion coefficient has not yet been measured. How would you make an approximate estimate of its value? What value would you then report?
Exercise E12.5	Interior wall insulation should insulate well, meaning low thermal conductivity, λ, but require as little heat as possible to warm up when the central heating system is turned on (if the wall absorbs heat the room stays cold). Use the λ–a chart of Figure 12.5 to find the materials that do this best (the contours will help).
Exercise E12.6	You notice that the ceramic coffee mugs in the office get too hot to hold about 10 seconds after pouring in the hot coffee. The wall thickness of the cup is 2 mm. (a) What, approximately, is the thermal diffusivity of the ceramic of the mug? (b) Given that the volume specific heat of solids, ρC_p, is more or less constant at 2×10^6 J/m³ · K, what approximately is the thermal conductivity of the cup material? (c) If the cup were made of a metal with a thermal diffusivity of 2×10^{-5} m²/s, how long could you hold it?
Exercise E12.7	A structural material is sought for a low-temperature device for use at $-20°C$ that requires high strength but low thermal conductivity. Use the λ–σ_y chart of Figure 12.6 to suggest two promising candidates (reject ceramics on the grounds that they are too brittle for structural use in this application).
Exercise E12.8	A material is needed for a small, super-efficient pressurized heat exchanger. The text explained that the index for this application is $M = \lambda\sigma_y$. Plot contours of this index onto a copy of the λ–σ_y chart of Figure 12.6 and hence find the two classes of materials that maximize M.

12.10 Exploring design with CES (use Level 2, Materials, for all selections)

Exercise E12.9 Use the 'Search' facility of CES to find materials for:

(a) Thermal insulation.
(b) Heat exchangers.

Exercise E12.10 The analysis of storage heaters formulated the design constraints for the heat-storage material, which can be in the form of a particle bed or a solid block with channels to pass the air to be heated. Use the selector to find the best materials. Here is a summary of the design requirements. List the top six candidates in ranked order.

Function	• Storage heater
Constraints	• Maximum service temperature >150°C
	• Non-flammable (a durability rating)
Objective	• Maximize specific heat/ material cost, C_p/C_m
Free variables	• Choice of material

Exercise E12.11 The requirements for a material for an automobile radiator, described in the text, are summarized in the table. Use CES to find appropriate materials to make them. List the top four candidates in ranked order.

Function	• Automobile heat exchanger
Constraints	• Elongation >10%
	• Maximum use temperature >180°C
	• Price <$5/kg
	• Durability in fresh water = very good
Objective	• Maximize thermal conductivity × yield strength, $\lambda\sigma_y$
	• Wall thickness, t
Free variables	• Choice of material

Exercise E12.12 A structural material is sought for a low-temperature device for use at −20°C that requires high tensile strength σ_{ts} but low thermal conductivity λ. For reasons of damage tolerance the fracture toughness K_{1c} must be greater than 15 MPa.m$^{1/2}$. Apply the constraint on K_{1c} using a 'Limit' stage, then make a chart with σ_{ts} on the x-axis and λ on the y-axis, and observe which materials best meet the requirements. Which are they?

Exercise E12.13   Interior wall insulation should insulate well, meaning low thermal conductivity, λ, but require as little heat as possible to warm up to the desired room temperature when the central heating system is turned on—that means low specific heat. The thickness of the insulation is almost always limited by the cavity space between the inner and outer wall, so it is the specific heat per unit volume, not per unit mass, that is important here. To be viable the material must have enough stiffness and strength to support its own weight and be easy to install—take that to mean a modulus $E > 0.01$ GPa and a strength $\sigma_y > 0.1$ MPa. Make a selection based on this information. List the materials you find that best meet the design requirements.

Exercise E12.14   Here is the gist of an email one of the authors received as this chapter was being written. 'We manufacture wood-burning stoves and fireplaces that are distributed all over Europe. We want to select the best materials for our various products. The important characteristics for us are: specific heat, thermal conductivity, density . . . and price, of course. What can your CES software suggest?'

Form your judgement about why these properties matter to them. Rank them, deciding which you would see as constraints and which as the objective. Consider whether there are perhaps other properties they have neglected. Then—given your starting assumptions of just how these stoves are used—use CES to make a selection. Justify your choice.

There is no 'right' answer to this question—it depends on the assumptions you make. The essence is in the last sentence of the last paragraph: justify your choice.

12.11   Exploring the science with CES Elements

Exercise E12.15   The text says that materials expand about 2% between 0 K and their melting point. Use CES Elements to explore the truth of this by making a bar chart of the expansion at the melting point, $\alpha T_m/10^6$ (the 10^6 is to correct for the units of α used in the database).

Exercise E12.16   When a solid vaporizes, the bonds between its atoms are broken. You might then expect that latent heat of vaporization, L_v, should be nearly the same as the cohesive energy, H_c, since it is the basic measure of the strength of the bonding. Plot one against the other, using CES Elements. How close are they?

Exercise E12.17   When a solid melts, some of the bonds between its atoms are broken, but not all—liquids still have a bulk modulus, for example. You might then expect that the latent heat of melting, L_m, should be less than the cohesive energy, H_c, since it is the basic measure of the strength of the bonding. Plot one against the other, using CES Elements (L_m is called the heat of fusion in the database). By what factor is L_m less than H_c? What does this tell you about cohesion in the liquid?

Exercise E12.18   The latent heat of melting (heat of fusion), L_m, of a material is said to be about equal to the heat required to heat it from absolute zero to its melting point, $C_p T_m$, where C_p is the specific heat and T_m is the absolute melting point. Make a chart with L_m on one axis and $C_p T_m$ on the other. To make the comparison right we have to change the units of L_m in making the chart to J/kg instead of kJ/mol. To do this multiply L_m by

$$\frac{10^6}{\text{Atomic weight in kg/kmol}}$$

using the 'Advanced' facility when defining axes in CES. Is the statement true?

Exercise E12.19   The claim was made in the text that the modulus E is roughly proportional to the absolute melting point T_m. If you use CES Elements to explore this correlation you will find that it is not, in fact, very good (try it). That is because T_m and E are measured in different units and, from a physical point of view, the comparison is meaningless. To make a proper comparison, we use instead $k_B T_m$ and $E\Omega$, where k_B is Boltzmann's (1.38×10^{-23} J/K) constant and Ωm^3/atom is the atomic volume. These two quantities are both energies, the first proportional to the thermal energy per atom at the melting point and the second proportional to the work to elastically stretch an atomic bond. It makes better sense, from a physical standpoint, to compare these.

Make a chart for the elements with $k_B T_m$ on the x-axis and $E\Omega$ on the y-axis to explore how good this correlation is. Correlations like these (if good) that apply right across the Periodic Table provide powerful tools for checking data, and for predicting one property (say, E) if the other (here, T_m) is known. Formulate an equation relating the two energies that could be used for these purposes.

Exercise E12.20   Above the Debye temperature, the specific heat is predicted to be $3R$, where R is in units of kJ/kmol.K. Make a plot for the elements with Debye temperature on the x-axis and specific heat in these units on the y-axis to explore this. You need to insert a conversion factor because of the units. Here it is, expressed in the units contained in the database:

Specific heat in kJ/kmol.K = Specific heat in J/kg.K
$\times$ Atomic weight in kg/kmol /1000.

Form this quantity, dividing the result by $R = 8.314$ kJ/kmol.K, and plot it against the Debye temperature. The result should be 3 except for materials with high Debye temperatures. Is it? Fit a curve by eye to the data. At roughly what temperature does the drop-off first begin?

Exercise E12.21   Explore the mean free path of phonons ℓ_m in the elements using equation (12.10) of the text. Inverting it gives

$$\ell_m = 3 \frac{\lambda}{\rho C_p c_o}$$

in which the speed of sound $c_o = \sqrt{E/\rho}$. Use the 'Advanced' facility when defining the axes in CES to make a bar chart of ℓ_m for the elements. Which materials have the longest values? Which have the shortest?

Chapter 13
Running hot: using materials at high temperatures

Shuttle flame. (Image courtesy of C. Michael Holoway, NASA Langley, USA.)

13.1 Introduction and synopsis

Material properties change with temperature. Some do so in a simple linear way, easy to allow for in design: the density, the modulus and the electrical conductivity are examples. But others, particularly the yield strength and the rates of oxidation and corrosion, change in more sudden ways that, if not allowed for, can lead to disaster.

This chapter explores the ways in which properties change with temperature and design methods to deal with the changes. To do this we must first understand *diffusion*—the intermixing of atoms in solids—and the ways it allows *creep* and *creep fracture*. This understanding lies behind procedures for high-temperature design with metals and ceramics. Polymers are a little more complicated in their behavior, but semi-empirical methods allow safe design with these too.

13.2 The temperature dependence of material properties

Maximum and minimum service temperatures

First, the simplest measure of tolerance to temperature: the *maximum* and *minimum service temperatures*, T_{max} and T_{min}. The former tells us the highest temperature at which the material can reasonably be used without oxidation, chemical change or excessive deflection or 'creep' becoming a problem (the *continuous use temperature*, or *CUT*, is a similar measure). The latter is the temperature below which the material becomes brittle or otherwise unsafe to use. These are empirical, with no universally accepted definitions. The minimum service temperature for carbon steels is the ductile-to-brittle-transition temperature—a temperature below which the fracture toughness falls steeply. For elastomers it is about $0.8\,T_g$, where T_g is the glass temperature; below T_g they cease to be rubbery and become hard and brittle.

Linear and nonlinear temperature dependence

Some properties depend on temperature T in a linear way, meaning that

$$P \approx P_o \left(1 + \beta \frac{T}{T_m}\right) \tag{13.1}$$

where P is the value of the property, P_o its low-temperature value and β is a constant. Figure 13.1 shows four examples: density, modulus, refractive index and—for metals—electrical resistivity. Thus, the density ρ and refractive index n decrease by about 6% on heating from cold to the melting point T_m ($\beta \approx -0.06$), the modulus E falls by a factor of 2 ($\beta \approx -0.5$), and the resistivity R increases by a factor of about 7 ($\beta \approx +6$). These changes cannot be neglected, but are easily accommodated by using the value of the property at the temperature of the design.

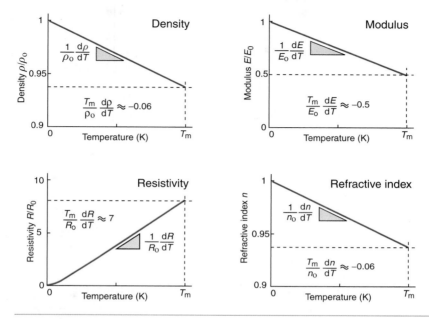

Figure 13.1   Linear dependence of properties on temperature: density, modulus, resistivity and refractive index.

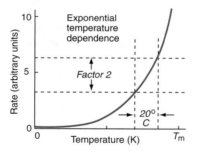

Figure 13.2   The exponential increase of rate of straining with temperature at constant load. The rate can double with a rise of temperature of only 20°C.

Other properties are less forgiving. Strength falls in a much more sudden way and the rate of *creep*—the main topic here—increases exponentially (Figure 13.2). This we need to explore in more detail.

Viscous flow
When a substance flows, its particles change neighbors; flow is a process of shear. Newton's law describes the flow rate in fluids under a shear stress τ:

$$\dot{\gamma} = \frac{\tau}{\eta}$$

where $\dot{\gamma}$ is the shear rate and η the viscosity. This is a linear law, like Hooke's law, with the modulus replaced by the viscosity and the strain by strain rate (units s^{-1}). Viscous flow occurs at constant volume (Poisson's ratio = 0.5) and this means that problems of viscous flow can be solved by taking the solution for elastic deformation and replacing the strain ε by the strain rate $\dot{\varepsilon}$ and, using equation (4.10), Young's modulus E by 3η. Thus, the rate at which a rod of a very viscous fluid, like tar, extends when pulled in tension is

$$\dot{\varepsilon} = \frac{\sigma}{3\eta} \tag{13.2}$$

The factor 1/3 appears because of the conversion from shear to normal stress and strain.

Creep

At room temperature, most metals and ceramics deform in a way that depends on stress but not on time. As the temperature is raised, loads that are too small to give permanent deformation at room temperature cause materials to *creep*: to undergo slow, continuous deformation with time, ending in fracture. It is usual to refer to the time-independent behavior as 'low-temperature' response, and the time-dependent flow as 'high-temperature' response. But what, in this context, is 'low' and what is 'high'? The melting point of tungsten, used for lamp filaments, is over 3000°C. Room temperature, for tungsten, is a very low temperature. The filament temperature in a tungsten lamp is about 2000°C. This, for tungsten, is a high temperature: the filament slowly sags over time under its own weight until the turns of the coil touch and the lamp burns out.

To design against creep we need to know how the strain rate $\dot{\varepsilon}$ or time to failure t_f depends on the stress σ and temperature T to which it is exposed. That requires *creep testing*.

Creep testing and creep curves

Creep is measured in the way shown in Figure 13.3. A specimen is loaded in tension or compression, usually at constant load, inside a furnace which is maintained at a constant temperature, T. The extension is measured as a function of time. Metals, polymers and ceramics all have creep curves with the general shape shown in the figure.

The *initial elastic* and the *primary creep* strains occur quickly and can be treated in much the way that elastic deflection is allowed for in a structure. Thereafter, the strain increases steadily with time in what is called the *secondary creep* or the *steady-state creep* regime. Plotting the log of the steady-state creep rate, $\dot{\varepsilon}_{ss}$, against the log of the stress, σ, at constant T, as in Figure 13.4(a), shows that

$$\dot{\varepsilon}_{ss} = B\sigma^n \tag{13.3}$$

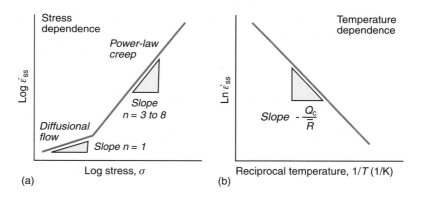

Figure 13.3   Creep testing and the creep curve, showing how strain ε increases with time t up to the fracture time t_f.

Figure 13.4   The stress and temperature dependence of the creep rate.

where n, the *creep exponent*, usually lies between 3 and 8 and for that reason this behavior is called *power-law creep*. At low σ there is a tail with slope $n \approx 1$ (the part of the curve labeled 'Diffusional flow' in Figure 13.4(a)). By plotting the natural logarithm (ln) of $\dot{\varepsilon}_{ss}$ against the reciprocal of the absolute temperature ($1/T$) at constant stress, as in Figure 13.4(b), we find that:

$$\dot{\varepsilon}_{ss} = C \exp - \left(\frac{Q_c}{RT} \right) \qquad (13.4)$$

Here $\bar{R}$ is the gas constant (8.31 J/mol/K) and Q_c is called the *activation energy for creep*, with units of J/mol. The creep rate $\dot{\varepsilon}_{ss}$ increases exponentially in the way suggested by Figure 13.2; increasing the temperature by as little as 20°C can double the creep rate. Combining these two findings gives

$$\dot{\varepsilon}_{ss} = C'\sigma^n \exp-\left(\frac{Q_c}{\overline{R}T}\right) \tag{13.5}$$

where C' is a constant. Written in this way the constant C' has weird units $(s^{-1} \cdot MPa^{-n})$ so it is more usual and sensible to write instead

$$\dot{\varepsilon}_{ss} = \dot{\varepsilon}_o\left(\frac{\sigma}{\sigma_o}\right)^n \exp-\left(\frac{Q_c}{\overline{R}T}\right) \tag{13.6}$$

The four constants $\dot{\varepsilon}_o$ (units: s^{-1}), σ_o (units: MPa), n and Q_c characterize the steady state creep of a material; if you know these, you can calculate the strain rate $\dot{\varepsilon}_{ss}$ at any temperature and stress using equation (13.6). They vary from material to material and have to be measured experimentally. We meet values for them in Section 13.5.

Sometimes creep is desirable. Extrusion, hot rolling, hot pressing and forging are carried out at temperatures at which power-law creep is the dominant mechanism of deformation; exploiting it reduces the pressure required for the operation. The change in forming pressure for a given change in temperature can be calculated from equation (13.6).

Creep damage and creep fracture
As creep continues, damage accumulates. It takes the form of voids or internal cracks that slowly expand and link, eating away the cross-section and causing

Figure 13.5 Creep fracture times. The scales are logarithmic. The data are typical of carbon steel.

the stress to rise. This makes the creep rate accelerate as shown in the tertiary stage of the creep curve of Figure 13.3. Since $\dot{\varepsilon} \propto \sigma^n$ with $n \approx 5$, the creep rate goes up even faster than the stress: an increase of stress of 10% gives an increase in creep rate of 60%.

Times to failure, t_f, are normally presented as *creep–rupture* diagrams (Figure 13.5). Their application is obvious: if you know the stress and temperature you can read off the life. If instead you wish to design for a certain life at a certain temperature, you can read off the design stress. Experiments show that

$$\dot{\varepsilon}_{ss}\, t_f = C \tag{13.7}$$

which is called the Monkman–Grant law. The Monkman–Grant constant, C, is typically 0.05–0.3. Knowing it, the creep life (meaning t_f) can be estimated from equation (13.6).

13.3 Charts for creep behavior

Melting point

Figure 13.6 shows melting points for metals, ceramics and polymers. Most metals and ceramics have high melting points and, because of this, they start to creep only at temperatures well above room temperature. *Lead*, however, has a melting point of 327°C (600 K), so room temperature is almost half its absolute

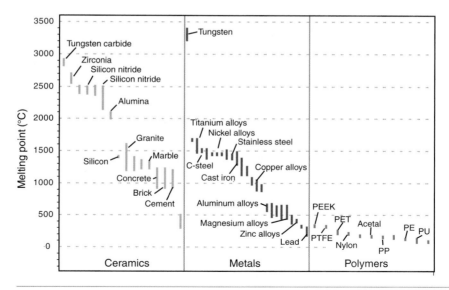

Figure 13.6   Melting points of ceramics, metals and polymers.

melting point and it creeps—a problem with lead roofs and cladding of old buildings. Crystalline polymers, most with melting points in the range 150–200°C, creep slowly if loaded at room temperature; glassy polymers, with T_g of typically 50–150°C, do the same. The point, then, is that the temperature at which materials start to creep depends on their melting points. As a general rule, it is found that creep starts when $T \approx 0.35 \, T_m$ for metals and $0.45 \, T_m$ for ceramics, although alloying can raise this temperature significantly.

Maximum service temperature and strength

Figure 13.7 charts the maximum service temperature T_{max} and room temperature strength σ_y. It shows that polymers and low melting metals like the alloys of zinc, magnesium and aluminum offer useful strength at room temperature but by 300°C they cease to be useful—indeed, few polymers have useful strength above 135°C. Titanium alloys and low-alloy steels have useful strength up to 600°C; above this temperature high-alloy stainless steels and more complex alloys based on nickel, iron and cobalt are needed. The highest temperatures require refractory metals like tungsten or technical ceramics such as silicon carbide (SiC) or alumina (Al_2O_3).

Figure 13.7   The strength and the maximum service temperature of materials. The strengths decrease with temperature in ways described in the text.

Creep strength at 950°C and density

Figure 13.8 is an example of a chart to guide selection of materials for load-bearing structures that will be exposed to high temperatures. It shows the creep strength at 950°C, $\sigma_{950°C}$. Strength at high temperatures (and this is a very high temperature), as we have said, is rate dependent, so to construct the chart we first have to choose an acceptable strain rate, one we can live with. Here 10^{-6}/s has been chosen. $\sigma_{950°C}$ is plotted against the density ρ. The chart is used in exactly the same way as the σ_y–ρ chart of Figure 6.6, allowing indices like

$$M_t = \frac{\sigma_{950°C}}{\rho}$$

to be plotted to identify materials for lightweight design at high temperature. Several such contours are shown.

The figure shows that titanium alloys, at this temperature, have strengths of only a few megapascals—about the same as lead at room temperature. Nickel- and iron-based super-alloys have useful strengths of 100 MPa or more. Only

Figure 13.8   A chart showing the strength of selected materials at a particularly high temperature—950°C and a strain rate of 10^{-6}/s—plotted against density. Software allows charts like this to be constructed for any chosen temperature. (The chart was made with a specialized high-temperature materials database that runs in the CES system.)

refractory metals like alloys of tungsten, technical ceramics like SiC and advanced (and very expensive) ceramic–ceramic composites offer high strength.

At room temperature we need only one strength–density chart. For high temperature design we need one that is constructed for the temperature and acceptable strain rate or life required by the design. It is here that computer-aided methods become particularly valuable, because they allow charts like Figure 13.8 to be constructed and manipulated for any desired set of operating conditions quickly and efficiently.

13.4 The science: diffusion and creep

Viscous flow and creep require the relative motion of atoms. How does this happen, and what is its rate? Think first of the unit step—one atom changing its position relative to those around it by what is called *diffusion*.

Diffusion

Diffusion is the spontaneous intermixing of atoms over time. In gases and liquids intermixing is familiar—the dispersion of smoke in still air, the dispersion of an ink drop in water—and understood as the random motion of the atoms or molecules (Brownian motion). But in crystalline solids atoms are confined to lattice sites—how then are they going to mix? In practice they do. If atoms of type A are plated onto a block of atoms of type B and the A–B couple is heated, the atoms interdiffuse at a rate described by Fick's law:[1]

$$J = -D \frac{dc}{dx} \tag{13.8}$$

Here J is the atom flux (the number of atoms of type A diffusing across unit area of surface per second), D is the diffusion coefficient and dc/dx is the gradient in concentration c of A atoms in the x direction.

Heat, as we have said, is atoms in motion. In a solid they vibrate about their mean position with a frequency ν (about 10^{13} per second) with an average energy, kinetic plus potential, of $k_B T$ in each mode of vibration, where k_B is Boltzmann's constant (1.38×10^{-23} J/atom·K). This is the average, but at any instant some atoms have less, some more. Statistical mechanics gives the distribution of energies. The Maxwell–Boltzmann equation describes the probability p that a given atom has an energy greater than a value q joules:

$$p = \exp - \left(\frac{q}{k_B T} \right) \tag{13.9}$$

---

[1] Adolph Eugen Fick (1829–1901), German physicist, physiologist and inventor of the contact lens. He formulated the law that describes the passage of gas through a membrane, but which also describes the heat-activated mixing of atoms in solids.

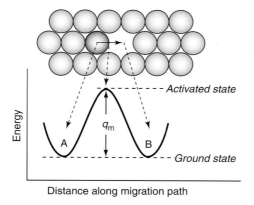

Distance along migration path

Figure 13.9   A diffusive jump. The energy graph shows how the energy of the red atom (which may be chemically the same as the green ones, or may not) changes as it jumps from site A to site B.

Crystals, as said in Chapter 4, contain vacancies—occasional empty atom sites. These provide a way for diffusive jumps to take place. Figure 13.9 shows an atom jumping into a vacancy. To make such a jump, the atom marked in red (though it is the same sort of atom as the rest) must break away from its original comfortable site at A, its *ground state*, and squeeze between neighbors, passing through an *activated state*, to drop into the vacant site at B where it is once again comfortable. There is an energy barrier, q_m, between the ground state and the activated state to overcome if the atom is to move. The probability p_m that a given atom has thermal energy this large or larger is just equation (13.9) with $q = q_m$.

So two things are needed for an atom to switch sites: enough thermal energy and an adjacent vacancy. A vacancy has an energy q_v, so—not surprisingly—the probability p_v that a given site be vacant is also given by equation (13.9), this time with $q = q_v$. Thus, the overall probability of an atom changing sites is

$$p = p_v p_m = \exp -\left(\frac{q_v + q_m}{k_B T}\right) = \exp -\left(\frac{q_d}{k_B T}\right) \qquad (13.10)$$

where q_d is called the *activation energy for self-diffusion*. If instead the red atom were chemically different from the green ones (so it is in solid solution), the process is known as interdiffusion. Its activation energy has the same origin.

Figure 13.10 illustrates how mixing occurs by interdiffusion. It shows a solid in which there is a concentration gradient dc/dx of red atoms: there are more in slice A immediately to the left of the central, shaded plane, than in slice B to its right. If atoms jump across this plane at random there will be a net flux of red atoms to the right because there are more on the left to jump, and a net flux of

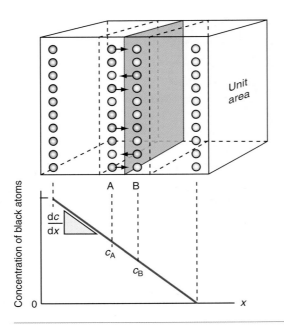

Figure 13.10   Diffusion in a concentration gradient.

white atoms in the opposite direction. The number of red atoms in slice A, per unit area, is $n_A = 2r_o c_A$ and that in slice B is $n_B = 2r_o c_B$, where $2r_o$, the atom size, is the thickness of the slices, and c_A and c_B are the concentration of red atoms on the two planes expressed as atom fractions. The difference is

$$n_A - n_B = 2r_o(c_A - c_B) = 4r_o^2 \frac{dc}{dx}$$

(since $c_A - c_B = 2r_o(dc/dx)$). The number of times per second that an atom on slice A oscillates towards B, or one on B towards A, is $\nu/6$, since there are six possible directions in which an atom can oscillate in three dimensions, only one of which is in the right direction. Thus, the net flux of red atoms from left to right is

$$J = -\frac{\nu}{6} \exp -\left(\frac{q_d}{k_B T}\right) 4r_o^2 \frac{dc}{dx}$$

It is usual to list the activation energy per mole, Q_d, rather than that per atom, q_d, so we write $Q_d = N_A q_d$ and $\bar{R} = N_A k_B$ (where N_A is Avogadro's number, 6.02×10^{23}), and assemble the terms $4\nu r_o^2/6$ into a single constant D_o to give

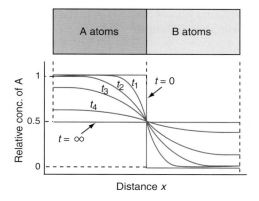

Figure 13.11   Intermixing by diffusion.

$$J = -D_o \exp -\left(\frac{Q_d}{\bar{R}T}\right)\frac{dc}{dx} \tag{13.11}$$

which has the form of Fick's law (equation (13.8)) with the diffusion coefficient D given by

$$D = D_o \exp -\left(\frac{Q_d}{\bar{R}T}\right) \tag{13.12}$$

(The CES Elements database contains data for D_o and Q_d for the elements.)

Fourier's law for heat flow (equation (12.2)) and Fick's law for the flow of matter (equation (13.8)) have identical forms. This means solutions to problems of heat flow are the same as those for matter flow if a is replaced by D and dT/dx is replaced by dc/dx. This includes solutions to transient problems. Figure 13.11 shows successive concentration profiles at times t_1, t_2, t_3, t_4 as red atoms on the left and green atoms on the right interdiffuse. This is a transient matter flow problem. Just as with transient heat flow (equation (12.15)), the mean distance x that one type of atom has penetrated the other is given by

$$x \approx \sqrt{Dt} \tag{13.13a}$$

There are two further rules of thumb, useful in making estimates of diffusion rates when data are not available and to give a feel for the upper limit for diffusion rate. The first is that the activation energy for diffusion, normalized by $\bar{R}T_m$, is approximately constant for metals:

$$\frac{Q_d}{\bar{R}T_m} \approx 18 \tag{13.13b}$$

The second is that the diffusion coefficient of metals, evaluated at their melting point, is also approximately constant:

$$D_{Tm} = D_o \exp -\left(\frac{Q_d}{\overline{R} T_m} \right) \approx 10^{-12} \text{ m}^2/\text{s} \qquad (13.13\text{c})$$

Diffusion in liquids and non-crystalline solids

The vacancies in a crystal can be thought of as *free volume*—free because it is able to move, as it does when an atom jumps. The free volume in a crystal is thus in the form of discrete units, all of the same size—a size into which an atom with enough thermal energy can jump. Liquids and non-crystalline solids, too, contain free volume, but because there is no lattice or regular structure it is dispersed randomly between all the atoms or molecules. Experiments show that the energy barrier to atom movement in liquids is not the main obstacle to diffusion or flow. There is actually plenty of free volume among neighbors of any given particle, but it is of little use for particle jumps except when, by chance, it comes together to make an atom-sized hole—a temporary vacancy. This is a fluctuation problem just like that of the Maxwell–Boltzmann problem, but a fluctuation of free volume rather than thermal energy, and with a rather similar solution: diffusion in liquids is described by

$$D = \frac{4}{6} r_o^2 \nu \exp -\left(\frac{v_a}{v_f} \right) \qquad (13.14)$$

where v_f is the average free volume per particle, v_a is the volume of the temporary vacancy, and r_o and ν have the same meaning as before.

Diffusion driven by other fields

So far we have thought of diffusion driven by a concentration gradient. Concentration c is a *field quantity*; it has discrete values at different points in space (the concentration field). The difference in c between two nearby points divided by the distance between them defines the local concentration gradient, dc/dx.

Diffusion can be driven by other field gradients. A stress gradient, as we shall see in a moment, drives diffusional flow and power-law creep. An electric field gradient can drive diffusion in non-conducting materials. Even a temperature gradient can drive diffusion of matter as well as diffusion of heat.

Diffusional flow

Liquids and hot glasses are wobbly structures, a bit like a bag loosely filled with beans. If atoms (or beans) move inside the structure, the shape of the whole structure changes in response. Where there is enough free volume, a stress favors the movements that change the shape in the direction of the stress, and opposes those that do the opposite, giving viscous flow. The higher the temperature, the more is the free volume and the faster the flow.

But how do jumping atoms change the shape of a crystal? A crystal is not like a bag of loose beans—the atoms have well-defined sites on which they sit. In the cluster of atoms in Figure 13.9 an atom has jumped but the shape of the cluster has not changed. At first sight, then, diffusion will not change the shape of a crystal. In fact it does, provided the material is *polycrystalline*—made up of many crystals meeting at grain boundaries. This is because the grain boundaries act as sources and sinks for vacancies. If a vacancy joins a boundary, an atom must leave it; repeat this many times and that face of the crystal is eaten away. If instead a vacancy leaves a boundary, an atom must join it and—repeated—that face grows. Figure 13.12 shows the consequences: the slow extension of the polycrystal in the direction of stress. It is driven by a stress gradient: the difference between the tensile stress σ on the horizontal boundaries from which vacancies flow and that on the others, essentially zero, to which they go. If the grain size is d the stress gradient is σ/d. The flux of atoms, and thus the rate at which each grain extends, $\delta d/\delta t$, is proportional to $D\sigma/d$. The strain rate, $\dot{\varepsilon}$, is the extension rate divided by the original grain size, d, giving

$$\dot{\varepsilon} = C\,\frac{D\sigma}{d^2} = C\,\frac{\sigma}{d^2}\,D_o\,\exp-\left(\frac{Q_d}{\bar{R}T}\right) \tag{13.15}$$

where C is a constant. This is a sort of viscous flow, linear in stress. The smaller the grain size, the faster it goes.

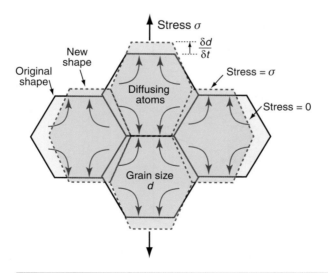

Figure 13.12 Deformation by diffusion alone, giving *diffusional flow*.

Dislocation climb and power-law creep

Plastic flow, as we saw in Chapter 5, is the result of the motion of dislocations. Their movement is resisted by dissolved solute atoms, precipitate particles, grain boundaries and other dislocations; the yield strength is the stress needed to force dislocations past or between them. Diffusion can unlock dislocations from obstacles in their path, making it easier for them to move. Figure 13.13 shows how it happens. Here a dislocation is obstructed by a particle. The glide force τb per unit length is balanced by the reaction f_o from the precipitate. But if the atoms of the extra half-plane diffuse away, thus eating a slot into the half-plane, the dislocation can continue to glide even though it now has a step in it. The stress gradient this time is less obvious—it is the difference between the local stress where the dislocation is pressed against the particle and that on the dislocation remote from it. The process is called 'climb' and, since it requires diffusion, it can occur at a measurable rate only when the temperature is above about $0.35T_M$.

Climb unlocks dislocations from the obstacles that pin them, allowing further slip. After a little slip, of course, the unlocked dislocations encounter the next obstacles, and the whole cycle repeats itself. This explains the *progressive, continuous*, nature of creep. The dependence on diffusion explains the dependence of creep rate on *temperature*, with

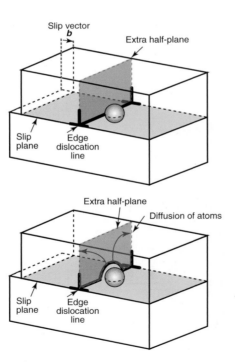

Figure 13.13   Climb of a dislocation: the extra half-plane is eaten away by diffusion, allowing the dislocation to pass the obstacle. The result is power-law creep.

$$\dot{\varepsilon}_{ss} = A \, \sigma^n \, \exp - \left(\frac{Q_c}{\bar{R}T} \right)$$
(13.16)

with $Q_c \approx Q_d$. The power-law dependence on *stress* σ is harder to explain. It arises partly because the stress gradient driving diffusion increases with σ and partly because the density of dislocations itself increases too.

Deformation mechanism diagrams

Let us now pull all this together. Materials can deform by dislocation plasticity (Chapter 6) or, if the temperature is high enough, by diffusional flow or power-law creep. If the stress and temperature are too low for any of these, the deformation is elastic. This competition between mechanisms is summarized in *deformation mechanism diagrams*, of which Figure 13.14 is an example. It shows the range of stress and temperature in which we expect to find each sort of deformation and the strain rate that any combination of them produces (the contours). Diagrams like these are available for many metals and ceramics, and are a useful summary of creep behavior, helpful in selecting a material for high-temperature applications—one appears in the examples at the end of this chapter.

Creep fracture

Diffusion, we have seen, gives creep. It also gives creep fracture. You might think that a creeping material would behave like toffee or chewing gum—that

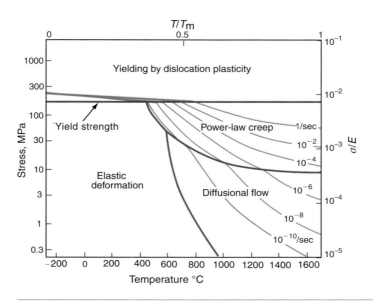

Figure 13.14 A deformation mechanism map, showing the regime in which each mechanism operates.

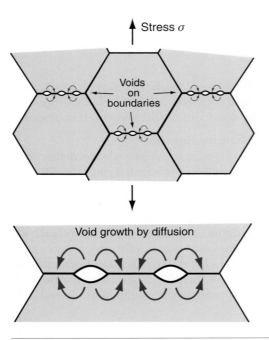

Figure 13.15   Creep fracture caused by the diffusive growth of voids on grain boundaries.

it would stretch a long way before breaking in two—but for crystalline materials this is very rare. Indeed, creep fracture (in tension) can happen at unexpectedly small strains, often only 2–5%, by the mechanism shown in Figure 13.15. Voids nucleate on grain boundaries that lie normal to the tensile stress. These are the boundaries to which atoms diffuse to give diffusional creep, coming from the boundaries that lie more nearly parallel to the stress. But if the tensile boundaries have voids on them, they act as sources of atoms too, and in doing so, they grow. The voids cannot support load, so the stress rises on the remaining intact bits of boundary, making the voids grow more and more quickly until finally (bang) they link.

Many engineering components (e.g. tie-bars in furnaces, super-heater tubes, high-temperature pressure vessels in chemical reaction plants) are expected to withstand moderate creep loads for long times (say 20 years) without failure. The loads or pressure they can safely carry are calculated by methods such as those we have just described. One would like to be able to test new materials for these applications without having to wait for 20 years to get the results. It is thus tempting to speed up the tests by increasing the load or the temperature to get observable creep in a short test time, and this is done. But there are risks. Tests carried out on the steep $n = 3\text{--}8$ branch of Figure 13.4(a), if extrapolated to lower stresses where $n \approx 1$, greatly underestimate the creep rate.

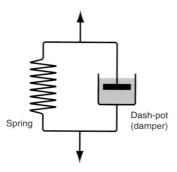

Figure 13.16   Visco-elastic behavior can be modeled as a spring (the elastic part) in parallel with a dash-pot (the viscous part).

Creep mechanisms: polymers

Creep in crystalline materials, as we have seen, is closely related to diffusion. The same is true of polymers, but because most of them are partly or wholly amorphous, diffusion is controlled by free volume (equation (13.14)). Free volume increases with temperature (its fractional change per degree is just the volumetric thermal expansion, 3α) and it does so most rapidly at the glass transition temperature T_g. Thus, polymers start to creep as the temperature approaches T_g and that, for most, is a low temperature: 50–150°C.

This means that the temperature range in which most polymers are used is that near T_g when they are neither simple elastic solids nor viscous liquids; they are *visco-elastic* solids. If we represent the elastic behavior by a spring and the viscous behavior by a dash-pot, then visco-elasticity (at its simplest) is described by a coupled spring and dash-pot as in Figure 13.16. Applying a load causes creep, but at an ever-decreasing rate because the spring takes up the tension. Releasing the load allows slow reverse creep, caused by the extended spring.

Real polymers require more elaborate systems of springs and dash-pots to describe them. This approach of *polymer rheology* can be developed to provide criteria for design, but these are complex. A simpler approach is to use graphical data for what is called the *creep modulus*, E_c, to provide an estimate of the deformation during the life of the structure. Figure 13.17 shows data for E_c as a function of temperature T and time t. Creep modulus data allow elastic solutions to be used for design against creep. The service temperature and design life are chosen, the resulting creep modulus is read from the graph, and this is used instead of Young's modulus in any of the solutions to design problems listed in Chapter 5.

13.5 Materials to resist creep

Metals and ceramics to resist creep

Diffusional flow is important when grains are small (as they often are in ceramics) and when the component is subject to high temperatures at low loads. Equation

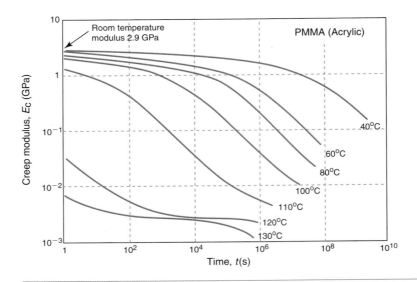

Figure 13.17   The creep modulus of PMMA.

(13.15) says that the way to avoid diffusional flow is to choose a material with a high melting temperature and arrange that it has a large grain size, d, so that diffusion distances are long. Single crystals are best of all; they have no grain boundaries to act as sinks and sources for vacancies, so diffusional creep is suppressed completely. This is the rationale behind the wide use of single-crystal turbine blades in jet engines.

That still leaves power-law creep. Materials that best resist power-law creep are those with high melting points, since diffusion and thus creep rates scale as T/T_m, and a microstructure that maximizes obstruction to dislocation motion through alloying to give a solid solution and precipitate particles. Current creep-resistant materials, known as *super-alloys*, are remarkably successful in this. Most are based on iron, nickel or cobalt, heavily alloyed with aluminum, chromium and tungsten. The chart of Figure 13.8 shows both effects: the high melting point *refractory metals* and the heavily alloyed *super-alloys*.

Many ceramics have high melting points, meeting the first criterion. They do not need alloying because their covalent bonding gives them a large lattice resistance. Typical among them are the technical ceramics alumina (Al_2O_3), silicon carbide (SiC) and silicon nitride (Si_3N_4)—they too appear in Figure 13.8. The lattice resistance pins down dislocations but it also gives the materials low fracture toughness, even at high temperature, making design to use them difficult.

Polymers to resist creep
The creep resistance of polymers scales with their glass temperature. T_g increases with the degree of cross-linking; heavily cross-linked polymers (e.g. epoxies)

with high T_g are therefore more creep resistant at room temperature than those that are not (like polyethylene). The viscosity of polymers above T_g increases with molecular weight, so the rate of creep is reduced by having a high molecular weight. Finally, crystalline or partly crystalline polymers (e.g. high-density polyethylene) are more creep resistant than those which are entirely glassy (e.g. low-density polyethylene).

The creep rate of polymers is reduced by filling them with powdered glass, silica (sand) or talc, roughly in proportion to the amount of filler. PTFE on saucepans and polypropylene used for automobile components are both strengthened in this way. Much better creep resistance is obtained with composites containing chopped or continuous fibers (GFRP and CFRP), because much of the load is now carried by the fibers that, being very strong, do not creep at all.

Selecting materials to resist creep

Classes of industrial applications tend to be associated with certain characteristic temperature ranges. There is the cryogenic range, between $-273°C$ and roughly room temperature, associated with the use of liquid gases like hydrogen, oxygen or nitrogen. Here the issue is not creep, but the avoidance of brittle fracture. There is the regime at and near room temperature (-20 to $+150°C$) associated with conventional mechanical and civil engineering: household appliances, sporting goods, aircraft structures and housing are examples. Above this is the range 150–400°C, associated with automobile engines and with food and industrial processing. Higher still are the regimes of steam turbines and superheaters (typically 400–650°C), and of gas turbines and chemical reactors (650–1000°C). Special applications (lamp filaments, rocket nozzles) require materials which withstand even higher temperatures, extending as high as 2800°C.

Materials have evolved to fill the needs of each of these temperature ranges (Figure 13.18). Certain polymers, and composites based on them, can be used in applications up to 250°C, and now compete with magnesium and aluminum alloys and with the much heavier cast irons and steels, traditionally used in those ranges. Temperatures above 400°C require special creep-resistant alloys: ferritic steels, titanium alloys (lighter, but more expensive) and certain stainless steels. Stainless steels and ferrous super-alloys really come into their own in the temperature range above this, where they are widely used in steam turbines and heat exchangers. Gas turbines require, in general, nickel-based or cobalt-based super-alloys. Above 1000°C, the refractory metals and ceramics become the only candidates. Materials used at high temperatures will, generally, perform perfectly well at lower temperatures too, but are not used there because of cost.

The chart of Figure 13.8 showed the performance of materials able to support load at 950°C. Charts like this are used in design in the same way as those for low-temperature properties.

T °C	Materials	Applications	*T* K
1200	Refractory metals: Mo, W, Ta Alloys of Nb, Mo, W, Ta Ceramics: Oxides Al_2O_3, MgO, etc. Nitrides, Si_3N_4, Carbides, SiC	Rocket nozzles Special furnaces Experimental turbines	1400
1000			
	Austenitic stainless steels Nichromes, nimonics Nickel based super-alloys Cobalt based super-alloys Iron based super-alloys	Gas turbines Chemical engineering Petrochemical reactors Furnace components Nuclear construction	1200
800			
			1000
600	Iron-based super-alloys Ferritic stainless steels Austenitic stainless steels Inconels and nimonics	Steam turbines Superheaters Heat exchangers	
	Low-alloy steels Titanium alloys (up to 450°C) Inconels and nimonics	Heat exchangers Steam turbines Gas turbine compressors	800
400			
			600
200	Fiber-reinforced polymers Copper alloys (up to 400°C) Nickel, monels and nickel-silvers PEEK, PEK, PI, PPD, PTFE and PES (up to 250°C)	Food processing Automotive (engine)	
			400
0	Most polymers (max temp: 60 to 150°C) Magnesium alloys (up to 150°C) Aluminum alloys (up to 150°C) Monels and steels	Civil construction Household appliances Automotive Aerospace	
			200
	Austenitic stainless steels Aluminum alloys	Rocket casings, pipework, etc. Liquid O_2 or N_2 equipment	
−200			
−273	Copper alloys Niobium alloys	Superconduction	0

Figure 13.18   Materials for each regime of temperature.

13.6 Design to cope with creep

Creep problems are of four types:

- Those in which limited creep strain can be accepted but creep rupture must be avoided, as in the creep of pipework or of lead roofs and cladding on buildings.
- Those in which creep strain is design limiting, as it is for blades in steam and gas turbines where clearances are critical.

Figure 13.19   A pressurized pipe.

- Those involving more complex problems of creep strain, loss of stiffness and risk of buckling—a potential problem with space-frames of supersonic air-craft and space vehicles.
- Those involving stress relaxation—the loss of tension in a pre-tightened bolt, for instance.

Predicting life of high temperature pipework

Chemical engineering and power generation plants have pipework that carries hot gases and liquids under high pressure. A little creep, expanding the pipe slightly, can be accepted; rupture, with violent release of hot, high-pressure fluid, cannot. We take, as an example, pipework in a steam-turbine power-generating station. Pipes in a typical 600 MW unit carry steam at 650°C and a pressure p of 15 MPa. The stress in the wall of a thin-walled pipe with a radius R and a wall thickness t carrying a pressure p, as in Figure 13.19, is

$$\sigma = \frac{pR}{t} \tag{13.17}$$

Suppose you have been asked to recommend a pipe required as a temporary fix while modifications are made to the plant. Space is constrained: the pipe cannot be more than 300 mm in diameter. The design life is 6 months. A little creep does not matter, but the pipe must not rupture. Type 304 stainless steel pipe with a diameter of 300 mm and a wall thickness of 10 mm is available. Will it function safely for the design life?

The stress in the pipe wall with these dimensions, from equation (13.17), is 225 MPa. Figure 13.20 shows stress–rupture data for Type 304 stainless steel. Enter 225 MPa at 650°C and read off the rupture life: about 7 hours. Not so good.

So how thick should the pipe be? To find out, reverse the reasoning. The design life is 6 months—4380 hours. A safety-critical component such as this one needs a safety factor, so double it: 8760 hours. Enter this and the temperature on the stress–rupture plot and read off the acceptable stress: 80 MPa. Put this into

Figure 13.20   Stress–rupture time data for Type 304 stainless steel.

equation (13.17) to calculate the wall thickness. To be safe the pipe wall must be at least 28 mm thick.

Turbine blades

Throughout the history of its development, the gas turbine has been limited in thrust and efficiency by the availability of materials that can withstand high stress at high temperatures. The origin of the stress is the centrifugal load carried by the rapidly spinning turbine disk and rotor blades. Where conditions are at their most extreme, in the first stage of the turbine, nickel- and cobalt-based super-alloys are currently used because of their unique combination of high-temperature strength, toughness and oxidation resistance. Typical of these is MAR-M200, an alloy based on nickel, strengthened by a solid solution of tungsten and cobalt and by precipitates of $Ni_3(Ti,A1)$, and containing chromium to improve its resistance to attack by gases.

When a turbine is running at a steady speed, centrifugal forces subject each rotor blade to an axial tension (equation (7.8)). If the blade has a constant cross-section, the tensile stress rises linearly from zero at its tip to a maximum at its root. As an example, a rotor of radius $r = 0.3$ m rotating at an angular velocity ω of 10 000 rpm (1000 radians/s) induces an axial stress ($R\omega^2\rho x$) of order 150 MPa. (Here ρ is the density of the alloy, about 8000 kg/m^3, and x the distance from the tip; a typical blade is about 80 mm long.) Typical stress and temperature profiles for a blade in a medium-duty engine are shown in Figure 13.21. They are plotted as a shaded box onto two deformation mechanism maps in Figure 13.22. If made of pure nickel (Figure 13.22(a)) the blade would deform by *power-law creep*, at a totally unacceptable rate. The strengthening methods

Figure 13.21 A turbine blade, showing the stress and temperature profiles.

Figure 13.22 (a) The stress–temperature profile of the blade plotted onto a deformation map for pure nickel. (b) The same profile plotted on a map for the alloy MAR-M200. The strain rates differ by a factor of almost 10^6.

used in MAR-M200 with a typical as-cast grain size of 0.1 mm (Figure 13.22(b)) reduce the rate of power-law creep by a factor of 10^6 and change the dominant mechanism of flow from power-law creep to diffusional flow. Further solution strengthening or precipitation hardening is now ineffective unless it slows this mechanism. A new strengthening method is needed: the obvious one is to increase the grain size or remove grain boundaries altogether by using a single crystal. This slows diffusional flow or stops it altogether, while leaving the other flow mechanisms unchanged. The power-law creep field expands and the rate of creep of the turbine blade falls to a negligible level.

The point to remember is that creep has contributions from several distinct mechanisms; the one that is dominant depends on the stress and temperature applied to the material. If one is suppressed, another will take its place. Strengthening methods are selective: a method that works well in one range of stress and temperature may be ineffective in another. A strengthening method should be regarded as a way of attacking a particular flow mechanism. Materials with good creep resistance combine strengthening mechanisms in order to attack them all; single-crystal MAR-M200 is a good example of this.

Thermal barrier coatings

The efficiency and power-to-weight ratio of advanced gas turbines, as already said, is limited by the burn temperature and this in turn is limited by the materials of which the rotor and stator blades are made. Heat enters the blade from the burning gas, as shown in Figure 13.23. Blades are cooled by pumping air through internal channels, leading to the temperature profile shown on the left. The surface temperature of the blade is set by a balance between the heat transfer coefficient between gas and blade (determining the heat in) and conduction within

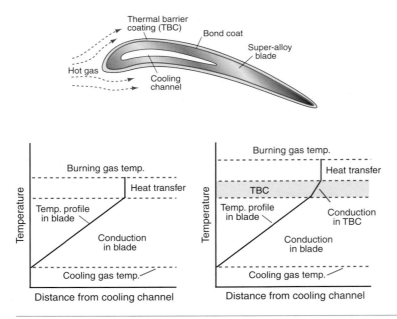

Figure 13.23   A cross-section of a turbine blade with a thermal barrier coating (TBC). The temperature profile for the uncoated blade is shown on the left, that for the coated blade on the right.

the blade to the cooling channel (determining the heat out). The temperature step at the surface is increased by bleeding a trickle of the cooling air through holes in the blade surface. This technology is already quite remarkable—air-cooled blades operate in a gas stream at a temperature that is above the melting point of the alloy of which they are made. How could the burn temperature be increased yet further?

Many ceramics have higher melting points than any metal and some have low thermal conductivity. Ceramics are not tough enough to make the whole blade, but coating the metal blade with a ceramic to form a *thermal barrier coating* (TBC) allows an increase in gas temperature with no increase in that of the blade, as shown on the right. How is the ceramic chosen? The first considerations are those of a low thermal conductivity, a maximum service temperature above that of the gas (providing some safety margin) and adequate strength. The α–λ chart (Figure 12.4) shows that the technical ceramic with by far the lowest thermal conductivity is zirconia (ZrO_2). The T_{max}–σ_f chart (Figure 13.7) confirms that it is usable up to very high temperature and has considerable strength. Zirconia looks like a good bet—and indeed it is this ceramic that is used for TBCs.

As always, it is not quite so simple. Other problems must be overcome to make a good TBC. It must stick to the blade, and that is not easy. To achieve it the blade surface is first plated with a thin *bond coat* (a complex Ni–Cr–Al–Y alloy); it is the glue, so to speak, between blade and coating, shown in Figure 13.23. Most ceramics have a lower expansion coefficient than the super-alloy of the blade (compare α for ZrO_2 and Ni in Figure 12.4) so when the blade heats up it expands more than the coating and we know what that means: thermal stresses and cracking. The problem is solved (amazingly) by arranging that the coating is *already* cracked on a fine scale, with all the cracks running perpendicular to its surface, making an array of interlocking columns like a microscopic Giant's Causeway. When the blade expands the columns separate very slightly, but not enough for hot gas to penetrate to any significant extent; its protective thermal qualities remain. Next time you fly, reflect on all this—the plane you are in almost certainly has coated blades.

Airframes

If you want to fly at speeds above Mach 1 (760 mph), aerodynamic heating becomes a problem. It is easiest to think of a static structure in a wind tunnel with a supersonic air stream flowing over it. In the boundary layer immediately adjacent to the structure, the velocity of the air is reduced to zero and its kinetic energy appears as heat, much of which is dumped into the skin of the structure. The surface temperature T_s can be calculated; it is approximately

$$T_s \approx (1 + 0.2M_a^2)\, T_o$$

Here M_a is the Mach number and T_o the ambient temperature. Supersonic flight is usually at altitudes above 35 000 ft, where $T_o = -50°C$. Figure 13.24 shows what T_s looks like.

This, not surprisingly, causes problems. First, creep causes a gradual change in dimension over time; wing deflection can increase, affecting aerodynamic

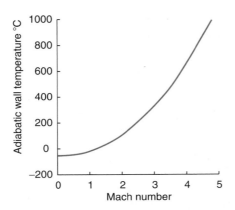

Figure 13.24 Adiabatic heating as a function of speed.

performance. Second, thermal stress caused by expansion can lead to further creep damage because it adds to aerodynamic loads. Finally, the drop in modulus, E, brings greater elastic deflection and changes the buckling and flutter (meaning vibration) characteristics.

In a large aircraft, the wing-tip displacement between ground and steady flight is about 0.5 m, corresponding to a strain in the wing-spar (which is loaded in bending) of about 0.1%. To avoid loss of aerodynamic quality, the creep strain over the life of the aircraft must be kept below this. Figure 13.25 illustrates how the strength of three potential structural materials influences the choice. Where the strength is flat with temperature (on the left) the material is a practical choice. The temperature at which it drops off limits the Mach number, plotted across the top. Aluminum is light, titanium twice as dense, super-alloys four times so, so increasing the Mach number requires an increase in weight. A heavier structure needs more power and therefore fuel and fuel, too, has weight—the structural weight of an aircraft designed to fly at Mach 3 is about three times that of one designed for Mach 1. Thus, it appears that there is a practical upper limit on speed. It has been estimated that sustained flight may be limited by the weight penalty to speeds below Mach 3.5.

Creep relaxation
Creep causes pre-tensioned components to relax over time: bolts in hot turbine casings must be regularly tightened; pipe connectors carrying hot fluids must be readjusted. It takes only tiny creep strain ε_{cr} to relax the stress—a fraction of the elastic strain ε_{el} caused by pre-tensioning is enough, and ε_{el} is seldom much greater than 10^{-3}. Figure 13.26 shows a bolt that is tightened onto a rigid component so that the initial stress in its shank is σ_i. The elastic strain is then

$$\varepsilon_{el} = \frac{\sigma_i}{E}$$

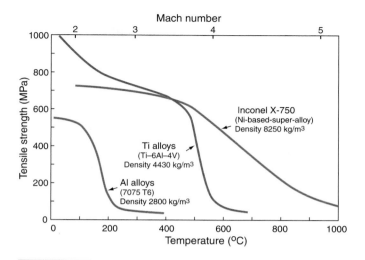

Figure 13.25   The strength of alloys as a function of temperature.

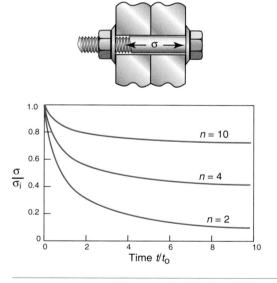

Figure 13.26   Stress relaxation by creep.

If creep strain ε_{cr} replaces part of ε_{el} the stress relaxes. At any time t

$$\varepsilon_{tot} = \varepsilon_{el} + \varepsilon_{cr}$$

This total strain ε_{tot} is fixed because the component on which the bolt is clamped is rigid. Differentiating with respect to time, inserting equation (13.3) for $\dot{\varepsilon}_{cr}$ gives

$$\dot{\varepsilon}_{tot} = \dot{\varepsilon}_{el} + \dot{\varepsilon}_{cr} = \frac{\dot{\sigma}}{E} + B\sigma^n = 0$$

The time t for the stress to relax from σ_i to σ, for $n > 1$, is found by integrating over time:

$$t = \frac{1}{(n-1)BE}\left(\frac{1}{\sigma^{n-1}} - \frac{1}{\sigma_i^{n-1}}\right)$$

Solving for σ/σ_i as a function of t gives

$$\frac{\sigma}{\sigma_i} = \frac{1}{(t/t_o + 1)^{1/(n-1)}} \quad \text{with} \quad t_o = (n-1)\,B\,E\,\sigma_i^{n-1}$$

The stress σ/σ_i is plotted as a function of t/t_o in Figure 13.26, for three values of n. The relaxation is most extensive when n is small. As it rises, the drop in stress becomes less until, as n approaches infinity, meaning rate-independent plasticity, it disappears. Since transient creep is neglected, these curves overestimate the relaxation time for the first tightening of the bolt, but improve for longer times.

13.7 Summary and conclusions

All material properties depend on temperature. Some, like the density and the modulus, change relatively little and in a predictable way, making compensation easy. *Transport properties*, meaning thermal conductivity and conductivity for matter flow (which we call diffusion), change in more complex ways. The last of these—diffusion—has a profound effect on mechanical properties when temperatures are high. To understand and use diffusion we need the idea of *thermal activation*—the ability of atoms to jump from one site to another, using thermal energy as the springboard. In crystals, atoms jump from vacancy to vacancy, but in glasses, with no fixed lattice sites, the jumps occur when enough free volume comes together to make an atom-sized hole. This atom motion allows the intermixing of atoms in a concentration gradient. External stress, too, drives a flux of atoms, causing the material to change shape in a way that allows the stress to do work.

Diffusion plays a fundamental role in the processing of materials, the subject of Chapters 18 and 19. It is also diffusion that causes creep and creep fracture. Their rates increase exponentially with temperature, introducing the first challenge for design: that of predicting the rates with useful accuracy. Exponential rates require precise data—small changes in activation energy give large changes in rates—and these data are hard to measure and are sensitive to slight changes of composition. And there is more than one mechanism of creep, compounding the problem. The mechanisms compete, the one giving the fastest rate winning. If you design using data and formulae for one, but under service conditions another is dominant, you are in trouble. Deformation mechanism maps help here, identifying both the mechanism and the approximate rate of creep.

A consequence of all this is that materials selection for high-temperature design, and the design itself, is largely based on empirical data rather than on modeling of the sort used for elastic and plastic design in Chapters 5, 7 and 10. Empirical data means plots, for individual alloys, of the creep strain rate and life as functions of temperature and stress. For polymers a different approach is used. Because their melting points and glass temperatures are low, they are visco-elastic at room temperature. Then design can be based on plots of the *creep modulus*—the equivalent of Young's modulus for a creeping material. The creep modulus at a given temperature and time is read off the plot and used in standard solutions for elastic problems (Chapter 5) to assess stress or deflection.

The chapter ends with examples of the use of some of these methods.

13.8 Further reading

Cottrell, A.H. (1964) *The Mechanical Properties of Matter*, Wiley, London, UK. Library of Congress Number 64-14262. (*Now 40 years old, inevitably somewhat dated, and now out of print, but still the best introduction to the mechanical properties of matter that I have ever found. Worth hunting for.*)

Finnie, I. and Heller, W.R. (1959) *Creep of Engineering Materials*, McGraw-Hill, New York, USA. Library of Congress No. 58-11171. (*Old and somewhat dated, but still an excellent introduction to creep in an engineering context.*)

Frost, H.J. and Ashby, M.F. (1982) *Deformation Mechanism Maps*, Pergamon Press, Oxford, UK. ISBN 0-08-029337-9. (*A monograph compiling deformation mechanism maps for a wide range of materials, with explanation of their construction.*)

Penny, R.K. and Marriott, D.L. (1971) *Design for Creep*, McGraw-Hill, New York, USA. ISBN 07-094157-2. (*A monograph bringing together the mechanics and the materials aspects of creep and creep fracture.*)

Waterman, N.A. and Ashby, M.F. (1991) *The Elsevier Materials Selector*, Elsevier Applied Science, Barking, UK. ISBN 1-85166-605-2. (*A three-volume compilation of data and charts for material properties, including creep and creep fracture.*)

13.9 Exercises

Exercise E13.1 The self-diffusion constants for aluminum are $D_o = 1.7 \times 10^{-4}\,m^2/s$ and $Q_d = 142\,kJ/mol$. What is the diffusion coefficient in aluminum at 400°C?

Exercise E13.2 A steel component is nickel plated to give corrosion protection. To increase the strength of the bond between the steel and the nickel, the component is heated for 4 hours at 1000°C. If the diffusion parameters for nickel in iron are $D_o = 1.9 \times 10^{-4}\,m^2/s$ and $Q_d = 284\,kJ/mol$, how far would you expect the nickel to diffuse into the steel in this time?

Exercise E13.3 The diffusion coefficient at the melting point for materials is approximately constant, with the value $D = 10^{-12}\,m^2/s$. What is the diffusion distance if a material is held for 12 hours at just below its melting temperature? This distance gives an idea of the maximum distance over which concentration gradients can be smoothed by diffusion.

Exercise E13.4 What are the requirements of a creep-resistant material? What materials would you consider for use at 550°C?

Exercise E13.5 Pipework with a radius of 20 mm and a wall thickness of 4 mm made of 2 1/4 Cr Mo steel contains a hot fluid under pressure. The pressure is 10 MPa at a temperature of 600°C. The table lists the creep constants for this steel. Calculate the creep rate of the pipe wall, assuming steady-state power-law creep.

Material	Reference strain rate, $\dot{\varepsilon}_o$ (1/s)	Reference stress, σ_o (MPa)	Rupture exponent, m	Activation energy, Q_c (kJ/mol)
2 1/4 Cr Mo steel	3.48×10^{10}	169	7.5	280

Exercise E13.6 There is concern that the pipework described in the previous exercise might rupture in less than the design life of 1 year. If the Monkman–Grant constant for 2 1/4 Cr Mo steel is 0.06, how long will it last before it ruptures?

Exercise E13.7 If the creep rate of a component made of 2 1/4 Cr Mo steel must not exceed 10^{-8}/second at 500°C, what is the greatest stress that it can safely carry? Use the data listed in the previous two examples to find out.

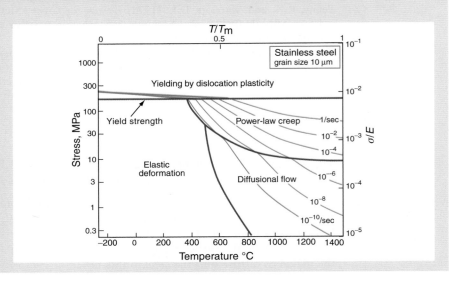

Exercise E13.8   A stainless steel suspension cable in a furnace is subjected to a stress of 100 MPa at 700°C. Its creep rate is found to be unacceptably high. By what mechanism is creep occurring? What action would you suggest to tackle the problem? The figure shows the deformation mechanism map for the material.

Exercise E13.9   The wall of a pipe of the same stainless steel as that of the previous exercise carries a stress of 3 MPa at the very high temperature of 1000°C. In this application it, too, creeps at a rate that is unacceptably high. By what mechanism is creep occurring? What action would you suggest to tackle the problem?

Exercise E13.10   It is proposed to make a shelf for a hot-air drying system from acrylic sheet. The shelf is simply supported, as in the diagram, and has a width $w = 500$ mm, a thickness $t = 8$ mm and a depth $b = 200$ mm. It must carry a distributed load of 50 N at 60°C with a design life of 800 hours (about a year) of continuous use. Use the creep modulus plotted in Figure 13.17 and the solution to the appropriate elastic problem (Chapter 5) to find out how much it will sag in that time.

13.10 Exploring design with CES (use Level 2 unless otherwise stated)

Exercise E13.11   Use the Search facility in CES to find:

(a) Materials for turbine blades.
(b) Materials for thermal barrier coatings.

Exercise E13.12   Use the CES software to make a bar chart for the glass temperatures of polymers like that for melting point in the text (Figure 13.6). What is the range of glass temperatures for polymers?

Exercise E13.13   The analysis of thermal barrier coatings formulated the design constraints. Use the selector to find the best material. Here is a summary of the constraints and the objective.

Function	• Thermal barrier coating
Constraints	• Maximum service temperature $>1300°C$
	• Adequate strength: $\sigma_y > 400\,MPa$
Objective	• Minimize thermal conductivity λ
Free variable	• Choice of material

Exercise E13.14   Find materials with $T_{max} > 500°C$ and the lowest possible thermal conductivity. Switch the database to Level 3 to get more detail. Report the three materials with the lowest thermal conductivity.

13.11 Exploring the science with CES Elements

Exercise E13.15   The claim is made in the text (equation (13.13b)) that the activation energy for self-diffusion, normalized by $\bar{R}T_m$, is approximately constant for metals:

$$\frac{Q_d}{\bar{R}T_m} \approx 18$$

Make a bar chart of this quantity and explore the degree to which it is true.

Exercise E13.16   The claim is made in the text (equation (13.13c)) that the diffusion coefficient of metals, evaluated at their melting point, is also approximately constant:

$$D_{Tm} = D_o \exp - \left(\frac{Q_d}{\bar{R}T_m} \right) \approx 10^{-12}\ m^2/s$$

Make a bar chart of this quantity and explore the degree to which it is true.

Exercise E13.17   What is a 'typical' value for the pre-exponential, D_o? Is it roughly constant for the elements? Make a chart with the activation energy for diffusion, Q_d, on the x-axis and the pre-exponential D_o on the y-axis to explore this.

Exercise E13.18   The diffusive jump shown in Figure 13.9 requires that the diffusing atom breaks its bonds in its starting position in order to jump into its final one. You might, then, expect that there would be at least an approximate proportionality between the activation energy for self-diffusion Q_d and the cohesive energy H_c of the material. Make a chart with H_c on the x-axis and Q_d on the y-axis. Report what you find.

Chapter 14
Conductors, insulators and dielectrics

Conductors and insulators. Copper, aluminum and silver (if you can afford it) make good conductors. Pyrex, porcelain and ceramics like alumina make good insulators – they are dielectrics. (Images courtesy of the Copper Development Association.)

14.1 Introduction and synopsis

York Minster, constructed between 1220 and 1400, is one of the great cathedrals of Europe. Many fine sermons have been preached from its pulpit, but when, in 1984, the Archbishop of York expressed his disbelief in the Virgin Birth, the Minster was struck by lightning and severely damaged. Some put this down to the wrath of God, but the reality is simpler: the cathedral was inadequately earthed. There is a message here: even when you are designing cathedrals, electrical properties matter.

This chapter is about the simplest of these: they relate to conduction, insulation and dielectric behavior. *Electrical conduction* (as in lightning conductors) and *insulation* (as in electric plug casings) are familiar properties. Dielectric behavior may be less so. A *dielectric* is an insulator. It is usual to use the word 'insulator' when referring to its inability to conduct electricity, and to use 'dielectric' when referring to its behavior in an electric field. Three properties are of importance here. The first, the *dielectric constant* (or *relative permittivity*), has to do with the way the material acquires a dipole moment (it *polarizes*) in an electric field. The second, the *dielectric loss factor*, measures the energy dissipated when radio-frequency waves pass through a material, the energy appearing as heat (the principle of microwave cooking). The third is the *dielectric breakdown potential*, and this takes us back to York Minster. Lightning is dielectric breakdown, and it can be as damaging on a small scale—in a piece of electrical equipment, for example—as on a large one.

Figure 14.1   The hierarchy of electrical behavior. The interesting ones are boxed in red, with examples of materials and applications. Their nature and origins are described in this chapter.

There are many different kinds of electrical behavior, all of them useful. Figure 14.1 gives an overview, with examples of materials and applications. The chapter introduces electrical properties and their physical origins, and examples of design for electrical performance.

14.2 Conductors, insulators and dielectrics

Resistivity and conductivity

The electrical resistance R (units: ohms,[1] symbol Ω) of a rod of material is the potential drop V (volts[2]) across it, divided by the current i (amps[3]) passing through it, as in Figure 14.2. This relationship is Ohm's law:

$$R = \frac{V}{i} \qquad (14.1)$$

The material property that determines resistance is the *electrical resistivity*, ρ_e. It is related to the resistance by

$$\rho_e = \frac{A}{L} R \qquad (14.2)$$

Figure 14.2   Electrical resistivity, ρ_e. Its value ranges from 1 to $10^{24} \mu\Omega.cm$.

---

[1] George Simon Ohm (1787–1854), dropped out of his studies at the University of Erlangen, preferring billiards to mathematics, but later became absorbed in the theory of electricity and magnetism.

[2] Alessandro Guiseppe Antonio Anastasio Volta (1745–1827), Italian physicist, inventor of the battery or, as it was then known, the Voltaic Pile. His face used to appear on the 10 000 lire note.

[3] André Marie Ampère (1775–1836), mathematician and physicist, who, it was said, had mastered all known mathematics by the age of 14. His life was not a happy one: father guillotined during the French Revolution, miserably married and with difficult children— yet he remained extensively productive up to his death.

where A is the section and L the length of a test rod of material—think of it as the resistance of a unit cube of the material. Its units in the metric system are Ω.m, but it is commonly reported in units of $\mu\Omega$.cm. It has an immense range, from a little more than 10^{-8} in units of Ω.m for good conductors (equivalent to $1\,\mu\Omega$.cm, which is why these units are still used) to more than $10^{16}\Omega$.m ($10^{24}\,\mu\Omega$.cm) for the best insulators. The electrical conductivity κ_e is simply the reciprocal of the resistivity. Its units are siemens per meter S/m or $(\Omega.\text{m})^{-1}$.

Dielectric properties

First, a reminder of what is meant by a field: it is a region of space in which objects experience forces if they have the right properties. Charge creates an *electric field*, E. The electric field strength between two oppositely charged plates separated by a distance t and with a potential difference V between them is

$$E = \frac{V}{t} \tag{14.3}$$

and is independent of position except near the edge of the plates.

Two conducting plates separated by a dielectric make a capacitor (Figure 14.3). Capacitors (sometimes called condensers) store charge. The charge Q (coulombs[4]) is directly proportional to the potential difference between the plates, V (volts):

$$Q = CV \tag{14.4}$$

Figure 14.3   Dielectric constant and dielectric breakdown. The capacitance of a condenser is proportional to the dielectric constant, and the maximum charge it can hold is limited by breakdown.

[4] Charles Augustin Coulomb (1736–1806), military physicist, laid the foundations of both the mathematical description of electrostatic forces and the laws of friction. Despite his appointment as *Intendant des Eaux et Fontaines de Paris*, he made no known contributions to hydrodynamics.

where C (farads[5]) is the capacitance. The capacitance of a parallel plate capacitor of area A, separated by empty space (or by air), is

$$C = \varepsilon_o \frac{A}{t} \tag{14.5}$$

where ε_o is the *permittivity* of *free* space (8.85×10^{-12} F/m, where F is farads). If the empty space is replaced by a dielectric, capacitance increases. This is because the dielectric *polarizes*. The field created by the polarization opposes the field E, reducing the voltage difference V needed to support the charge. Thus, the capacity of the condenser is increased to the new value

$$C = \varepsilon \frac{A}{t} \tag{14.6}$$

where ε is the *permittivity of the dielectric* with the same units as ε_o. It is usual to cite, not this, but the *relative permittivity* or *dielectric constant*, ε_r:

$$\varepsilon_r = \frac{C_{\text{with dielectric}}}{C_{\text{no dielectric}}} = \frac{\varepsilon}{\varepsilon_o} \tag{14.7}$$

making the capacitance

$$C = \varepsilon_r \varepsilon_o \frac{A}{t} \tag{14.8}$$

Being a ratio, ε_r is dimensionless. Its value for empty space and, for practical purposes, for most gases, is 1. Most dielectrics have values between 2 and 20, though low-density foams approach the value 1 because they are largely air. Ferroelectrics are special: they have values of ε_r as high as 20 000. More on them later.

Capacitance is one way to measure the dielectric constant of a material (Figure 14.3). The charge stored in the capacitor is measured by integrating the current that flows into it as the potential difference V is increased. The ratio Q/V is the capacitance. The dielectric constant ε_r is calculated from equation (14.7).

Small capacitors, with capacitances measured in microfarads (μF) or picofarads (pF), are used in R–C circuits to tune oscillations and give controlled time delays. The time constant for charging or discharging a capacitor in series with a resistor is

$$\tau = RC \tag{14.9}$$

---

[5] Michael Faraday (1791–1867), brilliant experimentalist both in physics and chemistry, discoverer of electromagnetic induction and inspiring lecturer at the Royal Institution, London.

where R is the resistance of the circuit. When charged, the energy stored in a capacitor is

$$\frac{1}{2}QV = \frac{1}{2}CV^2 \tag{14.10}$$

and this can be large: 'super-capacitors' with capacitances measured in farads store enough energy to power a hybrid car.

The *breakdown potential* or *dielectric strength* (units: MV/m) is the electrical potential gradient at which an insulator breaks down and a damaging surge of current flows through it. It is measured by increasing, at a uniform rate, a 60 Hz alternating potential applied across the faces of a plate of the material in a configuration like that of Figure 14.3 until breakdown occurs, typically at a potential gradient of between 1 and 100 MV/m.

The *loss tangent* and the *loss factor* take a little more explanation. We shall see in Section 14.4 that polarization involves the small displacement of charge (either of electrons or of ions) or of molecules that carry a dipole moment when an electric field is applied to the material. An oscillating field drives the charge between two alternative configurations. This charge motion is like an electric current that—if there were no losses—would be 90° out of phase with the voltage. In real dielectrics this current dissipates energy, just as a current in a resistor does, giving it a small phase shift, δ (Figure 14.4). The *loss tangent*, tan δ, also called the *dissipation factor*, D, is the tangent of the loss angle. The *power factor*, P_f, is the sine of the loss angle. When δ is small, as it is for the materials of interest here, all three are essentially equivalent:

$$P_f \approx D \approx \tan \delta \approx \sin \delta \tag{14.11}$$

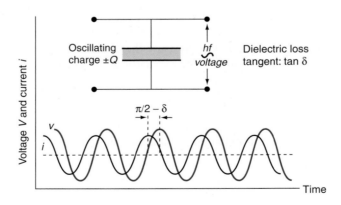

Figure 14.4   Dielectric loss. The greater the loss factor, the greater is the microwave coupling and heating.

More useful, for our purposes, is the *loss factor L*, which is the loss tangent times the dielectric constant:

$$L = \varepsilon_r \tan \delta \qquad (14.12)$$

It measures the energy dissipated in a dielectric when in an oscillating field. If you want to select materials to minimize or maximize dielectric loss, then the measure you want is L.

When a dielectric material is placed in a cyclic electric field of amplitude E and frequency f, power P is dissipated and the field is correspondingly attenuated. The power dissipated per unit volume, in W/m^3, is

$$P \approx f E^2 \varepsilon \tan \delta = f E^2 \varepsilon_o \varepsilon_r \tan \delta = f E^2 \varepsilon_o L \qquad (14.13)$$

where, as before, ε_r is the dielectric constant of the material and $\tan \delta$ is its loss tangent. This power appears as heat and is generated uniformly through the volume of the material. Thus, the higher the frequency or the field strength and the greater the loss factor $\varepsilon_r \tan \delta$, the greater is the heating and energy loss. This dielectric loss is exploited in processing—for example, in radio-frequency welding of polymers.

All dielectrics change shape in an electric field, a consequence of the small shift in charge that allows them to polarize; the effect is called *electrostriction*. Electrostriction is a one-sided relationship in that an electric field causes deformation, but deformation does not produce an electric field. *Piezo-electric* materials, by contrast, display a two-sided relationship between polarization and deformation: a field induces deformation and deformation induces charge differences between its surfaces, thus creating a field. The piezo-electric coefficient is the strain per unit of electric field, and although it is very small, it is a true linear effect, and this makes it useful: when you want to position or move a probe with nanoscale precision it is just what you need. *Pyro-electric* materials contain molecules with permanent dipole moments that, in a single crystal, are aligned, giving the crystal a permanent polarization. When the temperature is changed the polarization changes, creating surface charges or, if the surfaces are connected electrically, a pyro-electric current—the principle of intruder detection systems and of thermal imaging. *Ferro-electric* materials, too, have a natural dipole moment—they are polarized to start with, and the individual polarized molecules line up so that their dipole moments are parallel, like magnetic moments in a magnet. Their special feature is that the direction of polarization can be changed by applying an electric field, and the change causes a change of shape. Applications of all of these are described in Section 14.5.

14.3 Charts for electrical properties

All materials have an electrical resistivity and all dielectrics have a dielectric constant and loss factor. Property charts for these are useful because they allow

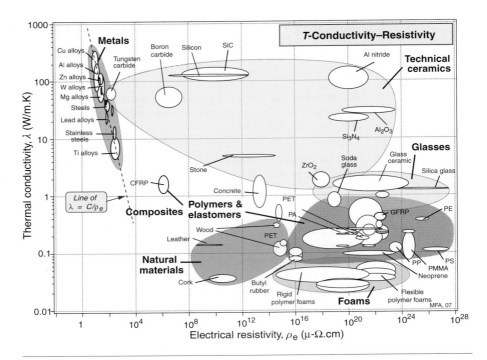

Figure 14.5 Electrical resistivity ρ_e and thermal conductivity λ.

comparison and selection from a large and varied population. Certain other properties are peculiar to only a handful of materials: useful piezo-electric and ferro-electric behavior are examples. Selection is then a case of a choice between a small number of candidates, easily explored by direct comparison. Property charts for these are not necessary.

Thermal conductivity and electrical resistivity

The first chart (Figure 14.5) shows the thermal conductivity λ and the electrical resistivity ρ_e. The first has a range of 10^5, the second a range of 10^{28}; no other material property has such a wide range. The chart is useful for selecting materials for applications in which these two properties are important. For metals, thermal and electrical conduction are linked because both depend on free electrons, giving the obvious correlation between the two at the upper left of the chart:

$$\lambda \approx \frac{1400}{\rho_e}$$

(λ in W/m.K, ρ_e in $\mu\Omega$.cm). It is known as the Wiedemann–Franz law.

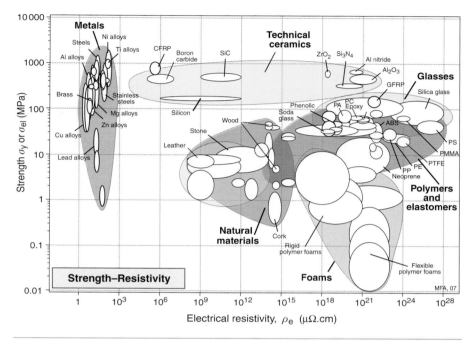

MFA, 07

Figure 14.6 Material property chart for electrical resistivity ρ_e and yield strength σ_y or elastic limit σ_{el}.

Strength and electrical resistivity

The second chart (Figure 14.6) shows electrical resistivity ρ_e and strength σ_{el}. Metals lie at the extreme left: copper and silver (not shown) have the lowest resistivities, with aluminum only just above. The strongest alloys have resistivities that are 10–100 times larger. Ceramics such as alumina, aluminum nitride and silica, at the upper right, have very high resistivities and are strong; they are used for making insulators for power lines and substrates for electronic circuitry. Below them lie polymers that are less strong, but are flexible, easily molded and excellent insulators; they are used for cable sheathing, insulation of switch gear and the like.

Strength and dielectric loss

The final chart (Figure 14.7) shows dielectric loss factor L and strength σ_y or σ_{el}. Materials that are transparent to microwaves (radar uses microwave frequencies) lie at the left: polymer foams, certain polymers (PP, PE, PTFE) and some ceramics have particularly low values. Materials that absorb microwaves, becoming hot in the process, lie at the right: polymers containing a polar group, like nylon and polyurethane, and natural materials like woods (woods are dried by microwave heating). Water, which has a polar molecule,

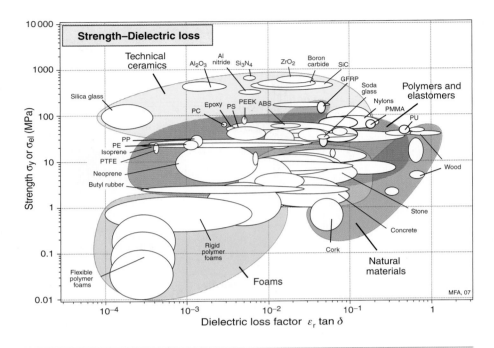

Figure 14.7 Material property chart for dielectric loss, $\varepsilon_r \tan \delta$ and yield strength σ_y or elastic limit σ_{el}.

has a particularly large loss factor, which is why most foods can be cooked in a microwave oven.

14.4 Drilling down: the origins and manipulation of electrical properties

Electrical conductivity

An electric field, E (volts/m), exerts a force Ee on a particle carrying a charge. Solids are made up of atoms containing electrons that carry a charge $-e$ and a nucleus containing protons, each with a positive charge $+e$. If charge carriers can move, the force Ee causes them to flow through the material—that is, it conducts. Metals are *electron conductors*, meaning that the charge carriers are the electrons. In ionic solids (which are composed of negatively and positively charged ions like $Na^+ Cl^-$) the diffusive motion of ions allows *ionic conduction*, but this is only possible at temperatures at which diffusion is rapid. Many materials have no mobile electrons, and at room temperature they are too cold

Isolated
atom

Crystal composed
of many atoms

Figure 14.8   When atoms are brought together to form crystals, the outermost electrons interact and the discrete energy levels split into bands.

to be ionic conductors. The charged particles they contain still feel a force in an electric field and it is enough to displace the charges slightly, but they are unable to move more than a tiny fraction of the atom spacing. These are insulators; the small displacement of charge gives them dielectric properties.

How is it that some materials have mobile electrons and some do not? To explain this we need two of the stranger results of quantum mechanics. Briefly, the electrons of an atom occupy discrete energy states or orbits, arranged in shells (designated 1, 2, 3, etc. from the innermost to the outermost); each shell is made up of sub-shells (designated s, p, d and f), each of which contains one, three, five or seven orbits respectively. The electrons fill the shells with the lowest energy, two electrons of opposite spin in each orbit; the *Pauli*[6] *exclusion principle* prohibits an energy state with more than two. When *n* atoms (a large number) are brought together to form a solid, the inner electrons remain the property of the atom on which they started, but the outer ones interact (Figure 14.8). Each atom now sits in the field created by the charges of its neighbors. This has the effect of decreasing slightly the energy levels of electrons spinning in a direction favored by the field of its neighbors and raising that of those with spins in the opposite direction, splitting each energy level. Thus, the discrete levels of an isolated atom broaden, in the solid, into *bands* of very closely spaced levels. The number of electrons per atom that have to be accommodated depends only on the atomic number of the atoms. These electrons fill the bands from the bottom, lowest energy, slot on up, until all are on board, so to speak. The topmost filled energy level is called the Fermi[7] level (more on this later). An electron in this level still has an energy that is lower than it would have if it

---

[6] Wolfgang Joseph Pauli (1900–1958), quantum theorist, conceiver of the neutrino and Nobel Prize winner. He was not, however, a happy man, requiring psychotherapy, which he received from none other than the great psychoanalyst Carl Jung.

[7] Enrico Fermi (1901–1954), devisor of the statistical laws known as Fermi statistics governing the behavior of electrons in solids. He was one of the leaders of the team of physicists on the Manhattan Project for the development of the atomic bomb.

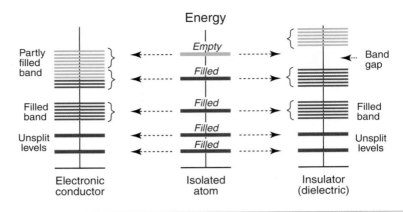

Figure 14.9 Conductors, on the left, have a partly filled outer band—electrons in it can move easily. Insulators, on the right, have an outer filled band, separated from the nearest unfilled band by a band gap. (Red bands are filled, blue bands are empty.)

were isolated in a vacuum far from the atoms. This energy difference is called, for historical reasons, the *work function* because it is the work that you have to do to remove an electron from the Fermi level to infinity. If you want to create an electron beam by pulling electrons out of a metal by using a high electric field (called a *field-emission electron gun*) you have to provide the work function to do it.

Whether the material is a conductor or an insulator depends on how full the bands are, and whether or not they overlap. In Figure 14.9 the central column describes an isolated atom and the outer ones illustrate the possibilities created by bringing atoms together into an array, with the energies spread into energy bands. Conductors like copper, shown on the left, have an unfilled outer band; there are many very closely spaced levels just above the last full one, and—when accelerated by a field—electrons can use these levels to move freely through the material. In insulators, shown on the right, the outermost band with electrons in it is full, and the nearest empty band is separated from it in energy by a wide *band gap*. Semiconductors, too, have a band gap, but it is narrower—narrow enough that thermal energy can pop a few electrons into the empty band, where they conduct. Deliberate doping (adding trace levels of impurities) creates new levels in the band gap, reducing the energy barrier to entering the empty states and thus allowing more carriers to become mobile.

Electrical resistance

If a field E exerts a force Ee on an electron, why does it not accelerate forever, giving a current that continuously increases with time? This is not what happens; instead, switching on a field causes a current that almost immediately

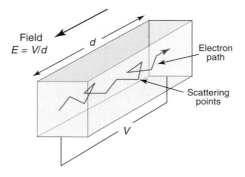

Figure 14.10   An electron, accelerated by the field E, is scattered by imperfection that create a resistance to its motion.

reaches a steady value. Referring back to equations (14.1) and (14.2), the current density i/A is proportional to the field E:

$$\frac{i}{A} = \frac{E}{\rho_e} = \kappa_e E \qquad (14.14)$$

where ρ_e is the resistivity and κ_e, its reciprocal, is the electrical conductivity.

Broadly speaking, the picture is this. Conduction electrons are free to move through the solid. Their thermal energy $k_B T$ (k_B = Boltzmann's constant, T = absolute temperature) causes them to move like gas atoms in all directions. In doing this they collide with *scattering centers*, bouncing off in a new direction. Impurity or solute atoms are particularly effective scattering centers (which is why alloys always have a higher resistivity than pure metals), but electrons are scattered also by imperfections such as dislocations and by the thermal vibration of the atoms themselves. When there is no field, there is no *net* transfer of charge in any direction even though all the conduction electrons are moving freely. A field imposes a drift velocity $v_d = \mu_e E$ on the electrons, where μ_e is the electron mobility and it is this that gives the current (Figure 14.10). The drift velocity is small compared with the thermal velocity; it is like a breeze in air—the thermal motion of the air molecules is far greater than the 'drift' that we feel as the breeze. The greater the number of scattering centers, the shorter is the mean free path, λ_{mfp}, of the electrons between collisions, and the slower, on average, they move. Just as with thermal conductivity, the electrical conductivity depends on mean free path, on the density of carriers (the number n_v of mobile electrons per unit volume) and the charge they carry. Thus, the current density, i/A, is given by

$$\frac{i}{A} = n_v e v_d = n_v e \mu_e E$$

Comparing this with equation (14.14) gives the conductivity:

$$\kappa_e = n_v\, e\, \mu_e \qquad (14.15)$$

Thus the conductivity is proportional to the density of free electrons and to the electron mobility, and this is directly proportional to the mean free path. The purer and more perfect the conductor, the higher is the conductivity.

Metals are strengthened (Chapter 7) by solid solution hardening, work hardening or precipitation hardening. All of these change the resistivity too. By 'zooming in' on part of the strength–resistivity chart, we can see how both properties are affected—Figure 14.11 shows this for two of the best conductors: copper and aluminum. Adding solute to either metal increases its solute atoms also act as scattering centers, increasing the electrical resistivity too. Dislocations add strength (by what we called *work hardening*) and they too scatter electrons a little, though not as much as solute. Precipitates offer the greatest gain in strength; their effect on resistivity varies, depending on their size and spacing compared to the electron mean free path and on the amount of residual solute left in the lattice. Precipitation hardening (with low residual solute) or work hardening are therefore the best ways to strengthen conductors. The two figures show that commercial conductor alloys have much greater strength, and only slightly greater resistivity than the pure metals.

The resistivity of metals increases with temperature because thermal vibration scatters electrons. Resistance decreases as temperature falls, which is why very-high-powered electromagnets are pre-cooled in liquid nitrogen. As absolute zero

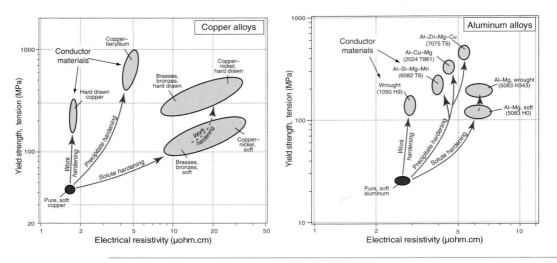

Figure 14.11 The best choice of material for a cable is one with high strength and low resistivity, but strengthening mechanisms increase resistivity. Work hardening and precipitation hardening do so less than solute hardening.

is approached most metals retain some resistivity, but a few, at between 0 and 4 K, suddenly lose all resistance and become superconducting. The resistivity of semiconductors, by contrast, decreases as temperature increases because, as explained earlier, thermal energy allows more carriers to cross the band gap.

Dielectric behavior

Dielectrics are insulators. So what happens to their electrons when a field E is applied? In zero field, the electrons and protons in most dielectrics are symmetrically distributed, so the material carries no net or dipole moment. A field exerts a force on a charge, pushing positive charges in the direction of the field and negative charges in the opposite direction. The effect is easiest to see in ionic crystals, since here neighboring ions carry opposite charges, as on the left of Figure 14.12. Switch on the field and the positive ions (charge $+q$) are pulled in the field direction, the negative ones (charge $-q$) in the reverse, until the restoring force of the inter-atomic bonds just balances the force due to the field at a displacement of Δx, as on the right of the figure. Two charges $\pm q$ separated by a distance Δx create a dipole with dipole moment, d, given by

$$d = q\Delta x \tag{14.16}$$

The polarization of the material, P, is the volume-average of all the dipole moments it contains:

$$P = \frac{\sum d}{\text{Volume}}$$

Even in materials that are not ionic, like silicon, a field produces a dipole moment because the nucleus of each atom is displaced a tiny distance in the direction of the field and its surrounding electrons are displaced in the opposite

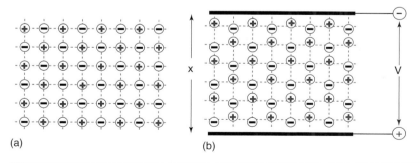

(a)   (b)

Figure 14.12   An ionic crystal in zero applied field, (a) and when a field V/x is applied (b). The electric field displaces charge, causing the material to acquire a dipole moment.

direction. The resulting dipole moment depends on the magnitude of the displacement and the number of charges per unit volume, and it is this that determines the dielectric constant. The bigger the shift, the bigger the dielectric constant. Thus, compounds with ionic bonds and polymers that contain polar groups like $-OH^-$ and $-NH^-$ (nylon, for example) have larger dielectric constants than those that do not.

Dielectric loss

Think now of polarization in an alternating electric field. When the upper plate of Figure 14.12(b) is negative, the displacements are in the direction shown in the figure. When its polarity is reversed, it is the negative ions that are displaced upwards, the positive ions downwards; in an oscillating field, the ions oscillate. If their oscillations were exactly in phase with the field, no energy would be lost, but this is never exactly true, and often the phase shift is considerable. Materials with high dielectric loss usually contain awkwardly shaped molecules that themselves have a dipole moment—the water molecule is an example (Figure 14.13). These respond to the oscillating field by rotating, but because of their shape they interfere with each other (you could think of it as molecular friction) and this dissipates energy that appears as heat—that is how microwave heating works. As equation (14.13) showed, the energy that is dissipated depends on the frequency of the electric field; generally speaking, the higher the frequency, the greater the power dissipated (because power is work per second, and the more times the molecules shuttle, the more energy is lost), but there are peaks at certain frequencies that are characteristic of the material structure.

Dielectric breakdown

In metals, as we have seen, even the smallest field causes electrons to flow. In insulators they can't, because of the band gap. But if, at some weak spot, one electron is torn free from its parent atom, the force Ee exerted by the field E accelerates it, giving it kinetic energy; it continues to accelerate until it collides with another atom. A sufficiently large field can give the electron so much kinetic energy that, in the collision, it kicks one or more new electrons out of the atom it hits, and they, in turn, are accelerated and gain energy. The result, sketched in Figure 14.14, is a cascade—an avalanche of charge. It is sufficiently violent that it can damage the material permanently.

The critical field strength to make this happen, called the *breakdown potential*, is hard to calculate: it is that at which the first electron breaks free at the

Figure 14.13   A water molecule. The asymmetry gives it a dipole moment.

weak spot—a defect in the material such as a tiny crack, void or inclusion that locally concentrates the field. The necessary fields are large, typically 1–15 MV/m. That sounds a lot, but such fields are found in two very different circumstances: when the voltage is very high or the distances are very small. In power transmission, the voltages are sufficiently high—20 000 volts or so—that breakdown can occur at the insulators that support the line, while in microcircuits and thin-film devices the distances between components are very small: a 1 V difference across a distance of $1 \mu m$ gives a field of 1 MV/m.

Piezo-electric materials

The word 'Quartz' on the face of your watch carries the assurance that its time-steps are set by the oscillations of a piezo-electric quartz crystal. Piezo-electric behavior is found in crystals in which the ions are unsymmetrically distributed (the crystal structure lacks a center of symmetry), so that each molecule carries a permanent dipole moment (Figure 14.15(a)). If you cut a little cube from such

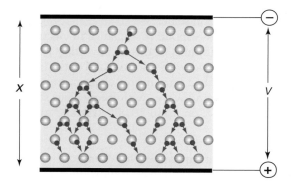

Figure 14.14   Breakdown involves a cascade of electrons like a lightning strike.

Figure 14.15   A piezo-electric material has unsymmetrically distributed charge, giving it a natural dipole moment. The surface charge associated with this is neutralized by pick-up of ions, but if it is deformed, as in (b), the dipole moment changes and the surfaces become charged. The inverse is also true: a field induces a change of shape, the basis of piezo-electric actuation, as in (c).

a material, with faces parallel to crystal planes, the faces would carry charge. This charge attracts ions and charged particles from the atmosphere just as a television screen does, giving an invisible surface layer that neutralizes the charge. If you now squeeze the crystal, its ions move relative to each other, the dipole moment changes and the charge on the surface changes too (Figure 14.15(b)). Given time, the newly appeared charge would attract neutralizing ions, but this does not happen immediately, giving a potential difference. This provides the basis of operation of electric microphones and pick-ups. The potential difference between the faces can be large—large enough to generate a spark across a narrow gap—it is the way that gas lighters work.

A strain, then, induces an electric field in a piezo-electric material. The inverse is also true: a field induces a strain. The field pulls the positive ions and pushes the negative ones, changing their spacing and so changing the shape of the crystal (Figure 14.15(c)). If a small strain produces a large field, then a large field will produce only a very small strain. But the strain is a linear function of field, allowing extremely precise, if small, displacements, used for positioning and actuation at the sub-micron scale.

Piezo-electric materials respond to a change in electric field faster than most materials respond to a stimulus of this or any other kind. Put them in a megahertz field and they respond with microsecond precision. That opens up many applications, some described later in this chapter. In particular, it opens up the world of ultrasonics—sound waves with frequencies starting at the upper limit of the human ear, 20 kHz, on up to 20 000 kHz and above.

Pyro-electric materials

Pyro-electricity—polarization caused by change of temperature—was known to (but not understood by) the Greek philosopher Theophrastus,[8] who noted that certain stone, when warmed or cooled, acquired the ability to pick up straw and dry leaves. It was not until the 20th century that this strange behavior was understood and exploited.

Some materials have a permanent dipole moment because their positive and negative ions balance electrically but are slightly out of line with each other. If a thin disk of one of these is cut so that its faces are parallel to the plane in which the misalignment happens, the disk has a dipole moment of its own. The unit cells of pyro-electric materials are like this. The dipole moment per unit volume of the material is called the spontaneous polarization P_s. This net dipole moment exists in the absence of an applied electric field and is equivalent to a layer of bound charge on each flat surface. Nearby free charges such as electrons or ions are attracted to the surfaces, neutralizing the charge. Imagine that conductive electrodes are then attached to the surfaces and connected through an ampmeter. If the temperature of

---

[8] Theophrastus (372–287 BC), successor to Aristotle, a teacher of science and author of a 10-volume history of plants which remained the most important contribution to botanical science for the following 1200 years.

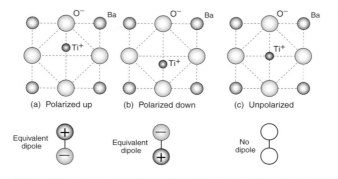

Figure 14.16 Ferro-electric materials have a permanent dipole moment that can switch: here the Ti^+ ion can flip from the upper to the lower position. Above the Curie temperature the asymmetry disappears.

the sample is constant, then so is P_s and no current flows through the circuit. An increase in temperature, however, causes expansion and that changes the net dipole moment and the polarization. Redistribution of free charges to compensate for the change in bound charge results in a current flow—the pyro-electric current—in the circuit. Cooling rather than heating reverses the sign of the current. Thus, the pyro-electric current only flows while the temperature is changing—it is the way that intruder alarms, automatic doors and safety lights are activated.

In an open circuit the free charges remain on the electrodes—and that has its uses too, described in the next section. Pyro-electric materials include minerals such as tourmaline (the one Theophrastus found), ceramics such as barium titanate, polymers such as polyvinylidene fluoride, and even biological materials, such as collagen.

Ferro-electric materials

Ferro-electrics are a special case of piezo-electric behavior. They too have an unsymmetric structure, but have the special ability to switch asymmetry. Barium titanate, $BaTiO_3$, shown schematically in Figure 14.16, is one of these. Below a critical temperature, the Curie[9] temperature (about 120°C for barium titanate), the titanium atom, instead of sitting at the center of the unit cell, is displaced up, down, to the left or to the right, as in (a) and (b). Above the Curie temperature the asymmetry disappears and with it the dipole moment, as in (c). In ferro-electrics, these dipoles spontaneously align so that large volumes of the material are polarized even when there is no applied field.

In the absence of an external field a ferro-electric divides itself up into *domains*—regions in which all the dipoles are aligned in one direction—separated

[9] Pierre Curie (1859–1906), French physicist, discoverer of the piezo-electric effect and of magnetic transformations; and husband of the yet more famous Marie Curie. He was killed in a street accident in Paris, a city with dangerous traffic even in 1906.

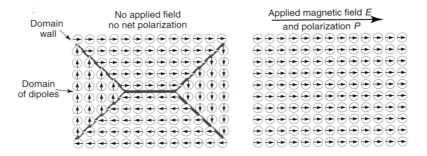

Figure 14.17   Ferro-electric domains arrange themselves so that their dipole moments cancel. In an electric field *E* they align, giving the crystal a net dipole moment, which can be large.

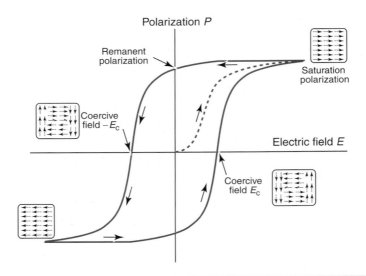

Figure 14.18   A ferro-electric hysteresis curve.

by *domain walls* at which the direction of polarization changes (Figure 14.17). The domains orient themselves so that the dipole moment of one more or less cancels those of its neighbors. If a field is applied the domain walls move so that those polarized parallel to the field grow and those polarized across or against it shrink, until the entire sample is polarized (or 'poled') in just one direction. Figure 14.18 shows how the polarization *P* changes as the field *E* is increased: *P* increases, reaching a maximum at the *saturation polarization*, P_s. If the field is now removed, a large part of the polarization remains (the *remanent polarization*), which is only removed by reversing the field to the value $-E_c$, the *coercive field*. The figure shows a complete cycle through full reverse polarization, ending up again with full forward poling. The little inserts show the domain structures round the cycle.

Ferro-electric materials have enormous dielectric constants. Those of normal materials with symmetric charge distributions lie in the range 2–20. Those for ferro-electrics can be as high as 20 000. It is this that allows their use to make super-capacitors that can store 1000 times more energy than conventional capacitors. Such is the energy density that super-capacitors now compete with batteries for energy storage.

14.5 Design: using the electrical properties of materials

Transmitting electrical power

The objective in power transmission is to minimize electrical loss, P_L (watts[10]):

$$P_L = i^2 R \qquad (14.17)$$

where i is the current (amps) and R the resistance of the transmission line, which, according to equation (14.2), is proportional to the material resistivity ρ_e. The best choice, for a given cross-section A, is that of a material with the lowest possible value of ρ_e; if the line is buried or supported along its entire length, the only other constraint is that of material cost. The chart of Figure 14.5 shows that, of the materials plotted on it, copper has the lowest resistivity, followed by aluminum. If instead the line is above ground and supported by widely spaced pylons, its strength and its density become important: high strength, to support self-weight and wind loads; low density to minimize self-weight.

Here we have a conflict: the materials with the lowest resistivity have low strength and those with high strength have high resistivities, for the reasons given in the last section. The aluminum and copper conductors used in overhead cables are those shown in Figure 14.11—precipitation hardened or work hardened to give strength with minimum loss of conductivity. Another solution is to create a hybrid cable—one in which two materials are combined, the first to give the conductivity, the second to give the strength. One such is sketched in Figure 14.19: a cable with a core of pure, high-conductivity copper or aluminum wrapped in a cage of high-strength carbon–steel wires. With a 50:50 mix the hybrid, when averaged, has roughly half the strength of the steel and twice the resistivity of the copper, a much better compromise than that offered by any single material. Many overhead cables use this approach, with a steel core surrounded by aluminum conductors; some use instead a combination of low- and high-strength aluminum alloys to reduce self-weight.

---

[10] James Watt (1736–1819), instrument maker and inventor of the condensing steam engine (the idea came to him whilst 'walking on a fine Sabbath afternoon'), which he doggedly developed. Unlike so many of the footnoted characters of this book, Watt, in his final years, was healthy, happy and famous.

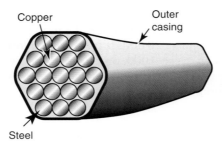

Figure 14.19   A hybrid cable made up of strands of copper reinforced with strands of steel.

Figure 14.20   An electric plug. The casing must be an insulator, be moldable and have
sufficient strength to tolerate working loads.

Electrical insulation

Electrical insulators keep electricity where it is supposed to be. The charts of
Figures 14.6 and 14.7 show that most polymers are excellent electrical insulators,
with resistivities above $10^{20}\,\mu\Omega.cm$. They are used for cable insulation, for casings
of plugs, connectors and switchgear. Many ceramics, too, are good electrical
insulators, notably alumina (Al_2O_3), silica (SiO_2) and glass (Pyrex); they are used
in applications in which rigidity and resistance to heat are important, such as substrates for electronic circuits and high-voltage insulators. Where electrical isolation is vital, the objective is to maximize resistivity, but as with conduction there
are usually additional constraints. Here is an example.

The electric plug (Figure 14.20) is perhaps the commonest of electrical products. It has a number of components, most performing more than one function.

Substrate

Connecting pins

Chips

Heat sink

Cooling fins

Figure 14.21   A heat sink. It must conduct heat but not electricity.

The most obvious are the casing and the pins, though there are many more (connectors, a cable clamp, fasteners and, in some plugs, a fuse). Power plugs have two or three pins and are of robust construction. System plugs, like that of the video port of a computer, have 25 or more pins and are miniaturized and delicate, placing more importance on the mechanical as well as the electrical properties of the materials of the pins and casing. The task here is to select a material for the casing of the plug.

The shape is complicated, the part will be made in very large numbers and pins have to be molded in. The only practical way to make it is by using a polymer. The designers set a required strength of at least 25 MPa and, of course, the electrical resistivity should be as high as possible. The chart of Figure 14.6 suggests ABS, PMMA and PC as suitable candidates—they are strong enough and have high resistivity. But there is a problem: all three are thermoplastics; they soften if they get hot. If circuit overload or poor contact generates heat, the plug will soften, distort and even melt, presenting hazards. So the choice has to be a thermoset, which will char rather than melt. The chart suggests Phenolic—and that is what plugs are really made of. The CES software allows much more sophisticated suggestions: try the Exercises at the end of this chapter to refine the choice.

Electrical insulation with thermal conduction: keeping microchips cool

As microchips shrink in scale and clock speeds rise, overheating becomes a problem. The Pentium chip of today's PCs already reaches 85°C, requiring forced cooling. Multiple-chip modules (MCMs) pack as many as 130 chips on to a single substrate, and there the problem is more severe. Heating is kept under control by attaching the chips to a heat sink (Figure 14.21), taking care that they are in good thermal contact. The heat sink now becomes a critical component, limiting further development of the electronics.

To drain heat away from the chips as fast as possible, the heat sink must have the highest possible thermal conductivity, λ. Metals like copper and aluminum have high λ, but they are good electrical conductors and this is bad because it allows electrical coupling and stray capacitance between chip and heat sink, slowing the information transfer rate (remember the time delay in a circuit is $\tau = RC$). To prevent this, the heat sink must be an excellent electrical insulator.

Figure 14.5 is the chart with the properties we want: thermal conductivity λ and electrical resistivity ρ_e. The materials at the top right have high thermal conductivity and very high resistivity—they are electrical insulators. The best choice of material for the heat sink, then, is aluminum nitride (AlN), alumina (Al_2O_3) or silicon carbide (SiC). It is these that are used for such heat sinks.

Using dielectric loss: microwave heating
When you cook a chicken in a microwave oven, you are using dielectric loss. Looking back at equation (14.13) we see that the dielectric loss, and thus the heating, is proportional to the loss factor $L = \varepsilon_r \tan \delta$, one axis of the chart of Figure 14.7. Chickens do not appear on the chart, but they *do* have a high loss factor, as do all foods that contain water. There are other ways of cooking chicken, but this one has the advantage that it is quick because microwaves absorb uniformly through the section, giving uniform heating. Microwave heating is not just useful in the kitchen; it is also widely used for industrial heating. Wood can be dried, adhesives cured, polymers molded and ceramics sintered using microwave heating. As equation (14.13) implies, the higher the frequency f of the electric field, the greater the heating, so it is common, in the industrial application of microwaves, to use gigahertz frequencies (0.3–3 GHz).

Using dielectric loss: stealth technology
Radar, developed during the Second World War, played a major role in protecting Britain (and others) from air attack. It works by transmitting bursts of radio-frequency waves that are reflected back by objects in their path. The radar antenna measures the time it takes for the reflection to arrive back, from which the distance and speed of the object are computed.

The metal body of an aircraft is an excellent radar reflector and its curved shape causes it to reflect in all directions—just what is needed for safe air-traffic control. But if your aim is to avoid detection, reflection is the last thing you want. Stealth technology combines three tricks to minimize it:

- A shape made up of flat planes that reflect the radar signal away from the detector unless it lies exactly normal to the surface.
- Non-metallic structural materials—usually composites—that are semi-transparent to microwaves (see Chapter 16 for how to achieve transparency).
- Surface coatings of radar-absorbing materials (called RAMs) that absorb microwaves rather than reflecting them. RAMs use dielectric loss. A surface layer of a material with a large value of loss factor $\varepsilon_r \tan \delta$, a ferro-electric for example, attenuates the incoming wave, diminishing the reflected signal.

Avoiding dielectric loss: materials for radomes
The function of a radome is to shield a microwave antenna from the adverse effects of wind and weather while having as little effect as possible on the signal it sends and receives (Figure 14.22). When trying to detect incoming signals that are weak to begin with, even a small attenuation of the signal as it passes through the radome decreases the sensitivity of the system. Yet the radome must withstand structural loads, loads caused by pressure difference between the inside and outside of the dome, and—in the case of supersonic flight—high

Radome

t

Δ*p*

2 *R*

Figure 14.22   A radome. The shell must be transparent to microwaves, requiring a low dielectric loss factor with high strength.

temperatures. So the problem here is the opposite of stealth coatings; it is how to design materials that do not attenuate electromagnetic waves. The answer is to seek those with exceptionally low values of $\varepsilon_r \tan \delta$. And since the radome carries structural loads, they should also be strong.

The chart of Figure 14.7 shows that some polymers have low loss factor, but they are not very strong. Reinforcing them with glass fibers, creating GFRPs, combines low loss with good strength. When, additionally, high temperatures are involved, the ceramics that appear on the upper left of Figure 14.7—silica (SiO_2), alumina (Al_2O_3) and silicon nitride (Si_3N_4)—are employed.

Piezo-electric actuation: ink-jet printers
Ink-jet printing is now widely used to make black and white or color prints. Many printers use thermal ('bubble-jet') technology, in which a tiny heating wire boils a minute volume of ink in the print-head, creating a vapor bubble. The sudden pressure this creates ejects a droplet to ink. The bubble collapses as the wire cools, sucking more ink from a reservoir into the print-head to replace that in the droplet. Very simple, but thermal technology gives little control of the droplet size or shape.

The alternative is the piezo print-head, one unit of which is sketched in Figure 14.23. An electric field applied to the disk of piezo-electric material transmits a kick to the vibration plate, creating a pressure pulse that spits a droplet of ink from the nozzle. On relaxation, more ink is drawn in to replace it. Like bubble-jet printers, piezo print-heads have 300–600 nozzles. The great advantage of piezo technology is the control of the shape and duration of the pulse delivered by the piezo disk. The droplets of ink are significantly smaller, and the control of droplet size gives greater resolution and color gradation.

Pyro-electric thermal imaging
Night vision, the location of survivors in disaster areas, the filming of wildlife and much else make use of thermal imaging. The technology is comparatively new—the

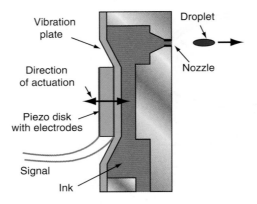

Figure 14.23   A piezo print-head.

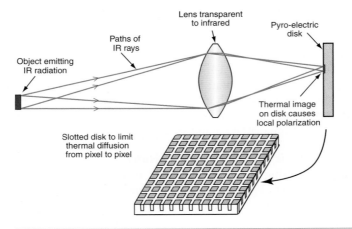

Figure 14.24   Thermal imaging.

first device for two-dimensional pyro-electric imaging was developed in the late 1970s. A typical set-up is sketched in Figure 14.24. A lens, made of a material transparent to infrared (such as germanium) focuses the IR radiation from an object on the left to form an image on the pyro-electric disk on the right. The focused IR rays warm points on the disk to different degrees, depending on the temperature and emissivity of the points on the object from which they came, and the change of temperature induces local polarization. The disk is an insulator, so the image creates a pattern of charge on the surface of the disk which can be 'read' with an electron beam.

The problem with early devices was that lateral heat diffusion in the disk during the time between successive electron-beam sweeps resulted in low resolution. The problem is overcome by reticulating the surface of the disk into 20 μm islands by ion milling (a way of cutting very thin channels with a high energy ion

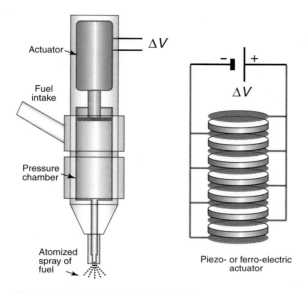

Figure 14.25   A diesel injector. The actuator of the injector must be very precise to meter the fuel accurately, and must adapt its timing and stroke to match the conditions of the engine. Piezo- or ferro-electric stacks, like that shown on the right, offer the best control.

beam), thermally isolating each island from its neighbors. Present-day imaging systems can detect a temperature change of about 0.2°C on the target disk.

Ferro-electric actuation: injectors for diesel engines

Fuel is injected into the combustion chamber of a diesel engine by micro-pumps, one for each cylinder. To meter the fuel accurately and adapt the quantity injected to the needs of the engine, the pump must allow precise proportional control. Until recently the standard actuator for this application was a solenoid (a coil with a moving ferro-magnetic core), but to exert sufficient force to inject fuel into the high pressure in the cylinder, the solenoid had to be large and the inertia of the core limited actuation speed. The answer is to use ferro-electric actuators (Figure 14.25). Most actuators of this sort are based on the compound lead zirconium titanate (PZT). When a potential gradient dV/dx is applied across the faces of a disk of PZT, it undergoes a strain ε that is proportional to the electric potential gradient:

$$\varepsilon = k\frac{dV}{dx} \tag{14.18}$$

where k is the *ferro-electric coefficient* of the material. In practice k is small, meaning that a high gradient is needed to get useful strain. The system voltage in cars or trucks is usually 12 or 24 V, so the only way to get a large gradient is to use very thin slices of PZT—typically 100 μm thick, giving a gradient dV/dx of 200 000 V per meter. The problem then is that the displacement (the strain ε

times the thickness of the disk, x) is small. The solution is to make a stack like that shown, expanded, on the right of Figure 14.25. The thin ($x = 100 \, \mu m$) slices of PZT are sandwiched between conducting electrodes, alternately charged, creating the gradient. A stack of 100 (still only 10 mm high) gives adequate displacements. These actuators are half the size and allow twice the pulse rate of solenoids, giving more precise control, performance and fuel economy.

14.6 Summary and conclusions

All materials contain charged particles: electrons, protons, ions, charged molecules and the like. An electric field E exerts a force qE on a charge q, pushing it in the direction of the field if q is positive and in the opposite direction if it is negative. If the particles can move freely, as electrons can in metals, an electric current flows through the material, which we refer to as conduction. If, instead, they cannot move freely, as is the case in insulators, the particles are still displaced slightly by the field, leading to dielectric polarization.

Materials that polarize easily have a high dielectric constant. If used as the insulating film of a capacitor, the polarization screens the electrodes from each other, increasing the capacitance. The slight displacement of charge has other consequences. One of these is electrostriction: the charge movement causes the material to change shape slightly. If the material is made up of ions and these are arranged in a way that lacks a center of symmetry, a further effect appears: that of piezo-electric response. The lack of symmetry means that each molecule carries a dipole moment, and within a single crystal these are aligned. An electric field causes a change of shape and a change of shape changes the molecular dipole moment, thus inducing a change in polarization, generating a field. A change of temperature, too, changes the molecular dipole moment, so materials that are piezo-electric are also pyro-electric. A few materials have a further property: that of molecular asymmetry that can switch direction because ions can occupy one of several equivalent sites in the crystal. These materials are ferro-electric; the molecular dipoles spontaneously align to form polarized domains. In the relaxed state the domains take up orientations such that the field of one is canceled by those of its neighbors, but when 'poled' by placing it in an electric field, the domains align, giving a large net polarization that is retained when the field is removed.

All these effects have practical utility, examples of which are described in the chapter.

14.7 Further reading

Braithwaite, N. and Weaver, G. (1990) *Electronic Materials*, The Open University and Butterworth-Heinemann, Oxford, UK. ISBN 0-408-02840-8. *(One of the excellent Open University texts that form part of their Materials program.)*

Jiles, D. (2001) *Introduction to the Electronic Properties of Materials*, 2nd edition, Nelson Thornes, Cheltenham, UK. ISBN 0-7487-6042-3. *(The author develops the*

electronic properties of materials, relating them to thermal and optical as well as the electrical behavior.)
Solymar, L. and Walsh, D. (2004) *Electrical Properties of Materials*, 7th edition, Oxford University Press, Oxford, UK. ISBN 0-19-926793-6. *(A mature introduction to electrical materials.)*

14.8 Exercises

Exercise E14.1	Why do metals conduct electricity? Why do insulators not conduct electricity, at least when cold?
Exercise E14.2	What is a dielectric? What are its important dielectric properties? What is meant by polarization? Why do some dielectrics absorb microwave radiation more than others?
Exercise E14.3	It is much easier to measure the electrical conductivity of a material than to measure its thermal conductivity. Use the Weidemann–Franz law Section 14.3 to find the thermal conductivities of: (a) An alloy that has an electrical resistivity ρ_e of $28\,\mu\Omega.\text{cm}$. (b) Tungsten of the kind used for lamp filaments that has an electrical conductivity κ_e of $9.9 \times 10^6\,\text{S/m}$ ($1\,\text{S/m}$ is $1/\Omega.\text{m}$).
Exercise E14.4	The metal zinc has free electron concentration per unit volume, n_v, of $1.3 \times 10^{29}/\text{m}^3$ and an electron mobility μ_e of $8 \times 10^{-4}\,\text{m}^2/\text{V.s}$. The charge carried by an electron, e, is 1.6×10^{-19} coulomb. Based on this information, what is the electrical conductivity of zinc? Handbooks list the measured resistivity of zinc as $5.9\,\mu\Omega.\text{cm}$. Is this consistent with your calculation? (Watch the units.)
Exercise E14.5	A power line is to carry $5\,\text{kA}$ at $11\,\text{kV}$ using pylons $460\,\text{m}$ apart. The dip d in a wire of weight m_l per unit length strung between pylons L apart at a tension T is given by $d = L^2 m_l/8T$. The maximum tension allowed is 0.8 of the yield stress, σ_y. If the maximum allowable dip is $6\,\text{m}$, which of the materials in the table could be used?

Material	Electrical resistivity, ρ_e (Ωm)	Yield stress, σ_y (MPa)	Density, ρ (kg/m³)
Aluminum	1.7×10^{-8}	102	2700
Copper	1.5×10^{-8}	300	8900
Carbon steel	55×10^{-8}	510	7800

Exercise E14.6	A material is required for a transmission line that gives the lowest full-life cost over a 20-year period. The total cost is the sum of the material cost and

the cost of the energy dissipated in joules. The cost of electricity C_E is 6×10^{-3} \$/MJ. Material prices are listed in the table. Derive an expression for the total cost per meter of cable in terms of the cross-sectional area A (which is a free parameter), the material and electrical costs and the material parameters. Show that the minimum cost occurs when the two contributions to the cost are equal. Hence derive a performance index for the material and decide on the best of the materials in the table.

Material	Electrical resistivity, ρ_e (Ωm)	Density, ρ (kg/m³)	Price, C_m (\$/kg)
Aluminum	1.7×10^{-8}	2700	1.6
Copper	1.5×10^{-8}	8900	5.2
Steel	55×10^{-8}	7800	0.5

Exercise E14.7   In the discussion of conductors, a 50:50 mix of copper and steel strands was suggested for transmission cables. Using the values in the table for resistivity and strength, calculate the effective values of both for the cable, assuming a rule of mixtures. Plot this on copies of the $\sigma_y - \rho_e$ chart of Figures 14.6 and 14.11 to explore its performance.

Material	Strength, σ_y (MPa)	Resistivity, ρ_e ($\mu\Omega$.cm)
High-strength steel, cold drawn	1700	22
High-conductivity copper, cold drawn	300	1.7

Low density, we said, was important for long-span transmission lines because the self-weight becomes significant. Suppose you were asked to design a hybrid power transmission cable with the lowest possible weight, what combination of materials would you choose? Use information from the charts for inspiration.

Exercise E14.8   Roughly 50% of all cork that is harvested in Portugal ends up as cork dust, a worthless by-product. As a materials expert, you are approached by an entrepreneur who has the idea of making useful products out of cork dust by compacting it and heating it, using microwave heating. The loss factor L of cork is 0.21. The entrepreneur reckons he needs a power density P of at least 2 kW per m³ for the process to be economic. If the maximum field E is limited to 10^2 V/m, what frequency f of microwaves will be needed?

14.9 Exploring design with CES (use Level 2 unless stated otherwise)

Exercise E14.9   Use the 'Search' facility of CES to search for:

(a) Electrical conductors.
(b) Heat sinks.

Exercise E14.10 The analysis of heat sinks in Chapter 3 formulated the following design constraints and objective. They are summarized in the table. Use the selector to find the material that best meets them.

Function	• Heat sink
Constraints	• Material must be good electrical insulator
	• Maximum operating temperature >200°C
Objective	• Maximize thermal conductivity
Free variable	• Choice of material

Exercise E14.11 The sheath of the cable shown in Figure 14.19 is molded over the bundled wires. The material of the sheath must be electrically insulating, flexible, water resistant and cheap. The table translates the requirements. Apply the constraints using a 'Tree' and a 'Limit' stage, then plot a bar chart of material price to find the three least expensive materials that meet all the constraints.

Function	• Cable sheathing
Constraints	• Able to be molded
	• Good electrical insulator
	• Durability in fresh water: very good
	• Young's modulus <3 GPa
Objective	• Minimize material price
Free variable	• Choice of material

Exercise E14.12 The pins of the plug of Figure 14.20 must conduct electricity well, be corrosion resistant so that contact remains good, and be hard enough to resist abrasion and wear. The table translates the requirements. Apply these and rank the promising candidates by material price.

Function	• Connecting pins for power plug
Constraints	• Good electrical insulator
	• Durability in fresh water: very good
	• Hardness >200 H_V
Objective	• Minimize material price
Free variable	• Choice of material

Exercise E14.13   The casing of the plug of Figure 14.20 performs both a mechanical and an electrical function. It must insulate, and it must also be rigid and resilient enough to tolerate service loads. (Resilient means it must be capable of some elastic distortion requiring a high yield strain, σ_y/E).

Function	• Plug casing
Constraints	• Able to be molded
	• Good electrical insulator
	• Flammability: non-flammable or self-extinguishing
	• Young's modulus > 2 GPa
	• Yield strain $\sigma_y/E > 0.01$
Objective	• Minimize material price
Free variable	• Choice of material

14.10   Exploring the science with CES Elements

Exercise E14.14   Make a chart with Atomic number on the x-axis (use a linear scale) and Electrical conductivity κ on the y-axis. Convert the values of resistivity ρ_e (which are in units of $\mu\Omega$.cm in the database) to those of conductivity in siemens/m (S/m); use $\kappa = 10^8/\rho_e$. This can be made using the 'Advanced' facility in the axis-choice dialog box. Does conductivity vary in a periodic way across the Periodic Table? Which three elements have the highest conductivities?

Exercise E14.15   Explore the Wiedemann–Franz relation for the elements. To do so, make a chart of electrical conductivity, constructed as in the previous example, on the x-axis and thermal conductivity on the y-axis. Limit the selection to metals only by using a 'Limit' stage—State at 300 K. Fit an equation to the resulting plot. If you were now given a new alloy with an electrical conductivity of 4×10^7 S/m, what would you estimate its thermal conductivity to be? (Relations such as this one provide ways of building 'intelligent checking' of data into material property databases.)

Exercise E14.16   Make a chart with atomic number on the x-axis (use a linear scale) and dielectric constant on the y-axis. Do you see any general trend? If so, how would you explain it?

Exercise E14.17   It is proposed to use the temperature dependence of electrical resistivity for sensing. Plot this quantity against relative cost to identify three cheap metals with high values of this temperature dependence.

Exercise E14.18   Equation (14.15) of the text derives the electrical conductivity as

$$\kappa_e = n_v \, e \, \mu_e$$

The CES Elements database contains data for the free electron concentration n_v and the electron mobility μ_e. You will find the electronic charge under 'Constants/Parameters' in the 'Advanced' facility for choosing the axes of the chart. Make a chart using the 'Advanced' facility with $\kappa_e = n_v \, e \, \mu_e$ on the y-axis and the reciprocal of the resistivity on the x-axis. Make sure you have made all the units self-consistent. How well does this equation describes the measured conductivities of metals?

Exercise E14.19   Explore superconductivity across the Periodic Table. Use CES Elements to plot superconducting transition temperature, T_c, against the atomic number (use linear scales). Which element has the highest transition temperature? What is T_c for this element?

Chapter 15
Magnetic materials

Magnet materials in action. The audio-frequency transformer on the left has a soft magnetic core. The DC motor on the right has a hard (permanent) magnetic stator and a soft magnetic rotor

15.1 Introduction and synopsis

Migrating birds, some think, navigate using the earth's magnetic field. This may be questionable, but what is beyond question is that sailors, for centuries, have navigated in this way, using a natural magnet, *lodestone*, to track it. Lodestone is a mineral, magnetite (Fe_3O_4), which sometimes occurs naturally in a magnetized state. Today we know lodestone as one of a family of *ferrites*, members of which can be found in every radio, television set and microwave oven. Ferrites are one of two families of magnetic material; the other is the *ferro-magnet* family, typified by iron but also including nickel, cobalt and alloys of all three. Placed in a magnetic field, these materials become magnetized, a phenomenon called *magnetic induction*; on removal, some, called soft magnets, lose their magnetization, others, the hard magnets, retain it.

Magnetic fields are created by moving electric charge—electric current in electromagnets, electron spin in atoms of magnetic materials. This chapter is about magnetic materials: how they are characterized, where their properties come from and how they are selected and used. It starts with definitions of magnetic properties and the way they are measured. As in other chapters, charts display them well, separating the materials that are good for one sort of application from those that are good for others. The chapter continues by drilling down to the origins of magnetic behavior, and concludes with a discussion of applications and the materials that best fill them.

15.2 Magnetic properties: definition and measurement

Magnetic fields in a vacuum

First, some definitions. When a current i passes through a long, empty coil of n turns and length L, as in Figure 15.1, a magnetic field is generated. The magnitude of the field, H, is given by Ampère's[1] law as

$$H = \frac{ni}{L} \tag{15.1}$$

and thus has units of amps/meter (A/m). The field has both magnitude and direction—it is a vector field.

Magnetic fields exert forces on a wire carrying an electric current. A current i flowing in a single loop of area S generates a dipole moment m, where

$$m = i\,S \tag{15.2}$$

---

[1] André Marie Ampère (1775–1836), French physicist, largely self-taught (and said by his fiancée to have 'no manners'), contributor to the theory of light, of electricity, of magnetism and to chemistry (he discovered fluorine). His father, a Justice of the Peace, was guillotined during the French Revolution.

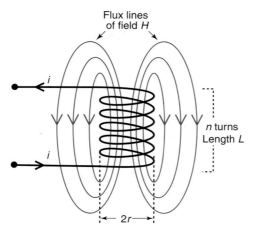

Figure 15.1   A solenoid creates a magnetic field *H*; the flux lines indicate the field strength.

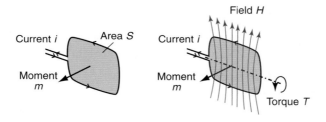

Figure 15.2   Definition of magnetic moment and moment-field interaction.

with units A.m², and it too is a vector with a direction normal to the plane of *S* (Figure 15.2). If the loop is placed at right angles to the field *H* it feels a torque *T* (units: newton.meter, or N.m) of

$$T = \mu_o \, m \, H \qquad (15.3)$$

where μ_o is called the *permeability of a vacuum*, $\mu_o = 4\pi \times 10^{-7}$ henry/meter (H/m). To link these we define a second measure of the magnetic field, one that relates directly to the torque it exerts on a unit magnetic moment. It is called the *magnetic induction* or *flux density*, *B*, and for a vacuum or non-magnetic materials it is

$$B = \mu_o \, H \qquad (15.4)$$

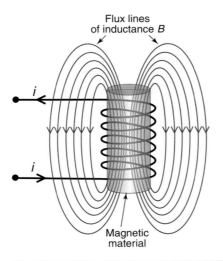

Flux lines
of inductance *B*

Magnetic
material

Figure 15.3   A magnetic material exposed to a field *H* becomes magnetized, concentrating
the flux lines.

Its units are *tesla*[2], so 1 tesla is $1\,HA/m^2$. A magnetic induction *B* of 1 tesla
exerts a torque of 1 N.m on a unit dipole at right angles to the field *H*.

Magnetic fields in materials

If the space inside the coil of Figure 15.2 is filled with a material, as in Figure
15.3, the induction within it changes. This is because its atoms respond to the
field by forming little magnetic dipoles in ways that are explained in Section 15.4.
The material acquires a macroscopic dipole moment or *magnetization*, *M* (its
units are A/m, like *H*). The induction becomes

$$B = \mu_o\,(H + M) \tag{15.5}$$

The simplicity of this equation is misleading, since it suggests that *M* and *H* are
independent; in reality *M* is the response of the material to *H*, so the two are
coupled. If the material of the core is ferro-magnetic, the response is a very strong
one and it is nonlinear, as we shall see in a moment. It is usual to rewrite equation
(15.5) in the form

$$B = \mu_R\,\mu_o\,H$$

where μ_R is called the *relative permeability*, and like the relative permittivity
(the dielectric constant) of Chapter 14, it is dimensionless. The magnetization,
M, is thus

$$M = (\mu_R - 1)\,H = \chi\,H \tag{15.6}$$

[2] Nikola Tesla (1856–1943), Serbian-American inventor, discoverer of rotating magnetic
fields, the basis of most alternating current machinery, inventor of the Tesla coil and of a
rudimentary flying machine (never built).

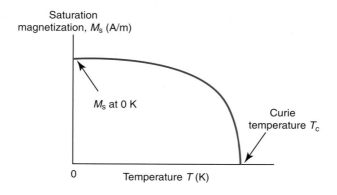

Figure 15.4   Saturation magnetization decreases with temperature, falling to zero at the Curie temperature, T_c.

where χ is the *magnetic susceptibility*. Neither μ_R nor χ are constants—they depend not only on the material but also on the magnitude of the field, H, for the reason just given.

Nearly all materials respond to a magnetic field by becoming magnetized, but most are paramagnetic with a response so faint that it is of no practical use. A few, however, contain atoms that have large dipole moments and have the ability to spontaneously magnetize—to align their dipoles in parallel—as electric dipoles do in ferro-electric materials. These are called *ferro-magnetic* and *ferri-magnetic* materials (the second one is called *ferrites* for short), and it is these that are of real practical use.

Magnetization decreases with increasing temperature. Just as with ferro-electrics, there is a temperature, the Curie temperature T_c, above which it disappears, as in Figure 15.4. Its value for the materials we shall meet here is well above room temperature (typically 300–500°C) but making magnets for use at really high temperatures is a problem.

Measuring magnetic properties

Magnetic properties are measured by plotting an *M–H* curve. It looks like Figure 15.5. If an increasing field H is applied to a previously demagnetized sample, starting at A on the figure, its magnetization increases, slowly at first and then faster, following the broken line, until it finally tails off to a maximum, the *saturation magnetization* M_s at the point B. If the field is now backed off, M does not retrace its original path, but retains some of its magnetization so that when H has reached zero, at the point C, some magnetization remains: it is called the *remanent magnetization* or *remanence* M_R and is usually only a little less than M_s. To decrease M further we must increase the field in the opposite direction until M finally passes through zero at the point D when the field is $-H_c$, the *coercive field*, a measure of the resistance to demagnetization. Some applications require H_c to be as high as possible, others as low a possible. Beyond

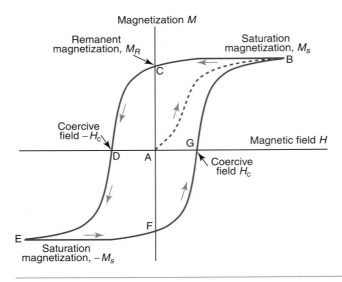

Figure 15.5   A hysteresis curve, showing the important magnetic properties.

point D the magnetization M starts to increase in the opposite direction, eventually reaching saturation again at the point E. If the field is now decreased again, M follows the curve through F and G back to full forward magnetic saturation again at B to form a closed $M–H$ circuit called the *hysteresis loop*.

Magnetic materials are characterized by the size and shape of their hysteresis loops. The initial segment AB is called the *initial magnetization curve* and its average slope (or sometimes its steepest slope) is the magnetic susceptibility, χ. The other key properties—the saturation magnetization M_s, the remanence M_R and the coercive field H_c—have already been defined. Each full cycle of the hysteresis loop dissipates an energy per unit volume equal to the area of the loop multiplied by μ_o, the permeability of a vacuum. This energy appears as heat (it is like magnetic friction). Many texts plot not the $M–H$ curve, but the curve of inductance B against H. Equation (15.5) says that B is proportional to $(M + H)$, and the value of M for any magnetic materials worthy of the name is very much larger than the H applied to it, so $B \approx \mu_o M$ and the $B–H$ curve of a ferro-magnetic material looks very like its $M–H$ curve (it's just that the M-axis has been scaled by μ_o).

There are several ways to measure the hysteresis curve, one of which is sketched in Figure 15.6. Here the material forms the core of what is, in effect, a transformer. The oscillating current through the primary input coil creates a field H that induces magnetization M in the material of the core, driving it round its hysteresis loop. The secondary coil picks up the inductance, from which the instantaneous state of the magnetization can be calculated, mapping out the loop.

Magnetic materials differ greatly in the shape and area of their hysteresis loop, the greatest difference being that between *soft magnets*, which have thin

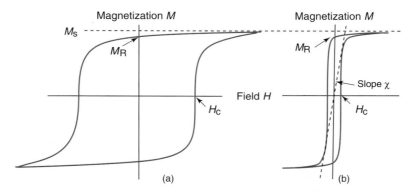

Figure 15.6   Measuring the hysteresis curve.

Figure 15.7   Hysteresis loops. (a) A fat loop typical of hard magnets. (b) A thin, square loop typical of soft magnets.

loops, and *hard magnets*, which have fat ones, as sketched in Figure 15.7. In fact the differences are much greater than this figure suggests: the coercive field H_c (which determines the width of the loop) of hard magnetic materials like Alnico is greater by a factor of about 10^5 than that of soft magnetic materials like silicon–iron. More on this in the next two sections.

15.3   Charts for magnetic properties

The remanence–coercive field chart
The differences between families of soft and hard magnetic materials are brought out by Figure 15.8. The axes are remanent magnetization M_r and coercive field

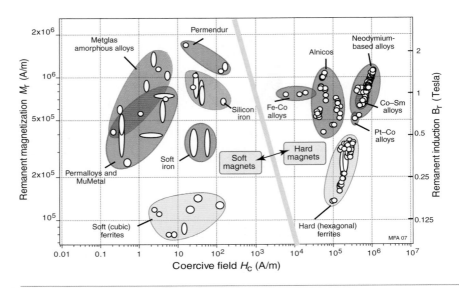

Figure 15.8 Remanent magnetization and coercive force. Soft magnetic materials lie on the left, hard magnetic materials on the right. The red and purple envelopes enclose electrically conducting materials, the yellow ones enclose insulators.

H_c. The saturation magnetization M_s, more relevant for soft magnets, is only slightly larger than M_r, so an M_s–H_c chart looks almost the same as this one. There are 12 district families, each enclosed in a colored envelope. The members of each family, shown as smaller ellipses, have unhelpful tradenames (such as 'H FerriteYBM-1A') so they are not individually labeled.[3] Soft magnets require high M_s and low H_c; they are the ones on the left, with the best near the top. Hard magnets must hold their magnetism, requiring a high H_c, and to be powerful they need a large M_r; they are the ones on the right, with the best again at the top.

In many applications the electrical resistivity is also important because of eddy-current losses, described later. Changing magnetic fields induces eddy currents in conductors but not in insulators. The red and purple ellipses on the chart enclose metallic ferro-magnetic materials, which are good conductors. The yellow ones describe ferrites, which are insulators.

The saturation magnetization–susceptibility chart

This is the chart for selecting soft magnetic materials (Figure 15.9). It shows the saturation magnetization, M_s—the ultimate degree to which a material can concentrate magnetic flux—plotted against its susceptibility, χ, which measures the ease with which it can be magnetized. Many texts use, instead, the saturation inductance B_s and maximum relative permeability, μ_R. They are shown on the

---

[3] This chart and the next were made with Level 3 of the CES Edu database, which contains data for 222 magnetic materials. Levels 1 and 2 do not include magnetic materials because of their specialized nature.

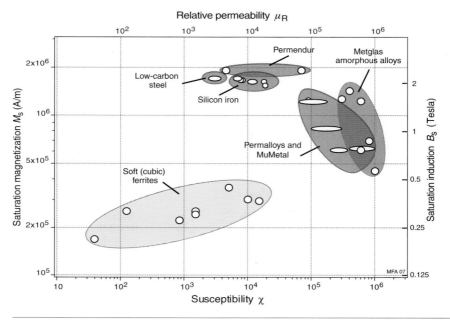

Figure 15.9 Saturation magnetization and susceptibility for soft magnetic materials. The red and purple envelopes enclose electrically conducting materials, the yellow ones enclose insulators.

other axes of the chart. As in the first chart, materials that are electrical conductors are enclosed in red or purple ellipses, those that are insulators yellow ones.

15.4 Drilling down: the physics and manipulation of magnetic properties

The classical picture of an atom is that of a nucleus around which swing electrons, as in Figure 15.10. Moving charge implies an electric current, and an electric current flowing in a loop creates a magnetic dipole, as in Figure 15.2. There is, therefore, a magnetic dipole associated with each orbiting electron. That is not all. Each electron has an additional moment of its own: its spin moment. A proper explanation of this requires quantum mechanics, but a way of envisaging its origin is to think of an electron not as a point charge but as slightly spread out and spinning on its own axis, again creating rotating charge and a dipole moment—and this turns out to be large. The total moment of the atom is the vector sum of the whole lot.

A simple atom like that of helium has two electrons per orbit and they configure themselves such that the moment of one exactly cancels the moment of

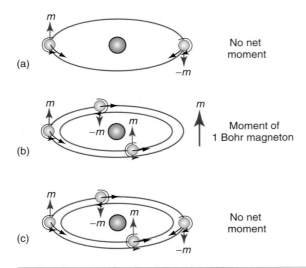

(a)

No net moment

(b)

Moment of 1 Bohr magneton

(c)

No net moment

Figure 15.10   Orbital and electron spins create a magnetic dipole. Even numbers of electrons filling energy levels in pairs have moments that cancel, as in (a) and (c). An unpaired electron gives the atom a permanent magnetic moment, as in (b).

the other, as in Figure 15.10(a) and (c), leaving no net moment. But now think of an atom with three, not two, electrons as in (b). The moments of two may cancel, but there remains the third, leaving the atom with a net moment represented by the red arrow at the right of Figure 15.10(b). Thus, atoms with electron moments that cancel are non-magnetic; those with electron moments that don't cancel carry a magnetic dipole. Simplifying a little, one unpaired electron gives a magnetic moment of 9.3×10^{-24} A.m^2, called a Bohr[4] magneton; two unpaired electrons give two Bohr magnetons, three gives three and so on.

Think now of the magnetic atoms assembled into a crystal. In most materials the atomic moments interact so weakly that thermal motion is enough to randomize their directions, as in Figure 15.11(a). Despite their magnetic atoms, the structure as a whole has no magnetic moment; these materials are *paramagnetic*. In a few materials, though, something quite different happens. The fields of neighboring atoms interact such that their energy is reduced if their magnetic moments line up. This drop in energy is called the *exchange energy* and it is strong enough that it beats the randomizing effect of thermal energy so long as the temperature is not too high (the shape of the Curie curve of Figure 15.4 shows how thermal energy overwhelms the exchange energy as the Curie temperature is approached). They may line up anti-parallel, head to tail so to speak, as in Figure 15.11(b), and there is still no net moment; such materials are called *anti-ferro-magnets*.

[4] Niels Henrik David Bohr (1885–1962), Danish theoretical physicist, elucidator of the structure of the atom, contributor to the Manhattan Project and campaigner for peace.

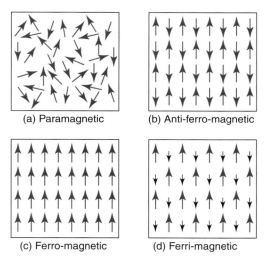

(a) Paramagnetic   (b) Anti-ferro-magnetic

(c) Ferro-magnetic   (d) Ferri-magnetic

Figure 15.11   Types of magnetic behavior.

But in a few elements, notably iron, nickel and cobalt, exactly the opposite happens: the moments spontaneously align so that—if all are parallel—the structure has a net moment that is the sum of those of all the atoms it contains. These materials are *ferro-magnetic*. Iron has three unpaired electrons per atom, nickel has two and cobalt just one, so the net moment if all the spins are aligned (the saturation magnetization M_s) is greatest for iron, less for nickel and still less for cobalt.

Compounds give a fourth possibility. The materials we have referred to as ferrites are oxides; one class of them has the formula MFe_2O_4, where M is also a magnetic atom, such as cobalt (Co). Both the iron and the cobalt atoms have dipoles but they differ in strength. They line up in the anti-parallel configuration, but because of the difference in moments the cancelation is incomplete, leaving a net moment M; these are *ferri-magnets*, ferrites for short. The partial cancelation and the smaller number of magnetic atoms per unit volume means they have lower saturation magnetization than, say, iron, but they have other advantages, notably that, being oxides, they are electrical insulators.

Domains

If atomic moments line up, shouldn't every piece of iron, nickel or cobalt be a permanent magnet? Magnetic materials they are; magnets, in general, they are not. Why not?

A uniformly magnetized rod creates a magnetic field, H, like that of a solenoid. The field has a potential energy

$$U = \frac{1}{2} \mu_o \int_V H^2 dV \tag{15.7}$$

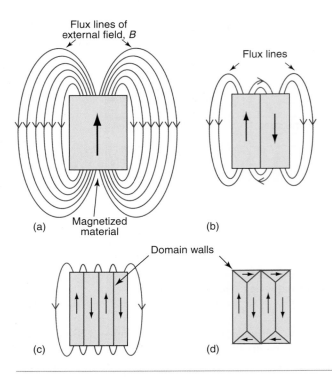

Figure 15.12   Domains allow a compromise: the cancelation of the external field while retaining magnetization of the material itself. The arrows show the direction of magnetization.

where the integral is carried out over the volume V within which the field exists. Working out this integral is not simple, but we don't need to do it. All we need to note is that the smaller is H and the smaller the volume V that it invades, the smaller is the energy. If the structure can arrange its moments to minimize its H or get rid of it entirely (remembering that the exchange energy wants neighboring atom moments to stay parallel) it will try to do so.

Figure 15.12 illustrates how this can be done. The material splits up into *domains*, within which the atomic moments are parallel but with a switch of direction between mating domains to minimize the external field. The domains meet at *domain walls*, regions a few atoms thick in which the moments swing from the orientation of one domain to that of the other. Splitting into parallel domains of opposite magnetization, as in (b) and (c), reduces the field a lot; adding caps magnetized perpendicular to both as in (d) kills it almost completely. The result is that most magnetic materials, unless manipulated in some way, adopt a domain structure with minimal external field—that is the same as saying that, while magnetic, they are not magnets.

How can they be magnetized? Placed in a magnetic field created, say, with a solenoid, the domains already aligned with the field have lower energy than those

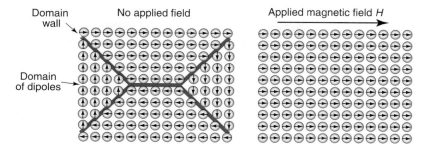

Figure 15.13   An applied field causes domain boundaries to move. At saturation the sample has become a single domain, as on the right. Domain walls move easily in soft magnetic materials. Hard magnetic materials are alloyed to prevent their motion.

aligned against it. The domain wall separating them feels a force, the *Lorentz*[5] *force*, pushing it in a direction to make the favorably oriented domains grow at the expense of the others. As they grow, the magnetization of the material increases, moving up the M–H curve, finally saturating at M_s, when the whole sample is one domain oriented parallel to the field, as in Figure 15.13.

The saturation magnetization, then, is just the sum of all the atomic moments contained in a unit volume of material when they are all aligned in the same direction. If the magnetic dipole per atom is $n_m m_B$ (where n_m is the number of unpaired electrons per atom and m_B is the Bohr magneton), then the saturation magnetization is

$$M_s = \mu_o \frac{n_m m_B}{\Omega} \qquad (15.8)$$

where Ω is the atomic volume. Iron has the highest because its atoms have the most unpaired electrons and they are packed close together. Anything that reduces either one reduces the saturation magnetization. Nickel and cobalt have lower saturation because their atomic moments are less; alloys of iron tend to be lower because the non-magnetic alloying atoms dilute the magnetic iron. Ferrites have much lower saturation both because of dilution by oxygen and because of the partial cancelation of atomic moments sketched in Figure 15.11.

There are, none the less, good reasons for alloying and making iron into ferrites. One relates to the coercive field, H_c. A good permanent magnet retains its magnetization even when placed in a reverse field, and this requires a high H_c. If domain walls move easily the material is a soft magnet with a low H_c; if the opposite, it is a hard one. To make a hard magnet it is necessary to pin the domain

---

[5] Hendrik Antoon Lorentz (1853–1928), Dutch mathematical physicist, friend and collaborator with Raleigh and Einstein, originator of key concepts in optics, electricity, relativity and hydraulics (he modeled and predicted the movement of water caused by the building of the Zuyderzee).

walls to stop them moving. Impurities, foreign inclusions, precipitates and porosity all interact with domain walls, tending to pin them in place (they act as obstacles to dislocation motion too, so magnetically hard materials are mechanically hard as well). And there are subtler barriers to domain-wall motion: *magnetic anisotropy* arising because certain directions of magnetization in the crystal are more favorable than others, and even the shape and size of the specimen itself—small, single-domain magnetic particles are particularly resistant to having their magnetism reversed.

The performance of a hard magnet is measured by its *energy product*, roughly proportional to the area of the magnetization loop. The higher the energy product, the more difficult it is to demagnetize it.

15.5 Materials selection for magnetic design

Selection of materials for their magnetic properties follows the new familiar pattern of translating design requirements followed by screening and ranking. In this case, however, the first screening step—'must be magnetic'—immediately isolates the special set of materials shown in the property charts of this chapter. The key selection issue is then whether a soft or hard magnetic response is required.

Soft magnetic devices

Electromagnets, transformers, electron lenses and the like have magnetic cores that must magnetize easily when the field switches on, be capable of creating a high flux density, yet lose their magnetism when the field is switched off. They do this by using soft, low H_c, magnetic materials, shown on the left of the chart of Figure 15.8. Soft magnetic materials are used to 'conduct' and focus magnetic flux. Figure 15.14 illustrates this. A current-carrying coil induces a magnetic field in the core. The field, instead of spreading in the way shown in Figure 15.3, is trapped in the extended ferro-magnetic circuit, as shown here. The higher the susceptibility χ, the greater is the magnetization and flux density induced by a given current loop, up to a limit set by its saturation magnetization M_s. The magnetic circuit conducts the flux from one coil to another, or from a coil to an

Figure 15.14   Soft magnets 'conduct' magnetic flux and are concentrated if the cross-section of the magnet is reduced.

air gap, where it is further enhanced by reducing the section of the core, where it is used to actuate or to focus an electron beam, as shown on the right of Figure 15.14. Thus, the first requirement of a good soft magnet is a high χ and a high M_s with a thin hysteresis loop like that sketched in Figure 15.15. The chart of Figure 15.9 shows these two properties. Permendur (Fe–Co–V alloys) and Metglass amorphous alloys are particularly good, but being expensive they are used only in small devices. Silicon–iron (Fe–1–4% Si) is much cheaper and easier to make in large sections; it is the staple material for large transformers.

Most soft magnets are used in AC devices, and then energy loss, proportional to the area of the hysteresis loop, becomes a consideration. Energy loss is of two types: the hysteresis loss already mentioned—it is proportional to switching frequency, f—and eddy current losses caused by the currents induced in the core itself by the AC field, proportional to f^2. In cores of transformers operating at low frequencies (meaning f up to 100 Hz or so) the hysteresis loss dominates, so for these we seek high M_s and a thin loop with a small area. Eddy-current losses are present but are suppressed when necessary by *laminating*: making the core from a stack of thin sheets interleaved with thin insulating layers that interrupt the eddy currents. At audio frequencies ($f < 50$ kHz) both types of loss become greater because they occur on every cycle and now there are more of them, requiring an even narrower loop and thinner laminates; here the more expensive Permalloys and Permendurs are used. At higher frequencies still (f in the MHz range) eddy-current loss, rising with f^2, becomes really serious and the only way to stop this is to use magnetic materials that are electrical insulators. That means ferrites (shown in yellow envelopes on both charts), even though they

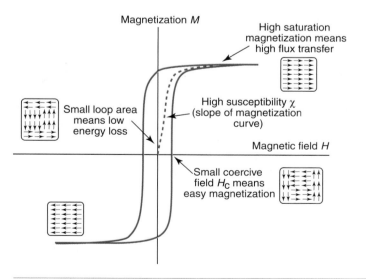

Figure 15.15   A characteristic hysteresis loop for a soft magnet.

Table 15.1 Materials for soft magnetic applications

Application	Frequency f	Material requirements	Material choice
Electromagnets	<1 Hz	High M_s, high χ	Silicon–iron Fe–Co alloys (Permendur)
Motors, low-frequency transformers	<100 Hz	High M_s, high χ	Silicon–iron AMA (amorphous Ni–Fe–P alloys)
Audio amplifiers, loudspeakers, microphones	<100 kHz	High M_s, very low H_c, high χ	Ni–Fe (Permalloy, Mumetal)
Microwave and UHF applications	<MHz	High M_s, very low H_c, electrical insulator	Cubic ferrites: MFe_2O_4 with M = Cu / Mn / Ni
Gigahertz devices	>100 MHz	Ultra low hysteresis, excellent insulator	Garnets

have a lower M_s. Above this ($f > 100$ MHz) only the most exotic ceramic magnets will work. Figure 15.15 summarizes the loop characteristics for soft magnets; Table 15.1 lists the choices, guided by the charts.

Hard magnetic devices

Many devices use permanent magnets: fridge doors, magnetic clutches, loudspeakers and earphones, DC motors, even frictionless bearings. For these, the key property is the remanence, M_R, since this is the maximum magnetization the material can offer without an imposed external field to keep it magnetized. High M_R, however, is not all. When a material is magnetized and the magnetizing field is removed, the field of the magnet itself tries to demagnetize it. It is this demagnetizing field that makes the material take up a domain structure that reduces or removes the net magnetization of the sample. A permanent magnet must be able to resist being demagnetized by it own field. A measure of its resistance is its coercive field, H_c, and this too must be large. A high M_R and a high H_c means a fat hysteresis loop like that of Figure 15.16. Ordinary iron and steel can become weakly magnetized with the irritating result that your screwdriver sticks to screws and your scissors pick up paper-clips when you wish they wouldn't, but neither are 'hard' in a magnetic sense. Here 'hard' means 'hard to demagnetize', and that is where high H_c comes in; it protects the magnet from itself.

Hard magnetic materials lie on the right of the $M_R – H_c$ chart of Figure 15.8. The workhorses of the permanent magnet business are the Alnicos and the hexagonal ferrites. An Alnico (there are many) is an alloy of aluminum, nickel, cobalt and iron—hence its name—with values of M_R as high as 10^6 A/m and a huge coercive field: 5×10^4 A/m. The hexagonal ferrites, too, have high M_R and H_c and, like the cubic soft-magnet ferrites, they are insulators. Being metallic, the Alnicos can be shaped by casting or by compacting and sintering powders; the ferrites can only be shaped by powder methods. Powders, of course, don't

Figure 15.16   A characteristic hysteresis loop for a hard magnet.

have to be sintered—you can bond them together in other ways, and these offer opportunities. Magnetic particles mixed with a resin can be molded to complex shapes. Bonding them with an elastomer gives magnets that are flexible or squashy—the magnetic seal of a fridge door is an example.

The chart of Figure 15.8 shows three families of hard magnets with coercive fiels that surpass those of Alnicos and hexagonal ferrites. These, the products of research over the last 30 years, provide the intense permanent-magnet fields that have allowed the miniaturization of earphones, microphones and motors. The oldest of these is the precious metal Pt–Co family but they are so expensive that they have now all but disappeared. They are outperformed by the cobalt–samarium family based on the inter-metallic compound $SmCo_5$, and they, in turn, are outperformed by the recent development of the neodymium–iron–boron family. None of these are cheap, but in the applications they best suit, the quantity of material is sufficiently small and the product value sufficiently high that this is not an obstacle. Table 15.2 lists the choices, guided by the charts.

Magnetic storage of information
Magnetic information storage requires hard magnets, but ones with an unusual loop shape: one that is rectangular (called 'square'). The squareness means that a unit of the material can flip in magnetization from $+M_R$ to $-M_R$ and back when exposed to fields above $+H_c$ or $-H_c$. The unit thus stores one binary bit of information. The information is read by moving the magnetized unit past a read head, where it induces an electric pulse. The choice of the word 'unit' was deliberately vague because it can take several forms: discrete particles embedded in a polymer tape or disk, or as a thin magnetic film that has the ability to

Table 15.2   Materials for hard magnetic applications

Application	Material requirements	Material choice
Large permanent magnets	High M_R, high H_c and high energy product $(MH)_{max}$; low cost	Alnicos Hexagonal ferrites
Small high-performance magnets	Acoustic, electronic and miniature mechanical applications	Cobalt–samarium Platinum–cobalt Neodymium–boron
Information storage	Thin, 'square' hysteresis loop	Elongate particles of Fe_2O_3, Cr_2O_3 or hexagonal ferrite

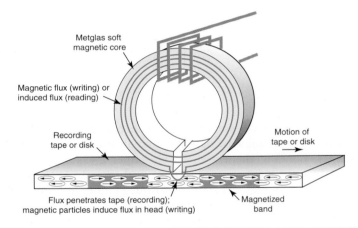

Figure 15.17   A magnetic read/write head. The gap in the soft magnetic head allows flux to escape, penetrating the tape and realigning the particles.

carry small stable domains called 'bubbles' that are magnetized in one direction, embedded in a background that is magnetized in another.

Figure 15.17 shows one sort of information storage system. A signal creates a magnetic flux in a soft magnetic core—often an amorphous metal alloy (AMA) because of its large susceptibility. The direction of the flux depends on the direction of the signal current; switching this reverses the direction of the flux. An air gap in the core allows field to escape, penetrating a tape or disk in which sub-micron particles of a hard magnetic material are embedded. When the signal is in one direction all particles in the band of tape or disk passing under the write-head are magnetized in one direction; when the signal is reversed, the direction of magnetization is reversed. The same head is used for reading. When the tape or disk is swept under the air gap, the stray field of the magnetized bands induces a flux in the soft iron core and a signal in the coil, which is amplified and processed to read it.

Why use fine particles? It is because domain walls are not stable in sufficiently small particles: each particle is a single domain. If the particle is elongated it prefers to be magnetized parallel to its long axis. Each particle then behaves like a little bar magnet with a north (N) and a south (S) pole at either end. The field from the write-head is enough to flip the direction, turning an N–S magnetization into one that is S–N, a binary response that is well suited to information storage. Rewritable tapes and disks use particles of Fe_2O_3 or of CrO_2, typically 0.1 μm long and with an aspect ratio of 10:1. Particles of hexagonal ferrites (MO) $(Fe_2O_3)_6$ have a higher coercive field, so a more powerful field is required to change their magnetization—but having done so, it is hard to erase it. This makes them the best choice for read-only applications like the identification strip of credit and swipe cards.

15.6 Summary and conclusions

The classical picture of an atom is that of a nucleus around which swing electrons in discrete orbits, each electron spinning at the same time on its own axis. Both spins create magnetic moments that, if parallel, add up, but if opposed, cancel to a greater or lesser degree. Most materials achieve near-perfect cancellation either within the atomic orbits or—if not—by stacking the atomic moments head to tail or randomizing them so that, when added, they cancel. A very few, most based on just three elements—Fe, Ni and Co—have atoms with residual moments and an inter-atomic interaction that causes them to line up to give a net magnetic moment or magnetization. Even these materials can find a way to screen their magnetization by segmenting themselves into domains: a ghetto-like arrangement in which atomic moments segregate into colonies or domains, each with a magnetization that is oriented such that it tends to cancel that of its neighbors. A strong magnetic field can override the segregation, creating a single unified domain in which all the atomic moments are parallel, and if the coercive field is large enough, they remain parallel even when the driving field is removed, giving a 'permanent' magnetization.

There are two sorts of characters in the world of magnetic materials. There are those that magnetize readily, requiring only slight urging from an applied field to do so. They transmit magnetic flux and require only a small reversal of the applied field to realign themselves with it. And there are those that, once magnetized, resist realignment; they give us permanent magnets. The charts of this chapter introduced the two, displaying the properties that most directly determine their choice for a given application.

15.7 Further reading

Braithwaite, N. and Weaver, G. (1990) *Electronic Materials*, The Open University and Butterworth-Heinemann, Oxford, UK. ISBN 0-408-02840-8. *(One of the excellent Open University texts that form part of their Materials program.)*

Campbell, P. (1994) *Permanent Magnetic Materials and their Applications*, Cambridge University Press, Cambridge, UK.

Douglas, W.D. (1995) Magnetically soft materials, in *ASM Metals Handbook*, 9th edition, Volume 2, *Properties and Selection of Non-ferrous Alloys and Special Purpose Materials*, pp 761–781, ASM, Metals Park, OH, USA.

Fiepke, J.W. (1995) Permanent magnet materials, in *ASM Metals Handbook*, 9th edition, Volume 2, *Properties and Selection of Non-ferrous Alloys and Special Purpose Materials*, pp. 782–803, ASM, Metals Park, OH, USA.

Jakubovics, J.P. (1994) *Magnetism and Magnetic Materials*, 2nd edition, The Institute of Materials, London, UK. ISBN 0-901716-54-5. *(A simple introduction to magnetic materials, short and readable.)*

15.8 Exercises

Exercise E15.1   Sketch an M_s–H curve for a ferro-magnetic material. Identify the important magnetic properties.

Exercise E15.2   Why are some elements ferro-magnetic when others are not?

Exercise E15.3   What is a ferrite? What are its characteristics?

Exercise E15.4   What is a Bohr magneton? A magnetic element has two unpaired electrons and an exchange interaction that causes them to align such that their magnetic fields are parallel. Its atomic volume, Ω, is 3.7×10^{-29} m^3. What would you expect its saturation magnetization, M_s, to be?

Exercise E15.5   A coil of 50 turns and length 10 mm carries a current of 0.01 amps. The core of the coil is made of a material with a susceptibility $\chi = 10^4$. What is the magnetization M and the induction B?

Exercise E15.6   An inductor core is required for a low-frequency harmonic filter. The requirement is for low loss and high saturation magnetization. Using the charts of Figures 15.8 and 15.9 as data sources, which class of magnetic material would you choose?

Exercise E15.7   A magnetic material is required for the core of a transformer that forms part of a radar speed camera. It operates at a microwave frequency of 500 kHz. Which class of material would you choose?

Exercise E15.8   A material is required for a flexible magnetic seal. It must be in the form of an elastomeric sheet and must be electrically insulating. How would you propose to make such a material?

Exercise E15.9   What are the characteristics required of materials for magnetic information storage?

15.9 Exploring design with CES

Open CES at Level 3—it opens in the 'Browse' mode. At the head of the Browse list are two pull-down menus reading 'Table' and 'Subset'. Open the 'Subset' menu and choose 'Magnetic'. Records for 233 magnetic materials are displayed, listing their magnetic properties and, where available, mechanical, thermal, electrical and other properties too.

To select, click on 'Select' in the main toolbar and choose Materials Universe > Magnetic materials in the dialog box. Now we can begin.

Exercise E15.10 Use the CES 'Search' facility to find materials for:

(a) Transformer cores.
(b) Electric motors.

Exercise E15.11 Find by browsing the records for:

(a) Cast Alnico 3. What is the value of its coercive force H_c?
(b) The amorphous alloy Metglas 2605-Co. What is the value of its coercive force H_c?

What do these values tell you about the potential applications of these two materials?

Exercise E15.12 Find by browsing the records for:

(a) Ferrite G (Ni–Zn ferrite). What are the values of its coercive force H_c and resistivity ρ_e?
(b) 2.5Si–Fe soft magnetic alloy. What are the values of its coercive force H_c and resistivity ρ_e?

If you were asked to choose one of these for a transformer core, what would be your first question?

Exercise E15.13 Make a bar chart of saturation induction B_s (the saturation magnetization $M_s = B_s/\mu_0$, so the two are proportional). Report the three materials with the highest values.

Exercise E15.14 Make a bar chart of coercive force, H_c. Report the four materials with the lowest value. What applications use them?

Exercise E15.15 A soft magnetic material is required for the laminated rotor of an AC induction electric motor. The material is to be rolled and further shaped by stamping, requiring an elongation of at least 40%. To keep hysteresis losses to a minimum it should have the lowest possible coercive force. Find the two materials that best meet these requirements, summarized below. Report them and their trade names.

Function	• Motor laminations
Constraints	• Soft magnetic material • Ductile (elongation > 40%)
Objective	• Minimize coercive force
Free variable	• Choice of material

Exercise E15.16 A magnetic material is required for the read/write head of a hard disk drive. It must have high hardness to resist wear and the lowest possible coercive force to give accurate read/write response.

(a) Use CES to identify possible candidates that meet these requirements, summarized in the table.
(b) Now add the further constraint that, for lower loss and high-frequency damping, the magnetic material must have an electrical resistivity above $10^8 \ \mu\Omega.\,\text{cm}$.

Function	• Magnetic read-head
Constraints	• Soft magnetic material • High hardness for wear resistance (Vickers hardness > 500 H_V)
Objective	• Minimize coercive force
Free variable	• Choice of material

15.10 Exploring the science with CES Elements

Exercise E15.17 Make a chart with atomic number on the *x*-axis and magneton moment per atom (Bohr magneton) on the *y*-axis (use linear scales) to identify the ferromagnetic elements. How many are there? Which have the highest magneton moment per atom?

Exercise E15.18 According to equation (15.8) of the text, the saturation magnetization is

$$M_s = \mu_o \frac{n_m m_B}{\Omega}$$

where n_m is the magnetic dipole per atom in units of Bohr magnetons, Ω is the atomic volume, m_B is the value of a Bohr magneton ($9.3 \times 10^{-24}\,\text{A/m}^2$) and μ_o the permeability of a vacuum ($\mu_o = 4\pi \times 10^{-7}\,\text{H/m}$). Make a chart with this combination of properties on the *x*-axis and the saturation magnetization M_s on the *y*-axis for the elements of the Periodic Table. How accurately is this equation obeyed?

Chapter 16
Materials for optical devices

Instruments using the ability of materials to reflect and refract. (Image of microscope courtesy of A–Z Microscope Corporation, California USA.)

16.1 Introduction and synopsis

It was at one time thought that the fact that light could travel through space—from the sun to earth, for instance—must mean that space was not really empty but filled with 'luminiferous ether'. It was not until the experiments of Michelson[1] and Morley in 1881 that it was realized that light did not need a 'material' for its propagation but could propagate through totally empty space at what is now seen as the ultimate velocity: 3×10^8 m/s.

When radiation strikes materials, things can happen. Materials interact with radiation by reflecting it, absorbing it, transmitting it and refracting it. This chapter is about these interactions, the materials that do them best and the ways we use them. The chapter opening page shows two: a reflecting telescope and a refracting microscope, each of which depend on the optical properties of materials.

16.2 The interaction of materials and radiation

Electromagnetic (e-m) radiation permeates the entire universe. Observe the sky with your eye and you see the visible spectrum, the range of wavelengths we call 'light' (0.40–0.77 μm). Observe it with a detector of X-rays or γ-rays and you see radiation with far shorter wavelengths (as short as 10^{-4} nm, one-thousandth the size of an atom). Observe it instead with a radio-telescope and you pick up radiation with wavelengths measured in millimeters, meters or even kilometers, known as radio and microwaves. The range of wavelengths of radiation is vast, spanning 18 orders of magnitude (Figure 16.1). The visible part of this spectrum is only a tiny part of it—but even that has entrancing variety, giving us colors ranging from deep purple through blue, green and yellow to deep red.

The intensity I of an e-m wave, proportional to the square of its amplitude, is a measure of the energy it carries. When radiation with intensity I_o strikes a material, a part I_R of it is reflected, a part I_A absorbed and a part I_T may be transmitted. Conservation of energy requires that

$$\frac{I_R}{I_o} + \frac{I_A}{I_o} + \frac{I_T}{I_o} = 1 \qquad (16.1)$$

The first term is called the *reflectivity* of the material, the second the *absorptivity* and the last the *transmittability* (all dimensionless). Each depends on the wavelength of the radiation, on the nature of the material and on the state of

---

[1] Albert A. Michelson (1852–1931), Prussian-American experimental physicist, who, with E.W. Morley, first demonstrated that the speed of light is independent of the earth's motion, a finding central to the establishment of the theory of relativity.

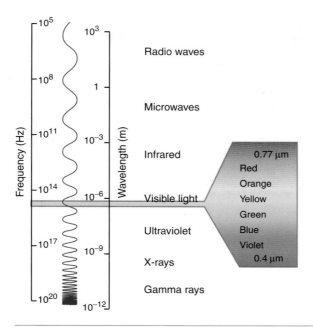

Figure 16.1   The spectrum of electromagnetic (e-m) waves. The visible spectrum lies between the wavelengths 0.4 and 0.77 μm.

its surfaces. They can be thought of as properties of the material in a given state of surface polish, smoothness or roughness.

In optics we are concerned with wavelengths in the visible spectrum. Materials that reflect or absorb all visible light, transmitting none, are called *opaque*, even though they may transmit in the near visible (infrared or ultraviolet). Those that transmit a little diffuse light are called *translucent*. Those that transmit light sufficiently well that you can see through them are called *transparent*; a subset of these that transmit almost perfectly, making them suitable for lenses, light-guides and optical fibers, are given the additional title of *optical quality*. Metals are opaque. To be transparent a material must be a dielectric.

Specular and diffuse reflection

Metals reflect almost all the light that strikes them; none is transmitted and little is absorbed. When light strikes a reflecting surface at an incident angle θ_1, part of it is reflected, leaving the surface with an angle of reflection θ_2 such that

$$\theta_1 = \theta_2$$

(16.2)

Specular surfaces are microscopically smooth and flat. A beam striking such a surface suffers specular reflection, meaning that it is reflected as a beam, as on

Figure 16.2   Optically flat, reflective surfaces give specular reflection, such that $\theta_1 = \theta_2$. The angles of incidence and reflection are always equal, but the rough surface gives diffuse reflection even though the angles of incidence and reflection are still, locally, equal.

the left of Figure 16.2. *Diffuse* surfaces are irregular; the law of reflection (equation (16.2)) still holds locally but the incident beam is reflected in many different directions because of the irregularities, as on the right of the figure.

Absorption

If radiation can penetrate a material, some is absorbed. The greater the thickness x through which the radiation passes, the greater the absorption. The intensity I, starting with the initial value I_0, decreases such that

$$I = I_o \exp - \beta x \qquad (16.3)$$

where β is the absorption coefficient, with dimensions of m^{-1} (or, more conveniently, mm^{-1}). The absorption coefficient depends on wavelength, with the result that white light passing through a material may emerge with a color corresponding to the wavelength that is least absorbed—that is why a thick slab of ice looks blue.

Transmission

By the time a beam of light has passed completely through a slab of material it has lost some intensity through reflection at the surface at which it entered, some in reflection at the surface at which it leaves, and some by absorption in between. Its intensity is

$$I = I_o \left(1 - \frac{I_R}{I_o} \right)^2 \exp - \beta x \qquad (16.4)$$

The term $(1 - I_R/I_o)$ occurs to the second power because intensity is lost through reflection at both surfaces.

Refraction

The velocity of light in a vacuum, $c_o = 3 \times 10^8$ m/s, is as fast as it ever goes. When it (or any other electromagnetic radiation) enters a material, it slows down. The *index of refraction*, n, is the ratio of its velocity in a vacuum, c_o, to that in the material, c:

$$n = \frac{c_o}{c} \tag{16.5}$$

This retardation makes a beam of light bend or *refract* when it enters a material of different refractive index. When a beam passes from a material 1 of refractive index n_1 into a material 2 of index n_2 with an angle of incidence θ_1, it deflects to an angle θ_2, such that

$$\frac{\sin \theta_1}{\sin \theta_2} = \frac{n_2}{n_1} \tag{16.6}$$

as in Figure 16.3(a); the equation is known as Snell's law[2]. The refractive index depends on wavelength, so each of the colors that make up white light is diffracted through a slightly different angle, producing a spectrum when light passes through a prism. When material 1 is vacuum or air, for which $n_1 = 1$, the equation reduces to

$$\frac{\sin \theta_1}{\sin \theta_2} = n_2$$

Equation (16.6) says that light passes from a material with index $n_1 = n$ into air with $n_2 = 1$, it is bent away from the normal to the surface, like that in Figure 16.3(b). If the incident angle, here θ_1, is slowly increased, the emerging beam tips down until it becomes parallel with the surface when $\theta_2 = 90°$ and $\sin \theta_2 = 1$. This occurs at an incident angle, from equation (16.6), of

$$\sin \theta_1 = \frac{1}{n} \tag{16.7}$$

For values of θ_1 greater than this, the equation predicts values of $\sin \theta_2$ that are greater than 1, and that can't be. Something strange has to happen, and it does: the ray is totally reflected back into the material, as in Figure 16.3(c). This *total*

[2] Willebrord Snell (1591–1626), Dutch astronomer, also known as Snell van Royen, or Snellius, who first derived the relationship between the different angles of light as it passes from one transparent medium to another. The lunar crater Snellius is named after him.

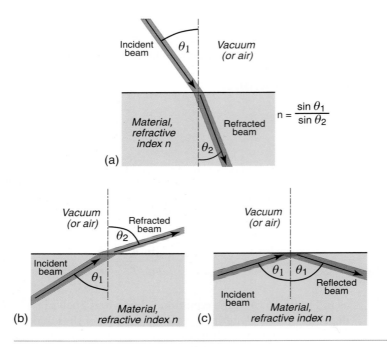

Figure 16.3   Refraction and total internal reflection.

internal reflection has many uses, one being to bend the path of light in prismatic binoculars and reflex cameras. Another is to trap light within an optical fiber, an application that has changed the way we communicate.

Reflection is related to refraction. When light traveling in a material of refractive index n_1 is incident normal to the surface of a second material with a refractive index n_2, the reflectivity is

$$R = \frac{I_R}{I_o} = \left(\frac{n_2 - n_1}{n_2 + n_1}\right)^2 \qquad (16.8a)$$

If the incident beam is in air, with refractive index 1, this becomes

$$R = \frac{I_R}{I_o} = \left(\frac{n - 1}{n + 1}\right)^2 \qquad (16.8b)$$

Thus, materials with high refractive index have high reflectivity.

Before leaving Snell and his law, ponder for a moment on the odd fact that the beam is bent at all. Light entering a dielectric, as we have said, slows down.

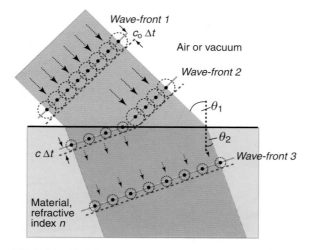

Figure 16.4   The Fresnel construction, explaining why a light beam is bent on entering a material of higher refractive index.

If that is so why does it not continue in a straight line, just more slowly? Why should it bend? To understand this we need Fresnel's[3] construction. We think of light as advancing via a series of wave-fronts. Every point on a wave-front acts as a source, so that, in a time Δt the front advances by $c\Delta t$, where c is the velocity of light in the medium through which it is passing ($c = c_0$ in vacuum or air)—as shown in Figure 16.4 with *wave-front 1* advancing $c_0\Delta t$. When the wave enters a medium of higher refractive index it slows down so that the advance of the wave-front within the medium is less than that outside, as shown with *wave-front* 2 advancing $c\Delta t$ in the figure. If the angle of incidence θ_1 is not zero, the wave-front enters the second medium progressively, causing it to bend, so that when it is fully in the material it is traveling in a new direction, characterized by the angle of refraction, θ_2. Simple geometry then gives equation (16.6).

16.3 Charts for optical properties

Refractive index and dielectric constant
Figure 16.5 shows the refractive index, n, of dielectrics, plotted against the dielectric constant ε_R (defined in Chapter 14). Those shown as white bubbles are transparent to visible light; those with red bubbles are not, but transmit

[3] Augustin Jean Fresnel (1788–1827), French physicist and engineer, known for his research on the wave theory of light and of diffraction and polarization.

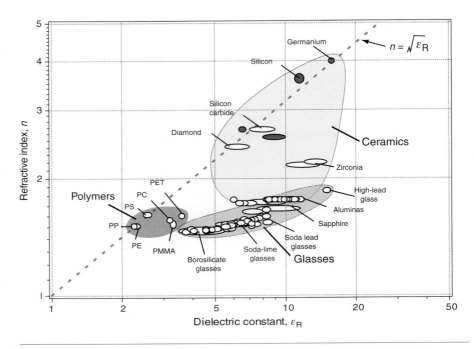

Figure 16.5   A chart of refractive index and dielectric constant, showing the approximate relationship $n = \sqrt{\varepsilon_R}$. (This chart and the next were made using Level 3 of the CES database.)

radiation of longer wavelengths (germanium, for instance, is used for lenses for infrared imaging). Note the wide and continuous range of refractive index of glasses, determined by their chemistry. This ability to control n by manipulating composition is central to the selection of materials for lenses and optical fibers. The relationship between n and ε_R is explored in Section 16.4.

Reflectivity and refractive index
The spectacular refraction and sparkle of diamond comes from its high refractive index, giving reflectance and extensive total internal reflection. Figure 16.6 plots these two properties. As in the previous chart, transparent materials are shown as white circles, those that are opaque as red. As the reflectance depends on n, the materials fall on a characteristic curve. High refractive index gives high reflectivity, and it is here that diamond ($n = 2.42$) excels; this, its unsurpassed hardness and durability ('diamonds are forever') and its scarcity create its unique desirability as jewelry. Costume jewelry uses cheap alternatives that mimic diamond in having high refractive index: zircon, $ZrSiO_4$ ($n = 1.98$) and cubic zirconia, ZrO_2 ($n = 2.17$). Plastic jewelry can't come close; as the chart shows, polystyrene (PS), with $n = 1.58$, has the highest refractive index among

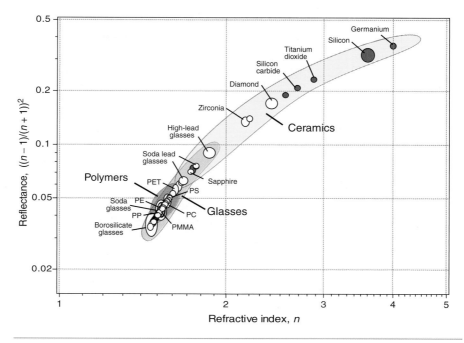

Figure 16.6 A chart of refractive index and reflectance. The materials with open circles are transparent to visible light; those with red circles are transparent to infrared and below, but not to visible radiation. Diamond is special because of its high refractive index, reflectance and transparency in the visible spectrum.

polymers (one reason it is used for the so-called 'jewel' cases in which CDs are packaged) but it is much lower than those of real jewels.

16.4 Drilling down: the physics and manipulation of optical properties

Light, like all radiation, is an electromagnetic (e-m) wave. The coupled fields are sketched in Figure 16.7. The electric part fluctuates with a frequency ν that determines where it lies in the spectrum of Figure 16.1. A fluctuating electric field induces a fluctuating magnetic field that is exactly $\pi/2$ out of phase with the electric one, because the induction is at its maximum when the electric field is changing most rapidly. A plane-polarized beam looks like this one: the electric and magnetic fields lie in fixed planes. Natural light is not polarized; then the wave also rotates so that the plane containing each wave continuously changes.

Many aspects of radiation are most easily understood by thinking of it as a wave. Others need a different picture—that of radiation as discrete packets of

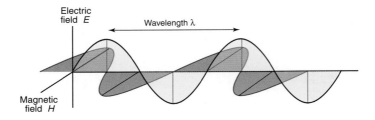

Figure 16.7   An electromagnetic wave. The electric component (gray) is $\pi/2$ out of phase with the magnetic component (red).

energy, *photons*. The idea of a wave that is also discrete energy units is not intuitive—it is another of the results of quantum theory that do not correspond to ordinary experience. The energy E_{ph} of a photon of radiation of frequency ν or wavelength λ is

$$E_{ph} = h\nu = \frac{hc}{\lambda} \qquad (16.9)$$

where h is Planck's[4] constant (6.626×10^{-34} J.s) and c is the speed of the radiation. Thus, radiation of a given frequency has photons of fixed energy, regardless of its intensity—an intense beam simply has more of them. This is the key to understanding reflection, absorption and transmission.

Why aren't metals transparent?

Recall from Chapter 14 that the electrons in materials circle their parent atom in orbits with discrete energy levels, and only two can occupy the same level. Metals have an enormous number of very closely spaced levels in their conduction band; the electrons in the metal only fill part of this number. Filling the levels in a metal is like pouring water into a container until it is part full—its surface is the Fermi[5] level; levels above it are empty. If you 'excite' the water—say, by shaking the container—some of it can slosh to a higher level. If you stop sloshing, it will return to its Fermi level.

Radiation excites electrons; and in metals there are plenty of empty levels in the conduction band into which they can be excited. But here quantum effects cut in. A photon with energy $h\nu$ can excite an electron only if there is an energy level that is exactly $h\nu$ above the Fermi level—and in metals there is. So all the photons of a light beam are captured by electrons of a metal, regardless of their wavelength. Figure 16.8 shows, on the left, what happens to just one.

---

[4] Max Karl Ernst Ludwig Planck (1858–1947), a central figure in the development of quantum theory; it was he who formulated equation (16.9).

[5] Enrico Fermi (1901–1954), devisor of the statistical laws known as Fermi statistics governing the behavior of electrons in solids. He was one of the leaders of the team of physicists on the Manhattan Project for the development of the atomic bomb.

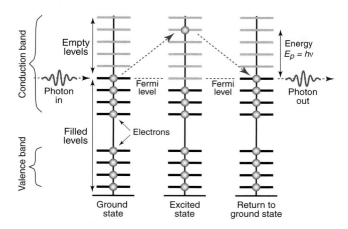

Figure 16.8   Metals absorb photons, capturing their energy by promoting an electron from the filled part of the conduction band into a higher, empty level. When the electron falls back, a photon is re-emitted.

What next? Shaken water settles back, and electrons do the same. In doing so it releases a photon with exactly the same energy that excited it in the first place, but in a random direction. Any photons moving into the material are immediately recaptured, so none make it more than about 0.01 μm (about 30 atom diameters) below the surface. All, ultimately, re-emerge from the metal surface—that is, they are reflected. Many metals—silver, aluminum and stainless steel are examples—reflect all wavelengths almost equally well, so if exposed to white light they appear silver. Others—copper, brass, bronze—reflect some colors better than others and appear colored because, in penetrating this tiny distance into the surface, some wavelengths are slightly absorbed.

Reflection by metals, then, has to do with electrons at the top of the conduction band. These same electrons provide electrical conduction. For this reason, the best metallic reflectors are the metals with the highest electrical conductivities—the reason that high-quality mirrors use silver and cheaper ones use aluminum.

How does light get through dielectrics?

If electrons snatch up photons, how is that some materials are transparent—light goes straight through them? Non-metals interact with radiation in a different way. Part may be reflected but much enters the material, inducing both dielectric and magnetic responses. Not surprisingly, then, the velocity of an e-m wave depends on the dielectric and magnetic properties of the material through which it travels. In a vacuum the velocity is

$$c_{\mathrm{o}} = \frac{1}{\sqrt{\varepsilon_{\mathrm{o}}\,\mu_{\mathrm{o}}}}$$ (16.10)

where ε_o is the electric permittivity of a vacuum and μ_o its magnetic permeability, defined in Chapters 14 and 15. Within a material its velocity is

$$c = \frac{1}{\sqrt{\varepsilon\,\mu}} \qquad (16.11)$$

where $\varepsilon = \varepsilon_R\,\varepsilon_o$ and $\mu = \mu_R\,\mu_o$ are the permittivity and permeability of the material. The refractive index, therefore, is

$$n = \frac{c_o}{c} = \sqrt{\varepsilon_R\,\mu_R} \qquad (16.12)$$

The relative permeability μ_R of most dielectrics is very close to unity—only the magnetic materials of Chapter 15 have larger values. Thus,

$$n \approx \sqrt{\varepsilon_R} \qquad (16.13)$$

The chart of Figure 16.5 has n as one axis and ε_R as the other. The diagonal line is a plot of this equation (the log scales make it a straight line). Polymers and very pure materials like diamond, silicon and germanium lie close to the line but the agreement with the rest is not so good. This is because refractive index and dielectric constant depend on frequency. Refractive index is usually measured optically, and that means optical frequencies, around 10^{15} Hz. Dielectric constants are more usually measured at radio frequencies or below—10^6 Hz or less. If the two are measured at the same frequency, equation (16.13) holds.

The reason that radiation of certain wavelengths can enter a dielectric is that its Fermi level lies at the top of the valence band, just below a band gap (Chapter 14 and Figure 16.9). The conduction band, with its vast number of empty levels, lies above it. To excite an electron across the gap requires a photon with an energy at least as great as the width of the gap, ΔE_{gap}. Thus, radiation with photon energy less than ΔE_{gap} cannot excite electrons—there are no energy states within the gap for the electron to be excited into. The radiation sees the material as transparent, offering no interaction of any sort, so it goes straight through.

Electrons are, however, excited by radiation with photons energies greater than ΔE_{gap} (i.e. higher frequency, shorter wavelength). These have enough energy to pop electrons into the conduction band, leaving a 'hole' in the valence band from which they came. When they jump back, filling the hole, they emit radiation of the same wavelength that first excited them, and for these wavelengths the material is not transparent (Figure 16.9, right-hand side). The critical frequency ν_{crit}, above which interaction starts, is given by

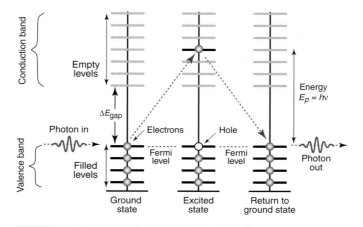

Figure 16.9   Dielectrics have a full band, separated from the empty conduction band by an energy gap. The material cannot capture photons with energy less than ΔE_{gap}, meaning that, for those frequencies, the material is transparent. Photons with energy greater than ΔE_{gap} are absorbed, as illustrated here.

$$h\nu_{crit} = \Delta E_{gap} \qquad (16.14)$$

The material is opaque to frequencies higher than this. Thus, bakelite is transparent to infrared light because its frequency is too low and its photons too feeble to kick electrons across the band gap, but the visible spectrum has higher frequencies with more energetic photons, exceeding the band gap energy; they are captured and reflected.

Although dielectrics can't absorb radiation with photons of energy less than that of the band gap, they are not all transparent. Most are polycrystalline and have a refractive index that depends on direction; then light is *scattered* as it passes from one crystal to another. Imperfections, particularly porosity, do the same. Scattering, sketched in Figure 16.10, makes the material appear translucent or even opaque, even though, when made as a perfect single crystal, it is completely transparent. Thus, sapphire (alumina, Al_2O_3), used for watch crystals and cockpit windows of aircraft, is transparent, but the polycrystalline, slightly porous form of the same material that is used for electronic substrates is translucent or opaque. Scattering explains why some polymers are translucent or opaque: their microstructure is a mix of crystalline and amorphous regions with different refractive indices. It explains, too, why some go white when you bend them: it is because light is scattered from internal microcracks—the crazes described in Chapter 6.

Color

If a material has a band gap with an energy ΔE_{gap} that lies within the visible spectrum, the wavelengths with energy greater than this are absorbed and those

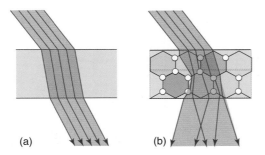

Figure 16.10   (a) A pure glass with no internal structure and a wide band gap is completely transparent. (b) Light entering a polycrystalline ceramic or a partly crystalline polymer suffers multiple refraction and is also scattered by porosity, making it translucent.

with energy that is less are not. The absorbed radiation is re-emitted when the excited electron drops back into a lower energy state, but this may not be the one it started from, so the photon it emits has a different wavelength than the one that was originally absorbed. The light emitted from the material is a mix of the transmitted and the re-emitted wavelengths, and it is this that gives it a characteristic color.

More specific control of color is possible by doping—the deliberate introduction of impurities that create a new energy level in the band gap, as in Figure 16.11. Radiation is absorbed as before but it is now re-emitted in two discrete steps as the electrons drop first into the dopant level, emitting a photon of frequency $\nu_1 = \Delta E_1/h$ and from there back into the valence band, emitting a second photon of energy $\nu_2 = \Delta E_2/h$. Particularly pure colors are created when glasses are doped with metal ions: copper gives blue; cobalt, violet; chromium, green; manganese, yellow.

Fluorescence, phosphorescence and electro luminescence

Electrons can be excited into higher energy levels by incident photons, provided they have sufficient energy. Energetic electrons, like those of the electron beam of a cathode ray tube, do the same. In most materials the time delay before they drop back into lower levels, re-emitting the energy they captured, is extremely short, but in some there is a delay. If, on dropping back, the photon they emit is in the visible spectrum, the effect is that of *luminescence*—the material continues to glow even when the incident beam is removed. When the time delay is fractions of seconds, it is called *fluorescence*, used in fluorescent lighting, where it is excited by ultraviolet from a gas discharge, and in TV tubes where it is excited by the scanning electron beam. When the time delay is longer it is called *phosphorescence*; it is no longer useful for creating moving images but is used instead for static displays like that of watch faces, where it is excited by electrons (β-particles) released by a mildly radioactive ingredient in the paint.

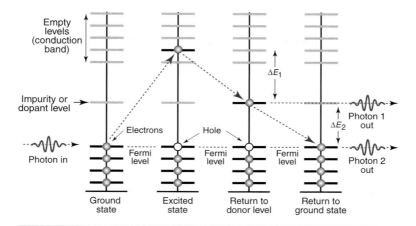

Figure 16.11 Impurities or dopants create energy levels in the band gap. Electron transitions to and from the dopant level emit photons of specific frequency and color.

Photo-conductivity

Dielectrics are true insulators only if there are no electrons in the conduction band, since if there are any, a field will accelerate them, giving an electric current. Dielectrics with a band gap that is sufficiently narrow that the photons of visible light excite electrons across it become conducting (though with high resistance) when exposed to light. The greater the intensity of light, the greater the conductivity. Photo light-meters use this effect; the meter is simply a bridge circuit measuring the resistance of a photo-conducting element such as cadmium sulfide.

16.5 Optical design

Using reflection and refraction

Telescopes, microscopes, cameras and car headlights all rely on the focusing of light. Reflecting telescopes and car headlights use metallized glass or plastic surfaces, ground or molded to a concave shape; the metal, commonly, is silver because of its high reflectivity across the entire optical spectrum.

Refracting telescopes, microscopes and cameras use lenses. Here the important property is the refractive index and its dependence on wavelength (since, if this is large, different wavelengths of light are brought to a focus in different planes), and the reflectivity (since reflected light is lost and does not contribute to the image). As the chart of Figure 16.5 shows, most elements and compounds have a fixed refractive index. Glasses are different: their refractive index can be tuned to any value between 1.5 and more than 2. Adding components with

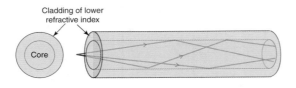

Figure 16.12   An optical fiber. The graded refractive index traps light by total internal reflection, making it follow the fiber even when it is curved.

light elements like sodium (as in soda-lime glass) gives a low refractive index; adding heavy elements like lead (as in lead glass or 'crystal') gives a high one.

Using total internal reflection
Optical fibers have revolutionized the digital transmission of information: almost all landlines now use these rather than copper wires as the 'conductor'. A single fiber consists of a core of pure glass contained in a cladding of another glass with a lower refractive index, as in Figure 16.12. A digitized signal is converted into optical pulses by a light-emitting diode and is fed into the fiber, where it is contained within the core, even when the fiber is bent, because any ray striking the core–cladding interface at a low angle suffers total internal reflection. The purity of the core—a silica glass—is so high that absorption is very small; occasional repeater stations are needed to receive, amplify and retransmit the signal to cover long distances.

16.6 Summary and conclusions

Materials interact with electromagnetic radiation in several ways: reflecting it, refracting it and absorbing it. This is because radiation behaves both as a wave of frequency ν and as a stream of discrete photons with energy $E_{ph} = h\nu$. Electrons in materials capture photons, grabbing their energy, provided there are energy levels exactly E_{ph} above their ground state for them to occupy. After a time delay that is usually very short, the electrons drop back down to their ground state, re-emitting their energy as new photons. In metals there are always usable energy levels so every photon hitting a metal is captured and thrown back out, giving reflection. In a dielectric, the atoms ignore low-frequency, low-energy photons and capture only those with energy above a critical level set by the band gap. Those that are not caught pass straight through; for these frequencies, the material is transparent. Doping introduces extra slots within the band gap for electrons to fall into as they release their energy, and in doing so they emit photons of specific frequency and color.

When radiation enters a dielectric, it slows down. A consequence of this is that a beam entering at an angle is bent—the phenomenon of refraction. It is this that allows light to be focused by lenses, reflected by prisms and trapped in fine, transparent fibers to transmit information.

Of all the transparent materials at our disposal, glasses offer the greatest range of refractive index and color. Glasses are based on amorphous silica, SiO_2. Silica is an extremely good solvent, allowing a wide range of other oxides to be dissolved in it over a wide range of concentrations. It is this that allows the optical properties of glasses to be adjusted and fine-tuned to match design needs.

16.7 Further reading

Callister, W.D., Jr (2003) *Materials science and engineering*, An *Introduction*, 6th edition, John Wiley, New York, USA. ISBN 0-471-13576-3. (*A well-established and comprehensive introduction to the science of materials.*)

Jiles, D. (2001) *Introduction to the Electronic Properties of Materials*, Nelson Thompson, Cheltenham, UK. ISBN 0-7487-6042-3. (*A refreshingly direct approach to the thermal, electrical, magnetic and optical properties of materials and their electronic origins.*)

16.8 Exercises

Exercise E16.1	Define refractive index. Give examples of devices that make use of refraction.
Exercise E16.2	Define reflectance. Give examples of devices that make use of reflection.
Exercise E16.3	The absorption coefficient of polyethylene for optical frequencies is $86.6\,m^{-1}$. How thick a slab of polyethylene is required to reduce the transmitted light intensity to one-half of its initial value?
Exercise E16.4	Why are metals good reflectors of radiation?
Exercise E16.5	It is proposed to replace soda glass windows of a greenhouse with polycarbonate (PC). Will the PC windows reflect more? The refractive index of the glass is 1.5 and that of PC 1.6.
Exercise 16.6	Waterford glass and Steuben glass, used for expensive ornamental and cut-glass objects, are high-lead glasses, meaning that they contain oxides of lead. Using the information shown on the chart of Figure 16.6, can you explain the choice of composition?
Exercise E16.7	An X-ray system has a beryllium window to transmit the beam. The absorption coefficient of beryllium for the wavelength of X-rays of interest here is $3.02 \times 10^2\,m^{-1}$. If the window is 2 mm thick, what fraction of the incident beam intensity will pass through the window? The rest of the equipment is shielded with 4 mm of lead, with absorption coefficient for X-rays of

$3.35 \times 10^6 \, \text{m}^{-1}$. What fraction of the intensity of the incident beam will escape through the casing?

Exercise E16.8 What principle and material would you choose to make a light-sensing switch for a greenhouse?

Exercise E16.9 What principle and material would you choose to make a heat-sensing switch to turn the lights off in the garage when no one is moving around in it?

Exercise E16.10 An optical fiber has a glass core with a refractive index of $n_1 = 1.48$, clad with a glass with refractive index $n_2 = 1.45$. What is the maximum angle that the incoming optical signal can deviate from the axis of the core while still remaining trapped in it?

16.9 Exploring design with CES (use Level 2 unless otherwise stated)

Exercise E16.11 Use the 'Search' facility to find materials for:
(a) Lenses.
(b) Mirrors.

Exercise E16.12 Explore the best choice of cheap polymer for a new line of injection-molded costume jewelry. Its transparency must be of optical quality, and it should have the highest possible refractive index. The table summarizes the requirements.

Function	• Plastic costume jewelry
Constraints	• Transparency of optical quality
	• Injection moldable
	• Price < $5/kg
Objective	• Maximize refractive index
Free variable	• Choice of material

Exercise E16.13 A material is required for the mirror backing of a precision reflecting telescope. It must have a modulus of at least 50 GPa so that it does not deflect under its own weight; it must be hard so that the surface (which will be silvered), once ground, does not distort; it must have the lowest possible thermal expansion to minimize thermal distortion and it must be able to be molded or cast to its initial shape before grinding. Use CES to find suitable candidates.

Function	• Mirror backing
Constraints	• Young's modulus >50 GPa
	• Processing: moldability 5
Objective	• Highest hardness with lowest thermal expansion coefficient
Free variable	• Choice of material

Exercise E16.14 Find materials that are optically clear, have a moldability rating of 5 and have excellent resistance to fresh and to salt water for use as contact lenses.

16.10 Exploring the science with CES Elements

Exercise E16.11 The Hagen–Rubens law—an empirical law—says that the reflectivity of metals, R, is given by

$$R \approx 1 - 0.02\sqrt{\rho_e}$$

(where the electrical resistivity ρ_e is in $\mu\Omega$.cm). Make a chart with R on the y-axis and electrical conductivity $1/\rho_e$ on the x-axis. Use it to find the three elements with the highest reflectivities.

Exercise E16.16 The text explained why the refractive index n was related to the dielectric constant ε_R by the equation

$$n = \sqrt{\varepsilon_R}$$

Make a chart with $\sqrt{\varepsilon_R}$ on the x-axis and n on the y-axis for the elements to see how accurate this relationship is. Which elements best obey it?

Chapter 17

Durability: oxidation, corrosion and degradation

Corrosion. (Image courtesy of Norbert Wodhnl ©
Norbert Wodhnl)

17.1 Introduction and synopsis

The Gospel according to St Matthew reminds us that we live in a world in which 'moth and rust doth corrupt'—a world of corrosion, degradation and decay (cover picture). Not a happy thought with which to start a chapter. But by understanding them we can, to a degree, control them. This chapter describes ways in which materials degrade or corrode, how they are measured and what can be done to slow them down.

Start with the first: the ways in which materials degrade. There are many. There is the damaging effect of radiation and heat, not in changing the properties in a reversible way as in the last chapter, but in causing chemical changes that destroy them irreversibly. There is attack by aggressive chemicals: acids, alkalis and, in the case of polymers, organic solvents. And there is aqueous corrosion, even in pure fresh water, which, provided oxygen is dissolved in it, can be remarkably aggressive.

Because many different properties are involved here, the exploration of the science is split into two, one delving into oxidation, flammability and photo-degradation, the other into corrosion by liquids. Each section ends with a description of ways in which degradation can be inhibited.

17.2 Oxidation, flammability and photo-degradation

The most stable state of most elements is as an oxide. For this reason the earth's crust is almost entirely made of simple or complex oxides: silicates like granite, aluminates like basalt, carbonates like limestone. Techniques of thermochemistry, electrochemistry and synthesis allows these to be refined into the materials we use in engineering, but they are not, in general, oxides. From the moment they are made to re-oxidize, some extremely slowly, others more quickly; and the hotter they are, the faster it happens. For safe high-temperature design we need to understand rates of oxidation.

Definition and measurement

Oxidation rates of metals are measured in the way sketched in Figure 17.1: a thin sheet of the material (to give a lot of surface) is held at temperature T for an increasing time t, and the gain or loss in weight, Δm, is measured. If the oxide adheres to the material the sample gains weight in a way that is either linear ($\Delta m \propto t$) or parabolic ($\Delta m \propto t^{1/2}$) in time t; if instead the oxide is volatile, the sample loses weight linearly with time ($\Delta m \sim -t$), as in Figure 17.2.

Polymers, too, oxidize, but in a more spectacular way.

Flammability

Ceramics and glasses do not burn. Metals, if dispersed as a fine powder, are potentially combustible, but in bulk flammability is not normally an issue, even for magnesium. Most polymers, on the other hand, are inherently flammable,

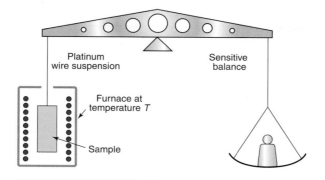

Figure 17.1   Measuring oxidation rates.

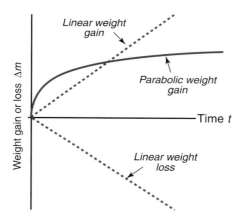

Figure 17.2   Oxidation rates: linear weight gain, parabolic gain and linear loss.

although to differing degrees: some burn spontaneously if ignited; others are self-extinguishing, burning only when directly exposed to flame.

There are several ways to characterize flammability. The most logical is the *limiting oxygen index (LOI)*: it is the oxygen concentration, in %, required to maintain steady burning. Fresh air has about 21% oxygen in it. A polymer with an oxygen index lower than this will burn freely in air; one with an oxygen index that is larger will extinguish itself unless a flame is played onto it—then it burns. Thus, a high oxygen index means resistance to self-sustained burning. Less logical, but much used, is the Underwriters Laboratory (UL) rating, in which a strip of polymer 1.6 mm thick is held horizontally (H) or vertically (V) and ignited. Its response is recorded as a code such as HB, meaning 'horizontal burn'. The scales are compared in Table 17.1.

Table 17.1   Flammability ratings compared

Oxygen index (LOI)	UL 94 (1.6 mm) rating	Flammability
Up to 16	Unrated	Flammable
16–20	HB	
21–24	HB	
25–29	V-2	Self-extinguishing
30–43	V-0	
Over 44	V-0 or better (e.g. 5 V)	

Photo-degradation

You don't have to set fire to a polymer for it to oxidize. Polymers and elastomers age when exposed to light (particularly UV) and oxygen, causing loss of strength, stiffness and toughness, discoloration and loss of gloss. This is countered by additives: *antioxidants*, *light stabilizers* and *fluorescent whitening agents*. Some of these are so universal that the 'standard' grade of the polymer already contains enough of one or another of them to give it acceptable resistance; PP, ABS, PS, PET, PMMA, PC, nylons and PU all need UV protection.

17.3 Oxidation mechanisms

Mechanisms of oxidation

The rate of oxidation of most metals at room temperature is too slow to be an engineering problem; indeed, it can be beneficial in protecting metals from corrosion of other sorts. But heat them up and the rate of oxidation increases, bringing problems. The driving force for a material to oxidize is its *free energy of oxidation*—the energy released when it reacts with oxygen—but a big driving force does not necessarily mean rapid oxidation. The rate of oxidation is determined by the *kinetics* (rate) of the oxidation reaction, and that has to do with the nature of the oxide. When any metal (with the exception of gold, platinum and a few others that are even more expensive) is exposed to air, an ultra-thin surface film of oxide forms on it immediately, following the oxidation reaction

$$M(metal) + O(oxygen) \rightarrow MO(oxide) + energy \qquad (17.1)$$

The film now coats the surface, separating the metal beneath from the oxygen. If the reaction is to go further, oxygen must get through the film.

Figure 17.3 Oxidation mechanisms. (a) Growth by metal diffusion and electron conduction (b) Growth by diffusion of oxygen and holes.

The weight gain shown in Figure 17.2 reveals two different types of behavior. For some metals the weight gain is linear, and this implies that the oxidation is progressing at a constant rate:

$$\frac{dm}{dt} = k_\ell \quad \text{giving} \quad \Delta m = k_\ell t \tag{17.2}$$

where k_ℓ is the *linear kinetic constant*. This is because the oxide film cracks (and, when thick, spalls off) and does not protect the underlying metal. Some metals behave better than this. The film that develops on their surfaces is compact, coherent and strongly bonded to the metal. For these the weight gain is parabolic, slowing up with time, and this implies an oxidation rate with the form

$$\frac{d(\Delta m)}{dt} = \frac{k_p}{\Delta m} \quad \text{giving} \quad \Delta m^2 = k_p t \tag{17.3}$$

where k_p is the *parabolic kinetic constant*. The film, once formed, separates the metal from the oxygen. To react further, either oxygen atoms must diffuse inward through the film to reach the metal or metal atoms must diffuse outward through the film to reach the oxygen. The driving force is the free energy of oxidation, but the rate of oxidation is limited by the rate of diffusion, and the thicker the film, the longer this takes.

Figure 17.3 shows a growing oxide film. The reaction creating the oxide MO

$$M + O = MO$$

goes in two steps. The metal first forms an ion, M^{2+} say , releasing electrons:

$$M = M^{2+} + 2e$$

The electrons are then absorbed by oxygen to give an oxygen ion:

$$O + 2e = O^{2-}$$

The problem is that the first of these reactions occurs at the metal side of the oxide, whereas the oxygen is on the other side. Either the metal ions and the electrons must diffuse out to meet the oxygen or the oxygen and electron holes (described in Chapter 14) must diffuse in to find the metal. If the film is an electrical insulator, as many oxides are, electrons cannot move through it, so it is the oxygen that must diffuse in. The concentration gradient of oxygen is that in the gas C_o divided by the film thickness, x. The rate of growth of the film is proportional to the flux of atoms diffusing through the film, giving

$$\frac{dx}{dt} \propto D\frac{C_o}{x} = K_o\left(\exp-\frac{Q_d}{RT}\right)\frac{C_o}{x} \qquad (17.4)$$

where D is the diffusion coefficient, K_o is a kinetic constant and Q_d is the activation energy for oxygen diffusion. Integrating gives

$$x^2 = k_p t$$

This has the same form as equation (17.3), with

$$k_p \propto \exp-\frac{Q_d}{RT} \qquad (17.5)$$

This explains why oxidation rates rise steeply with rising temperature, and why the growth is parabolic. The most protective films are those with low diffusion coefficients, and this means high melting points. This is why the Al_2O_3 oxide film on aluminum, the Cr_2O_3 film on stainless steel and chrome plate, and the SiO_2 film on high silicon cast iron are so protective.

Not all oxides grow with parabolic kinetics, as we have seen. Those that show linear weight gain do so because the oxide that forms is not compact, but has cracks or spalls off because of excessive volume change, leaving the fresh surface continually exposed to oxygen. Linear weight loss, the other behavior, occurs when the oxide is volatile, and simply evaporates as it forms.

17.4 Making materials that resist oxidation

Elements for heaters, furnace components, power generation and chemical engineering plants all require materials that can be used in air at high temperatures. If it is a pure metal that you want it has to be platinum (and, indeed, some special furnaces have platinum windings) because the oxide of platinum is less stable than the metal itself. If you want something more affordable, it has to be an alloy.

Oxides, of course, are stable at high temperature—they are already oxidized. One way to provide high-temperature protection is to coat metals like cast irons, steels or nickel alloys with an oxide coating. Stoves are protected by enameling (a glass coating, largely SiO_2); turbine blades are coated with plasma-sprayed TBCs based on the oxide zirconia (ZrO_2) (Chapter 13). But coatings of this sort are expensive, and, if damaged, they cease to protect.

There is another way to give oxidation resistance to metals, and it is one that repairs itself if damaged. The oxides of chromium (Cr_2O_3), of aluminum (Al_2O_3), of titanium (TiO_2) and of silicon (SiO_2) have very high melting points, so the diffusion of either the metal or oxygen through them is very slow. They are also electrical insulators so electrons cannot move through them either. The mechanism of oxidation requires both diffusion and conduction, so the rate constant k_p is very small: the oxide stops growing when it is a few molecules thick. These oxides adhere well and are very protective; it is they that make these otherwise reactive metals so passive. The films can be artificially thickened by *anodizing*—an electrochemical process that promotes the oxidation reaction thereby increasing their protective power. Anodized films accept a wide range of colored dyes, giving them a decorative as well as a protective function.

If enough chromium, aluminum or silicon can be dissolved in a metal like iron or nickel—'enough' means 12–20%—a similar protective oxide grows on the alloy. And if the oxide is damaged, more chromium, aluminum or silicon immediately oxidizes, repairing the damage. Stainless steels (typical composition Fe–18% Cr–8% Ni), widely used high-temperature equipment and nichromes (nickel with 10–30% chromium), used for heating elements, derive their oxidation resistance in this way; so, too, do the aluminum bronzes (copper with 10% aluminum) and high-silicon cast irons (iron with 16–18% silicon). To see just how important this way of suppressing oxidation has become, look back for a moment at Figure 13.8. It is the table guiding material choice at low and high temperatures. All the metals above the 400°C mark rely on alloy methods for their protection.

Flammability: how do polymers burn and how do you stop them?

Combustion is an exothermic reaction in which hydrocarbons are oxidized to CO_2 and H_2O. That sounds simple, but it isn't. Combustion is a *gas-phase reaction*; the polymer or its decomposition products must become gaseous for a fire to begin. When you light a candle you are melting the wax and raising it to the temperature at which it *pyrolyzes* (400–800°C), forming gaseous hydrocarbon decomposition products. These gases react in the flame to produce heat, which melts and pyrolyzes more wax, keeping the reaction going.

A fully developed fire results when an ignition source like a spark or cigarette ignites combustible material such as paper. Its heat radiates out, causing other combustible materials (particularly polymers and fabrics) to decompose into a

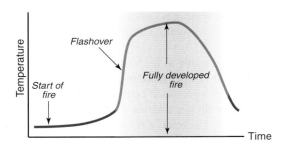

Figure 17.4   The temperature rise during a fire.

flammable gas mix. Flashover (Figure 17.4) occurs when these gases ignite, instantly spreading the fire over the entire area and producing temperatures of over 1000°C.

Combustion involves the reaction of *free radicals*. At high temperatures hydrogen, together with one electron from its covalent bond (symbol $H^\bullet$), is freed from a carbon atom of the polymer molecule, leaving a highly reactive free radical $R^\bullet$ in the gas. The hydrogen radical reacts with oxygen and the hydrocarbon radical to give CO_2 and heat, releasing the $H^\bullet$ again to propagate the reaction further.

Flame retardants work in one of two ways. Some scavenge the free radicals, tying them up harmlessly and so retarding or snuffing out the combustion reaction. These are usually compounds containing chlorine or bromine; they, too, pyrolyze to give free radicals $Cl^\bullet$ and $Br^\bullet$ that attach themselves to the hydrocarbon radical, removing it from the reaction. Others work by creating a protective layer of water vapor between the solid polymer and the gaseous decomposition products, limiting heat transfer, cooling it and reducing pyrolization. Typical among these is the addition of $Mg(OH)_2$ that decomposes at about 300°C, releasing H_2O and leaving inert MgO.

What causes photo-degradation?

As we have seen, commercial polymers are long-chain, high molecular weight hydrocarbons. When exposed to radiation chemical reactions are triggered that change their chemical composition and molecular weight. These reactions, called *photo-oxidation* or *photo-degradation*, also create free radicals. They trigger a change in the physical and optical properties of the polymer, making it brittle and, if transparent, turning it white or gray. Once this starts it sets off a chain reaction that accelerates degradation unless stabilizers are used to interrupt the oxidation cycle. Heat, too, can trigger degradation in an oxygen-containing atmosphere.

Ultraviolet absorbers such as benzophenone or benzotriazole work by preferentially absorbing UV radiation, following the grandly named Beer–Lambert law, which states the obvious: that the amount of UV radiation absorbed increases with increase in the sample thickness and stabilizer concentration. In practice, high concentrations of absorbers and sufficient thickness of the polymer are required before enough absorption takes place to effectively retard photo-degradation. *Hindered amine stabilizers* (HALS) are a more efficient way to stabilize against light-induced degradation of most polymers. They do not absorb UV radiation, but act to inhibit degradation of the polymer; their efficiency and longevity are due to a cyclic process wherein the HALS are regenerated rather than consumed. Low concentrations are enough to give good stabilization.

17.5 Corrosion: acids, alkalis, water and organic solvents

Acids, alkalis and aqueous liquids

Strong acids like H_2SO_4 (sulfuric acid), HNO_3 (nitric acid) and HCl (hydrochloric acid) are widely used in the chemical industry. So too are strong alkalis like NaOH (caustic soda). Household products contain some of these in dilute form, but in the home it is organic acids like vinegar (acetic acid) and milder alkalis like washing soda (sodium carbonate) that are more commonly found.

Dunk a metal into an acid and you can expect trouble. The sulfates, nitrates, chlorides and acetates of most metals are more stable than the metal itself. Zinc, for instance, reacts with sulfuric acid, releasing hydrogen. The question is not whether the metal will be attacked, but how fast. As with oxidation, this depends on the nature of the corrosion products and on any surface coating the metal may have. The oxide films on aluminum, titanium and chromium protect them from some acids provided the oxide itself is not damaged.

Alkalis, too, attack some metals. Zinc, tin, lead and aluminum react with NaOH, for instance, to give zincates, stannites, plumbates and aluminates (as anyone who has heated washing soda in an aluminum pan will have discovered).

Water is nearly as bad. Even when pure, water causes corrosion if oxygen is available. The rusting of steel alone presents enormous economic and reliability problems. Corrosion, however, can be beautiful: the turquoise patina of copper on roofs and the deep browns of bronze statues are created by corrosion.

Organic solvents

Organic liquids are ubiquitous: we depend on them as fuels and lubricants, for cooking, for removing stains, as face cream, as nail varnish and much more. Any material in service will encounter them. Metals, ceramics and glasses are largely immune to them, and some polymers can tolerate some organic liquids without problems. But not all.

Resistance to organic solvents, usually given as a ranking on a five-point scale from *Very bad* to *Very good*, gives only the broadest indication of susceptibility. The identification of solvents that are incompatible with a given polymer is largely based on experience, trial and error, and intuition guided by such rules as 'like dissolves like' and various definitions of solvent 'strength'.

17.6 Drilling down: mechanisms of corrosion

Ions in solution and pH

Pure water, H_2O, dissociates a little to give a hydrogen ion, H^+, and a hydroxyl ion, OH^-:

$$H_2O \leftrightarrow H^+ + OH^- \tag{17.6}$$

The product of the concentrations of the two ions is constant: increase one and the other falls. This is known as the Law of Mass Action:

$$[H^+].[OH^-] = \text{constant} \tag{17.7}$$

where the square brackets mean 'molar concentration'. In pure water there are equal numbers of the two types of ion, $[H^+] = [OH^-]$, and the value of the constant, when measured, is 10^{-14}. Thus, the molar concentration of both ion types is 10^{-7}; one H_2O molecule in 10^7 is ionized. The pH of the ionized water is defined as the negative of the log of the hydrogen ion concentration:

$$pH = -\log_{10}[H^+] \tag{17.8}$$

so, for pure water, pH = 7.

Acids dissociate in water to give H^+ ions. Sulfuric acid, for instance, dissociates:

$$H_2SO_4 \leftrightarrow 2H^+ + SO_4^{2-} \tag{17.9}$$

This pushes up the concentration $[H^+]$ and, because of equation (17.8), it pulls down the concentration $[OH^-]$; weak acids have a pH of 4–6; strong ones a pH down to 0. Alkalis do the opposite. Sodium hydroxide, for example, dissociates when dissolved in water, to give OH^- ions:

$$NaOH \leftrightarrow Na^+ + OH^- \tag{17.10}$$

Weak alkalis have a pH of 8–10, strong ones a pH up to 13 (Figure 17.5).

Corrosion by acids and alkalis is an electrochemical reaction. One-half of this is the dissociation reaction of a metal M into a metal ion, M^{z+}, releasing electrons e^-:

$$M \rightarrow M^{z+} + ze^-$$

where z, an integer of 1, 2 or 3, is the valence of the metal. Acidic environments, with high $[H^+]$ (and thus low pH) stimulate this reaction; thus, a metal such as copper, in sulfuric acid solution, reacts rapidly:

$$\left.\begin{array}{l} Cu \rightarrow Cu^{2+} + 2e \\ H_2SO_4 \leftrightarrow 2H^+ + SO_4^{2-} \end{array}\right\} \quad Cu^{2+} + SO_4^{2-} \rightarrow CuSO_4 \tag{17.11}$$

Figure 17.5   The scales of pH and pOH.

Some metals are resistant to attack by some acids. This is because the reaction product, here $CuSO_4$, forms a protective surface layer; thus, lead-lined containers are used to process sulfuric acid because lead sulfate is protective.

Alkalis, too, cause corrosion via an electrochemical reaction. Zinc, for instance, is attacked by caustic soda via the steps:

$$\left.\begin{array}{l} Zn \rightarrow Zn^{2+} + 2e \\[6pt] 2NaOH \leftrightarrow 2Na^+ + 2OH^- \end{array}\right\} \quad Zn^{2+} + 2Na^+ + 2OH^- \rightarrow Na_2ZnO_2 + H_2$$

Fresh and impure water

Corrosion, as we have just seen, is the degradation of a metal by an electrochemical reaction with its environment. Figure 17.6 illustrates in more detail the idea of an electrochemical reaction. If a metal is placed in a conducting solution like salt water, it dissociates into ions, releasing electrons, as the iron is shown doing in the figure, via the reaction:

$$Fe \leftrightarrow Fe^{2+} + 2e^- \tag{17.12}$$

The electrons accumulate on the iron, giving it a negative charge that grows until the electrostatic attraction starts to pull the Fe^{2+} ions back onto the metal surface, stifling further dissociation. At this point the iron has a potential (relative to a standard, the *hydrogen standard*) of -0.44 volts. Each metal has its

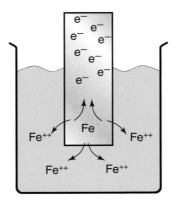

Figure 17.6 Ionization.

own characteristic potential (called the *standard reduction potential*), as plotted in Figure 17.7. The extra electrons enter the band structure of the metal just above the Fermi level, so the energy associated with the Fermi level determines how strongly they are held. The Fermi level also establishes the work function (Chapter 14), so you might expect that the two are related. This relationship is explored in Exercise E17.12 at the end of this chapter.

If two metals are connected together in a cell, like the iron and copper samples in Figure 17.8, a potential difference equal to their separation in Figure 17.7 appears between them. The corrosion potential of iron, -0.44 V, differs from that of copper, $+0.34$ V, by 0.78 volts, so if no current flows in the connection the voltmeter will register this difference. If a current is now allowed, electrons flow from the iron (the *anode*) to the copper (the *cathode*); the iron ionizes (that is, it corrodes), following the anodic reaction of equation (17.12) and—if the solution were one containing copper sulfate—copper ions, Cu^{2+}, plate out onto the copper following the cathodic reaction:

$$Cu^{2+} + 2e^- \leftrightarrow Cu \qquad (17.13)$$

Suppose now that the liquid is not a copper sulfate solution, but just water (Figure 17.9). Water dissolves oxygen, so unless it is specially degassed and protected from air, there is oxygen in solution. The iron and the copper still dissociate until their corrosion potential difference is established but now, if the current is allowed to flow, there is no reservoir of copper ions to plate out. The iron still corrodes but the cathodic reaction has changed; it is now the *hydrolysis reaction*

$$H_2O + O + 2e^- \leftrightarrow 2OH^- \qquad (17.14)$$

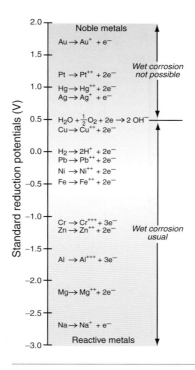

Figure 17.7   Standard reduction potentials of metals.

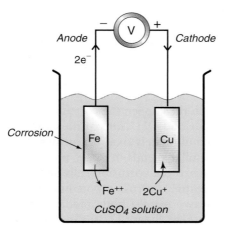

Figure 17.8   A bi-metal corrosion cell. The corrosion potential is the potential to which the metal falls relative to a hydrogen standard.

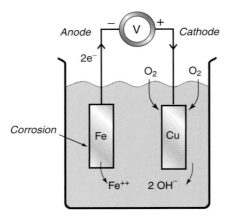

Figure 17.9   A bit-metal cell containing pure water in which oxygen can dissolve.

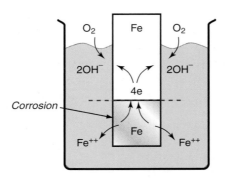

Figure 17.10   A corrosion cell created by differential access to oxygen.

While oxygen can reach the copper, the corrosion reaction continues, creating Fe^{2+} ions at the anode and OH^- ions at the cathode. They react to form insoluble $Fe(OH)_2$, which ultimately oxidizes further to $Fe_2O_3 \cdot H_2O$, rust.

Thus, connecting dissimilar metals in either pure water or water with dissolved salts is a bad thing to do: corrosion cells appear that eat up the metal with the lower (more negative) corrosion potential. Worse news is to come: it is not necessary to have two metals: both anodic and cathodic reactions can take place on the *same* surface. Figure 17.10 shows how this happens. Here an iron sample is immersed in water with access to air. The part of the sample nearest the water surface has an easy supply of oxygen; that further away does not. On the remoter part the ionization reaction of equation (17.12) takes place, corroding the iron and releasing electrons that flow up the sample to the near-surface part where oxygen is plentiful, where they enable the hydrolysis reaction of equation (17.14). The hydrolysis reaction has a corrosion potential of +0.81 volts—it is shown in Figure 17.7—and the difference between this and that of iron, −0.44 volts, drives the corrosion. If the sample could be cut in two along the broken line in Figure 17.10 and a tiny voltmeter inserted, it would register the difference: 1.23 volts.

This differential oxidation corrosion is one of the most usual and most difficult to prevent: where there is water and a region with access to oxygen and one that is starved of it, a cell is set up. Only metals above the hydrolysis reaction potential of +0.81 volts in Figure 17.7 are immune.

Selective corrosion
Often, wet corrosion, instead of being uniform, occurs selectively, and when it does it can lead to failure more rapidly than the uniform rate would suggest. Stress and corrosion acting together ('stress corrosion' and 'corrosion fatigue') frequently lead to localized attack, as do local changes in microstructure such as those at a welded joint. Briefly, the localized mechanisms are these.

Intergranular corrosion occurs because grain boundaries have chemical properties that differ from those of the grain because the disregistry there gives atoms there higher energy. *Pitting corrosion* is preferential attack that can occur at breaks in the natural oxide film on metals, or at precipitated compounds in certain alloys. *Galvanic attack* at the microstructural level appears in alloys with a two-phase microstructure, made up of grains alternately of one composition and of another. The two compositions will, in general, lie at slightly different points on the reduction potential scale. If immersed in a conducting solution, thousands of tiny corrosion cells appear, causing the phase that lies lower in reduction potential to be eaten away.

Stress corrosion cracking is accelerated corrosion, localized at cracks in loaded components. The stress breaks the protective oxide film and the elastic energy at and near the crack tip stimulates attack there. The result is that cracks grow under a stress intensity K_{scc} that is far below K_{1c}. Examples are brass in ammonia, mild steel in caustic soda, and some aluminum and titanium alloys in salt water. *Corrosion fatigue* refers to the accelerated rate at which fatigue cracks grow in a corrosive environment. The endurance limit of some steels in salt water is reduced by a factor as large as 4. The rate of fatigue crack growth, too, increases to a level that is larger than the sum of the rates of corrosion and of fatigue acting together.

17.7 Fighting corrosion

In fighting corrosion, there are a number of dos and don'ts. First the don'ts.

Two different metals, connected electrically and immersed in water with almost anything dissolved in it, create a corrosion cell like that of Figure 17.8.

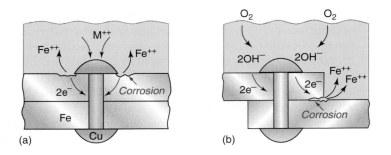

Figure 17.11   (a) Electrochemical corrosion where two different metals are in electrical contact. (b) Crevice corrosion caused by differential access of oxygen.

The metal that lies lower on the scale of reduction potential is the one that is attacked. The rate of attack can be large and localized at and near the contact. Thus, riveting steel with copper rivets risks severe corrosion of the steel around the rivet head (Figure 17.11(a)). Attaching an aluminum body shell to a steel auto chassis risks the same fate, as incautious car-makers have found. The answer is to avoid bi-metal couples if water is around or, when this is impossible, to insulate them electrically from each other. This, too, is not always possible; then one of the dos (below) is needed.

The second don't derives from the mechanism shown in Figure 17.10: if possible, avoid differential oxygen access. Figure 17.11(b) shows a riveted lap joint (the rivet, of course, made from the same metal as the plate). If water can get under the plate edge, it will. The water–air surface has free access to oxygen, but the metal between the plates does not. The result is corrosion of the joint surface, just where it is least wanted. The answer here is to put sealant in the joint before riveting. This is not just a joint problem—differential aeration is very hard to avoid. It is the reason that the legs of jetties and piers rust just below the water line, and why some edges of the tank shown as the cover picture are more severely rusted than others.

Now the dos. If attaching a metal to one lower in reduction potential causes the lower one to corrode, it follows that the upper one is protected. So connecting buried steel pipework to zinc plates sets up a cell that eats the zinc but spares the pipes (Figure 17.12(a)). The expensive bronze propellers of large ships carry no surface protection (the conditions are too violent for it to stay in place). Copper and bronze lie above steel in the reduction potential table and electrical isolation is impossible. Bronze connected to steel in salt water is a recipe for disaster. The solution is to shift the disaster elsewhere by attaching zinc plates to the hull around the propeller shaft. The zinc, in both examples, acts as a *sacrificial anode*, protecting both the bronze and the steel. The sacrificial bit is real—the zinc is consumed—so the protection only lasts as long as the zinc, requiring that it be replaced regularly.

(a)   (b)

Figure 17.12   (a) Protection of steel pipes by a zinc sacrificial anode. (b) Galvanized steel plate; the zinc protects the steel even when scratched.

Plating can work in the same way. *Galvanized iron* is steel sheet with a thin coating of zinc. The zinc acts as a sacrificial anode (Figure 17.12(b)): if scratched, the exposed steel does not rust because the zinc protects it even when it does not cover it. Other platings look good but do not protect so well. A plating of copper sets up a cell the works in the opposite direction, eating the iron at a scratch rather than the copper. Chromium plating looks as if it should protect, since chromium is below iron in the table, but the chromium protects itself so well with its oxide film that it becomes passive and no cell is set up. It is this passivity that makes the stainless alloys described in Section 17.3 good at resisting aqueous corrosion as well. Stainless steel is not just oxidation resistant, it is corrosion resistant too, and the same is true of the other alloys listed there.

When tricks like these cannot be used to prevent corrosion, the answer is to separate the reactive metal from the corrosive medium with a polymer film. Painting with *solvent-based paints* was, until recently, the most widely used method, but the evaporating solvent is generally toxic; it is a VOC (a volatile organic compound) and, for health reasons, their use is discouraged. *Water-based paints* overcome the problem but do not yet produce quite as good a paint film. *Polymer powder coating* works by plasma-spraying the polymer or dipping the component when hot into a fluidized bed of polymer powder, causing a thick film of polymer to melt onto the surface. While intact, paint films and polymer powder coats work well, but as soon as the coating is damaged the protection ceases unless the underlying metal has a sub-coat of something—like zinc—that protects it.

Organic solvents and the solubility parameter

Organic solvents attack certain polymers. The nearest that this can be made scientific is via what are called *solubility parameters*. Polymer chains are bonded strongly (covalent bonding) along their length, but only weakly (van der Waals bonding) between chains. To enter a polymer the solvent, like a virus, must trick these bonds into allowing them in—that is where the 'like dissolves like', mentioned earlier, comes from. Once in—like viruses—they do things: mostly bad, a few, good.

The bad: like UV radiation, aggressive solvents cause discoloration, reduce strength, induce brittleness and trigger crazing (the whitening of the polymer because of many tiny crack-like expansion cavities). The good: certain organic solvents act as plasticizers, reducing the glass temperature but not the strength, converting a rigid polymer into a flexible, leather-like material. Artificial leather—plasticized PVC—is an example. Many plasticizers are phthalates but increasing concern for their potential toxicity generates pressure to use alternatives. Increasingly they are replaced by biochemical plasticizers—vegetable oils such as soybean oil or linseed oil—although these are more expensive.

Which organic solvents are the worst? Not an easy question. Amyl acetate (nail varnish) is particularly aggressive, as is chloroform (anesthetics), but this is no answer. To get a full answer you have to turn to the more specialized CES OPS database (which runs in the same system as that of the one used to make the charts and for the exercises in this book); it gives detailed rankings of a large number of individual polymers in a large number of different organic (and other) gases and liquids.

17.8 Summary and conclusions

Corrosion is like cancer: it will kill you unless something else (like wear) kills you first. Almost as much money must have been invested in corrosion research as in research on cancer, such is its damaging effect on the economy. This research has revealed what goes on at an atomic scale when materials react with oxygen, acids, alkalis, aqueous solutions, aerated water and organic solvents.

All involve electrochemical reactions or reactions involving free radicals. In oxidation of metals, the metal ionizes, releasing electrons; the electrons combine with oxygen molecules to give oxygen ions, that in turn combine with metal ions to form the oxide. As the oxide film grows, ions and electrons must diffuse through it to enable further growth; its resistance to this diffusion, and its ability to adhere to the metal surface, determine how protective it is. Polymers oxidize through the action of free radicals. Radiation, including UV light, can generate the free radicals, as does heat. Polymers are protected by doping them with UV filters and with additions that absorb free radicals.

In corrosion, the electrochemical reactions occur in the corrosive fluid rather than in the solid. A metal ionizes when placed in water. The extent of ionization determines its reduction potential. Those with the smallest (most negative) reduction potential are the most vulnerable to attack and will corrode spontaneously, even in pure water if oxygen is present. If a metal with a low reduction potential is in electrical contact with one of higher potential, the first corrodes and the second is protected.

In designing materials to resist corrosion we rely on the ability of a few of them—chromium, aluminum, silicon and titanium—to form a protective oxide film that adheres strongly to the surface. By using these as alloying elements, other metals can be given the same protection. Alternatively, metals can be protected by coating their surfaces with a corrosion-resistant film of polymer or ceramic.

17.9 Further reading

Bradford, S.A. (1993) *Corrosion Control*, Van Nostrand Reinhold, New York, USA. ISBN 0-442-01088-5. *(A text emphasizing practical ways of preventing or dealing with corrosion.)*

Fontanta, M.G. (1986) *Corrosion Engineering*, 3rd edition, McGraw-Hill, New York, USA. ISBN 0-07-021463-8. *(A text focusing on the practicalities of corrosion engineering rather than the science, with numerous examples and data.)*

Schweitzer, P.A. (1995) *Corrosion Resistance Tables*, 4th edition, Volumes 1–3, Marcel Dekker, New York, USA. *(The ultimate compilation of corrosion data in numerous environments. Not bedtime reading.)*

Schweitzer, P.A. (1998) *Encyclopedia of Corrosion Technology*, Marcel Dekker, New York, USA. ISBN 0-8247-0137-2. *(A curious compilation, organized alphabetically, that mixes definitions and terminology with tables of data.)*

Tretheway, K.R. and Chamberlain, J. (1995) *Corrosion for Science and Engineering*, 2nd edition, Longman Scientific and Technical, Harlow, UK. ISBN 0-582-238692. *(An unusually readable introduction to corrosion science, filled with little benchtop experiments to illustrate principles.)*

Waterman, N.A. and Ashby, M.F. (1991) *Elsevier Materials Selector*, Elsevier, Oxford, UK. ISBN 1-85-166-605-2 and, in the CRC edition, ISBN 0-8493-7790-0. *(A three-volume compilation of materials data for design, with extensive tables and guidelines for the oxidation and corrosion characteristics of metals and polymers.)*

17.10 Exercises

Exercise E17.1	By what mechanisms do metals oxidize? What determines the rate of oxidation?
Exercise E17.2	The oxidation kinetics of titanium to TiO_2 is limited by oxygen diffusion, with an activation energy Q_d of 275 kJ/mol. If the oxide film grows to a thickness of 0.08 μm after 1 hour at 800°C, how thick a film would you expect if it had been grown at 1000°C for 30 minutes?
Exercise E17.3	What is meant by the standard reduction potential? A copper and a platinum electrode are immersed in a bath of dilute copper sulfate. What potential difference would you expect to measure between them? If they are connected so that a current can flow, which one will corrode?

17.11 Exploring design with CES (use Level 2 unless otherwise stated)

Exercise E17.4 Find, by browsing or searching, the record for the nickel–chromium alloys called nichromes. What are their main applications?

Exercise E17.5 Find, by browsing or searching, the record for polymer powder coating (remember to search in the Process Universe, not the Materials Universe). What are the three ways of applying a polymer powder coating?

Exercise E17.6 Use a 'Limit' stage, applied to the Level 2 Surface Treatment data table, to find surface treatment processes that impart resistance to gaseous corrosion.

Exercise E17.7 Use a 'Limit' stage, applied to the Level 2 Surface Treatment data table to find surface treatment processes that impart resistance to aqueous corrosion.

Exercise E17.8 Use the CES 'Search' facility to find materials for food processing equipment.

Exercise E17.9 Plastic cases for electrical plugs and switch-gear should not be made of flammable materials. Use the 'Select' facility in CES to find polymers that are non-flammable or self-extinguishing.

Exercise E17.10 A vat is required to hold hot caustic soda, NaOH, a strong alkali. Use the 'Select' facility in CES to find metals that resist strong alkalis very well.

Exercise E17.11 Pipework is required for a gherkin-pickling plant to carry vinegar (a weak acid) at 100°C from one vat to another. The liquid is under pressure, requiring a material with a strength of at least 100 MPa, and for ease of installation it must be able to be bent, requiring a ductility of at least 10%. Find the four cheapest materials that meet the constraints, summarized below.

Function	• Pipework for hot acetic acid
Constraints	• Durability in weak acid = very good
	• Maximum operating temperature > 100°C
	• Yield strength > 100 MPa
	• Elongation > 10%
Objective	• Minimize material price
Free variable	• Choice of material

17.12   Exploring the science with CES Elements

Exercise E17.12   The work function (Chapter 14) is the energy required to pluck an electron from the top of the Fermi level of a crystal and drag it away until it is isolated in vacuum. The standard reduction potential of Figure 17.7 involves electrons dropping into energy levels just above the Fermi level. You might suspect that the two were, in some way, related. Make a graph with standard reduction potential on the x-axis and work function on the y-axis to find out if they are. What is your conclusion?

Chapter 18

Heat, beat, stick and polish: manufacturing processes

A 1000 tonne forging press. (Image courtesy of the Fu Sheng Group, Sporting Goods Division.)

18.1 Introduction and synopsis

Materials and processing are inseparable. Ores, minerals and oil are processed to create the stuff of engineering in clean, usable form. These are shaped into components, finished and joined to make products. Chapter 3 defined a hierarchy for these processes; in this chapter we explore their technical capabilities and methods for choosing which to use. Increasingly, legislation and public opinion require that materials are recycled at the end of product life. New processes are emerging to extract usable material from what used to be regarded as waste.

The strategy for choosing a process parallels that for materials: *screen out* those that cannot do the job, *rank* those that can (usually seeking those with the lowest cost) and explore the *documented knowledge* of the top-ranked candidates. Many technical characteristics of a process can be described by numeric and non-numeric data: the size of part it can produce, the surface finish of which it is capable and, critically, the material families or classes it can handle and the shapes it can make. Constraints on these provide the inputs for screening, giving a short list of processes that meet the requirements of the design.

There is another aspect to this. It relates to the influence of processing on the properties of the materials themselves. Materials, of course, have properties before they are shaped, joined or finished and processing is used to enhance properties such as strength. But the act of processing can also damage properties by introducing porosity, cracks and defects or by changing the microstructure in a detrimental way. These interactions, central to the choice and control of processes, are explored further in Chapter 19. For now, it is sufficient to recognize that success in manufacturing often relies on accumulated know-how and expertise, resident, in the past, in the heads of long-established employees but now captured as design guidelines, best practice guides and software providing the information needed for the documentation stage.

18.2 Process selection in design

The selection strategy

The taxonomy of manufacturing processes was presented in Chapter 2. There, it was established that processes fall into three broad families: *shaping*, *joining* and *finishing* (Figures 2.4 and 2.5). Each is characterized by a set of attributes listed, in part, in Figures 2.6 and 2.7.

The strategy for materials selection was outlined in Chapter 3. The procedure, whether dealing with original design or refining an existing one, was to translate the design requirements into a set of constraints and objectives. The constraints are used to screen and the objectives to rank, delivering a preferred short-list of materials that best meet both, finally seeking documentation to guide the final choice. The strategy for selecting processes follows a parallel

Figure 18.1   The selection strategy applied to processes. As for materials, the four steps are translation, screening, ranking and supporting documentation. All can be implemented in software.

path, shown in Figure 18.1. It uses constraints on process attributes to screen, together with an appropriate objective to rank, culminating as before with a search for information documenting the details of the top-ranked candidates.

Translation

As we saw in earlier chapters, the *function* of a component dictates the initial choice of material and shape. This choice exerts constraints on the choice of processes. It is helpful to think of two types of constraint: *technical*—can the process do the job at all? And *quality*—can it do so sufficiently well? One technical constraint applies across all process families: it is the compatibility of material and process. Shape, too, is a general constraint, although its influence on the choice of shaping, joining and finishing processes differs. Quality includes precision and surface finish, together with avoidance of defects and achieving the target for material properties. The usual *objective* in processing is to minimize cost. The *free variables* are largely limited to choosing the process itself and its operating parameters (such as temperatures, flow rates and so on). Table 18.1 summarizes the outcome of the translation stage. The case studies of Section 18.9 illustrate its use for each process family.

Table 18.1   Function, constraints, objective and free variables

Function	• What does the process do?
Constraints	• What technical and quality limits must be met?
Objective	• Minimize cost
Free variables	• Choice of process and process operating conditions

Screening

Translation, as with materials, leads to constraints for process attributes. The screening step applies these, eliminating processes that cannot meet the constraints. Some process attributes are simple numeric ranges—the size or mass of component the process can handle, the precision or the surface smoothness it can achieve. Others are non-numeric—lists of materials to which the process can be applied, for example. Requirements such as 'made of magnesium and weighing about 3 kg' are easily compared with the process attributes to eliminate those that cannot shape magnesium or cannot handle a component as large as 3 kg.

Ranking

Ranking, it will be remembered, is based on *objectives*. The most obvious objective in selecting a process is that of minimizing cost. In certain demanding applications it may be replaced by the objective of maximizing quality regardless of cost, though more usually it is a trade-off between the two that is sought.

Documentation

Screening and ranking do not cope adequately with the less tractable issues (such as corrosion behavior); they are best explored through a documentation search. This is just as true for processes. Old hands who have managed a process for decades carry this information in their heads and can develop an almost mystical ability to detect problems that defy immediate scientific explanation. But jobs for life are rare today; experts move, taking their heads with them. Expertise is stored more accessibly in compilations of design guidelines, best practice guides, case studies and failure analyses. The most important technical expertise relates to productivity and quality. Machines all have optimum ranges of operating conditions under which they work best and produce products with uncompromised quality. Failure to operate within this window can lead to manufacturing defects, such as excessive porosity, cracking or residual stress. This in turn leads to scrap and lost productivity, and, if passed on to the user, may cause premature failure. At the same time material properties depend, to a greater or lesser extent, on process history through its influence on microstructure. Processes are chosen not just to make shapes but to fine-tune properties.

So, documentation and knowledge of how processes work (or don't work) is critical in design. There is only so far that screening and ranking can take us in finding the right process.

18.3 Process attributes: material compatibility

The main goal in defining process attributes is to identify the characteristics that discriminate between processes, to enable their selection by screening. Each of the three process families—shaping, joining and surface treatment—has it own set of characterizing attributes, described here and in the next three sections. It is also of interest to explore the physical limits to these for a given process, and to examine how the properties of the material being processed limit the rate of the process itself.

One process attribute applies to all three families—compatibility with material—so we examine this first.

Material–process compatibility
Figure 18.2 shows a material–process compatibility matrix. Shaping processes are at the top, with compatible combinations marked by dots the color of which identifies the material family. Its use for screening is straightforward—specify the material and some processes are immediately eliminated (or the reverse, as a screening step in material selection). The diagonal spread of the dots in the matrix reveals that each material class—metals, polymers, etc.—has its own set of process routes. This largely reflects the underlying process physics. Shaping is usually eased by melting or softening the material, and the melting temperatures of the material classes are quite distinct (Figure 13.6). There are some overlaps—powder methods are compatible with both metals and ceramics, molding with both polymers and glasses. Machining (when used for shaping) is compatible with almost all families. Joining processes using adhesives and fasteners are very versatile and can be used with most materials, whereas welding methods are material specific. Finishing processes are primarily used for the harder materials, particularly metals—polymers are molded to shape and rarely treated further except for decorative purposes. We will see why later.

Coupling of material selection and process selection
The choice of material and of process are clearly linked. The compatibility matrix shows that the choice of material isolates a sub-domain of processes. The process choice feeds back to material selection. If, say, *aluminum alloys* were chosen in the initial material selection stage and *casting* in the initial process selection stage, then only the subset of aluminum alloys that are *castable* can be used. If the component is then to be *age hardened* (a heat-treatment process), material choice is further limited to *heat-treatable aluminum casting alloys*. So we must continually iterate between process and material, checking that inconsistent combinations are avoided. The key point is that neither material nor process choice can be carried too far without giving thought to the other. Once past initial screening, it is essentially a co-selection procedure.

Figure 18.2   The material–process compatibility matrix for shaping, joining and finishing processes color coded by material. Ignore the boxes on first reading; they refer to case studies to come.

18.4 Shaping processes: attributes and origins

Mass and section thickness

There are limits to the size of component that a process can make. Figure 18.3 shows the limits. The color coding for material compatibility has been retained, using more than one color when the process can treat more than one material family. Size can be measured by volume or by mass, but since the range of either one covers many orders of magnitude, while densities only span a factor of about 10, it doesn't make much difference which we use—big things are heavy,

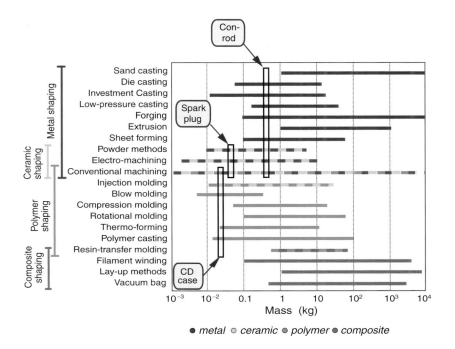

Figure 18.3   The process–mass range bar chart for shaping processes. Ignore, on first reading, the boxes; they refer to case studies to come. The colors refer to the material, as in Figure 18.2. Multiple colors indicate compatibility with more than one family.

whatever they are made of. Most processes span a mass range of about a factor of 1000 or so. Note that this attribute is most discriminating at the extremes; the vast majority of components are in the 0.1–10 kg range, for which virtually any process will work.

It is worth noting what the ranges represent. Each bar spans the size range of which the process is capable without undue technical difficulty. All can be stretched to smaller or larger extremes but at the penalty of extra cost because the equipment is no longer standard. During screening, therefore, it is important to recognize 'near misses'—processes which narrowly failed, but which could, if needed, be reconsidered and used.

Figure 18.4 shows a second bar chart: that for the ranges of section thickness of which each shaping process is capable. It is the lower end of the ranges—the minimum section thickness—where the physics of the process imposes limits. Their origins are the subject of the next sub-section.

Physical limits to size and section thickness

Casting and molding both rely on material flow in the liquid or semi-liquid state. Lower limits on section thickness are imposed by the physics of flow.

Figure 18.4   The process–section thickness bar chart for shaping processes color coded by material, annotated with the target values for the case studies.

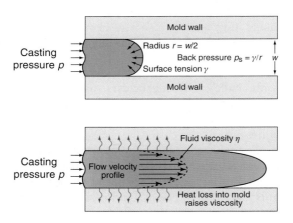

Figure 18.5.   Flow of liquid metal or polymer into thin sections is opposed by surface tension (a) and by viscous forces (b). Loss of heat into the mold increases viscosity and may cause premature solidification.

Viscosity and surface tension oppose flow through narrow channels, and heat loss from the large surface area of thin sections cools the flowing material, raising the viscosity before the channel is filled (Figure 18.5). Polymer and glass viscosities increase steadily as temperature drops. Pure metals solidify at a fixed

Tension caused
by differential
shrinkage

More uniform
cross-section
avoids problem

Figure 18.6   (a) Differential shrinkage causes internal stress, distortion and cavitation in thick sections or at changes of section. (b) Good design keeps section thickness as uniform as possible, avoiding the problem.

temperature, with a step increase in viscosity, but for alloys this happens over a range of temperature, known as the 'mushy zone', in which the alloy is part liquid, part solid. The width of this zone can vary from a few degrees centigrade to several hundred—so metal flow in castings depends on alloy composition. In general, higher pressure die-casting and molding methods enable thinner sections to be made, but the equipment costs more and the faster, more turbulent flow can entrap more porosity and cause damage to the molds.

Upper limits to size and section in casting and molding are set by problems of shrinkage. The outer layer of a casting or molding cools and solidifies first, giving it a rigid skin. When the interior subsequently solidifies, the change in volume can distort the product or crack the skin, or cause internal cavitation. Problems of this sort are most severe where there are changes of section, since the constraint introduces tensile stresses that cause *hot tearing*—cracking caused by constrained thermal contraction. Different compositions have different susceptibilities to hot tearing—another example of coupling between material, process and design detail.

Much of the documentation for casting and molding processes concerns guidance on designing both the component shape and the mold geometry to achieve the desired cross-sections while avoiding defects. Even when the component shape is fixed, there is freedom to choose where the material inlets are built into the mold (the 'runners') and where the air and excess material will escape (the 'risers'). Figure 18.6 shows an example of good practice in designing the cross-sectional shape of a polymer molding.

Powder methods for metals and ceramics too depend on flow. Filling a mold with powder uses free flow under gravity plus vibration to bed the powder down and achieve uniform filling. This may be followed by compression ('cold compaction'). Once full of powder, the mold is heated to allow densification by sintering or, if the pressure is maintained, by hot isostatic pressing (HIPing).

Metal shaping by deformation—hot or cold rolling, forging or extrusion—also involves flow. Solid metals flow by plastic deformation or by creep—Chapter 13 described how the flow rate depends on stress and temperature. Much forming

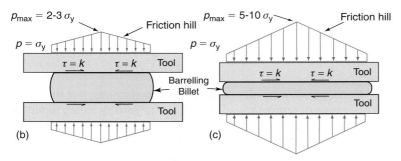

Figure 18.7 The influence of friction and aspect ratio on open die forging. (a) Uniaxial compression with very low friction. (b) With sticking friction the contact pressure rises in a 'friction hill' causing barrelling. (c) The greater the aspect ratio, the greater the pressure rise and the barrelling.

is done hot because the hot yield stress is lower than that at room temperature and work hardening, which drives the yield strength up during cold deformation, is absent. The thinness that can be rolled, forged or extruded is limited by plastic flow in much the same way that the thinness in casting is limited by viscosity: the thinner the section, the greater the required roll-pressure or forging force.

Figure 18.7 illustrates the problems involved in forging or rolling very thin sections. Friction changes the pressure distribution on the die and under the rolls. When they are well lubricated, as in (a), the loading is almost uniaxial and the material flows at its yield stress σ_y. With friction, as in (b), the metal shears at the die interface and the pressure ramps up because the friction resists the lateral spreading, giving a 'friction hill'. The area under the pressure distribution is the total forming load, so friction increases the load. The greater the aspect ratio of the section (width/thickness), the higher the maximum pressure needed to cause yielding, as in (c). This illustrates the fundamental limit of friction on section thickness—very thin sections simply stick to the tools and will not yield, even with very large pressures. Friction not only increases the load and limits the aspect ratio that can be formed, it also produces distortion in shape—'barrelling'—shown in both (b) and (c). A positive aspect of the high contact pressures in forging is the repeatability with which the metal can be forced to take up small features machined on the face of the tooling, exploited in making coins.

Very small or very thin objects—pushing the limits

So how do you make very small things? The obvious way to do it is to cut them out of bigger ones. That means machining: turning, shaping, drilling and milling. The upper limit on size is set by the budget available to buy the machine and the electricity bill for its use. Lower limits—now we are thinking of milligrams and fractions of a millimeter—are set by the stiffness of the machine and the material itself.

One continuous shaping process, rolling, is able to produce sections as thin as $10\,\mu m$—the thickness of aluminum kitchen foil. The foil only yields and thins just as it enters and leaves the roll bite—in between, the huge aspect ratio and frictional constraint mean that the high pressure actually flattens the steel rolls elastically. The process is helped by 'pack rolling'—feeding two sheets through together, with lubricant in between so that they can be peeled apart as they exit the rolls. This is why kitchen foil is shiny on one side (where contact was with the roll) and matt on the other.

Suppose you want to go still smaller. Flow won't work, and the sections are so thin that they simply bend if you try to machine them. Now you need chemistry. Conventional chemistry—etching, chemical milling, electro-discharge milling—allow features on the same scale as that of precision machining. To get really small, we need the screening, printing and etching techniques used to fabricate electronic devices. These are used to make MEMS (micro-electro-mechanical systems) — tiny sensors, actuators, gears and even engines. At present the range of materials is limited by this silicon-based technology, but some commercial devices, such as the inertial triggers for airbags, are made in this way.

Shape

A key attribute of a process is the families of shapes it can make. It is also one of the most difficult to characterize. A classification scheme is illustrated in Figure 18.8, with three generic classes of shape, each sub-divided in two. The merit of this approach is that processes map onto specific shapes. Figure 18.9 shows the matrix of viable combinations.

The prismatic shapes shown on the left of Figure 18.8, made by rolling, extrusion or drawing, have a special feature: they can be made in continuous lengths. The other shapes cannot—they are discrete, and the processes that make them are called batch processes. The distinction between continuous and discrete (or batch) processes is a useful one, bearing on the cost of manufacture. Batch processes require repeated cycles of setting-up, which can be time-consuming and costly. On the other hand, they are often net-shape or nearly so, meaning that little further machining is needed to finish the component. Casting and molding processes are near net-shape. Forging, on the other hand, is limited by the strains and shape changes by solid-state plasticity, starting

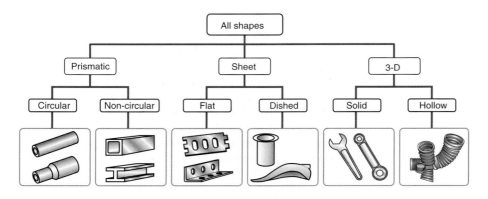

Figure 18.8   The classification of shape. Sub-division into more than six classes is difficult, and does not give useful further discrimination.

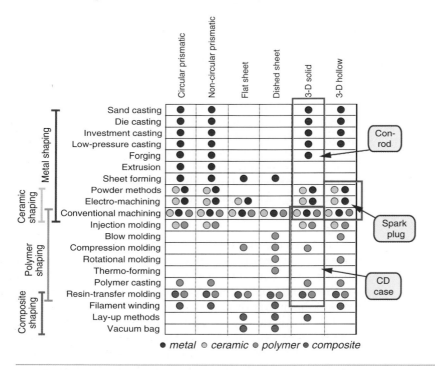

Figure 18.9   The shape–process compatibility matrix. A dot indicates that the pair forms a viable combination color coded as before.

from standard stock (bar, rod or sheet) to create a 3-D shape in several stages requiring a sequence of dies.

Continuous processes are well suited to long, prismatic products such as railway track or standard stock material such as tubes, plate and sheet. Cylindrical

Table 18.2 Typical levels of finish required in different applications, and suitable processes

Roughness (µm)	Typical application	Process
$R = 0.01$	Mirrors	Lapping
$R = 0.1$	High-quality bearings	Precision grind or lap
$R = 0.2$–0.5	Cylinders, pistons, cams, bearings	Precision grinding
$R = 0.5$–2	Gears, ordinary machine parts	Precision machining
$R = 2$–10	Light-loaded bearings, non-critical components	Machining
$R = 3$–100	Non-bearing surfaces	Unfinished castings

Figure 18.10   A section through a surface, showing its irregular surface (artistically exaggerated in the vertical direction). The irregularity is measured by the RMS roughness, R.

rolls produce sheets. Shaped rolls make more complex profiles—rail track is one of these. Extrusion is a particularly versatile continuous process, since complex prismatic profiles that include internal channels and longitudinal features such as ribs and stiffeners can be manufactured in one step.

Tolerance and roughness
We think of the precision and surface finish of a component as aspects of its *quality*. They are measured by the *tolerance*, *T*, and the *surface roughness*, *R*. When the dimensions of a component are specified the surface quality is specified as well, though not necessarily over the entire surface. Surface quality is critical in contacting surfaces such as the faces of flanges that must mate to form a seal or sliders running in grooves. It is also important for resistance to fatigue crack initiation and for aesthetic reasons. The tolerance T on a dimension y is specified as $y = 100 \pm 0.1$ mm, or as $y = 50^{+0.01}_{-0.001}$ mm, indicating that there is more freedom to oversize than to undersize. Surface roughness, R, is specified as an upper limit, e.g. $R < 100 \, \mu\text{m}$. The typical surface finish required in various products is shown in Table 18.2. The table also indicates typical processes that can achieve these levels of finish.

Surface roughness is a measure of the irregularities of the surface (Figure 18.10). It is defined as the root-mean-square (RMS) amplitude of the surface profile:

$$R^2 = \frac{1}{L} \int_0^L y^2(x) \, dx \qquad (18.1)$$

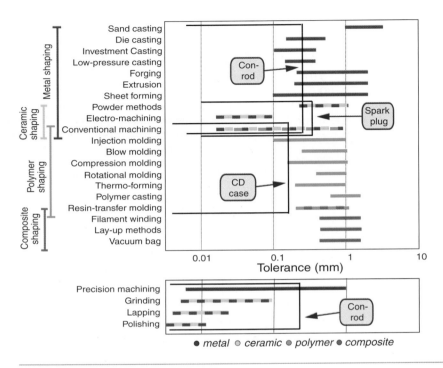

Figure 18.11 The process—tolerance bar chart for shaping processes and some finishing processes, enabling process chains to be selected. The chart is color coded by material and annotated with the target values for the case studies.

It used to be measured by dragging a light, sharp stylus over the surface in the x-direction while recording the vertical profile $y(x)$, like playing a gramophone record. *Optical profilometry*, which is faster and more accurate, has now replaced the stylus method. It scans a laser over the surface using interferometry to map surface irregularity. The tolerance T is obviously greater than $2R$; indeed, since R is the root-mean-square roughness, the peak roughness, and hence absolute lower limit for tolerance, is more like $5R$. Real processes give tolerances that range from $10R$ to $1000R$.

Figures 18.11 and 18.12 show the characteristic ranges of tolerance and roughness of which processes are capable, retaining the color coding for material family. Data for finishing processes are added below the shaping processes. Sand casting gives rough surfaces; casting into metal dies gives a smoother one. No shaping processes for metals, however, do better than $T = 0.1$ mm and $R = 0.5$ μm. Machining, capable of high dimensional accuracy and surface finish, is commonly used after casting or deformation processing to bring the tolerance or finish up to the desired level, creating a *process chain*. Metals and ceramics can be surface-ground and lapped to a high precision and

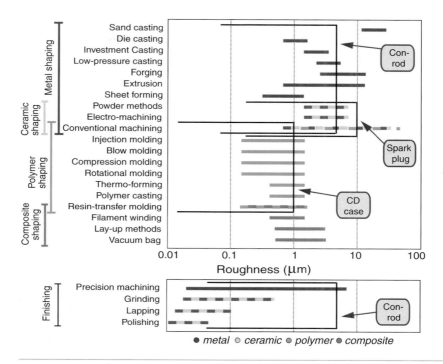

● *metal* ○ *ceramic* ● *polymer* ● *composite*

Figure 18.12 The process–roughness bar chart for shaping processes and some finishing processes, enabling process chains to be selected.

smoothness: a large telescope reflector has a tolerance approaching 5 μm and a roughness of about 1/100 of this over a dimension of a meter or more. But precision and finish carry a cost: processing costs increase exponentially as the requirements for both are made more severe. It is an expensive mistake to over-specify precision and finish.

Molded polymers inherit the finish of the molds and thus can be very smooth (Figure 18.12); machining to improve the finish is rarely necessary. Tolerances better than ±0.2 mm are seldom possible (Figure 18.11) because internal stresses left by molding cause distortion and because polymers creep in service.

18.5 Joining processes: attributes and origins

The design requirements to be met in choosing a joining process differ from those for choosing a shaping process. Even material compatibility is more involved than for shaping, in that joints are often to be made between dissimilar materials.

Material compatibility

Processes for joining metals, polymers, ceramics and glasses differ. Adhesives will bond to some materials but not to others; methods for welding polymers differ from those for welding metals; and specific metals often require specific types of welds. The material–process matrix (Figure 18.2) includes four classes of joining process.

When the joint is between dissimilar materials, further considerations arise, both in manufacture and in service. The process must obviously be compatible with both materials. Adhesives and fasteners generally allow joints between different materials; many welding processes do not. Practicality in service is a different issue. If two materials are joined in such a way that they are in electrical contact, a corrosion couple appears if close to water or a conducting solution. This can be avoided by inserting an insulating interlayer between the surfaces. Thermal-expansion mismatch gives internal stresses in the joint if the temperature changes, with risk of damage. Identifying good practice in joining dissimilar materials is part of the documentation step.

Joint geometry and mode of loading

Important joining-specific considerations are the geometry of the joint, the thickness of material they can handle and the way the joint will be loaded. Figures 18.13 and 18.14 illustrate standard joint geometries and modes of loading. The choice of geometry depends on the shapes and thicknesses of the parts to be joined, and the mode of loading—it is a coupled problem. Adhesive joints support shear but are poor in peeling—think of peeling off sticky tape. Adhesives need a good working area—lap joints are fine, butt joints are not. Rivets and staples, too, are well adapted for shear loading of lap joints but are less good in tension. Welds and threaded fasteners are more adaptable, but here, too, matching choice of process to geometry and loading is important. Screening on each in isolation correctly keeps options open, but the details of the coupling must be researched at the documentation stage.

When making joints, sheet thickness is an issue. Joining processes like stapling, riveting and sewing can be used with thin sections but not with thick. The thickness that can be welded is restricted, to varying degrees, by heat transfer considerations. Some processes are better adapted than others to joining greatly unequal sections: adhesives can bond a very thin layer to a very thick one, while riveting and sewing work best for sections that are nearly equal. In practice the thickness, too, is coupled to geometry.

Secondary functionalities and manufacturing constraints

Most joints carry load, but that is not all. Joints may need to conduct or insulate against the conduction of heat or electricity, or they may be required to operate at an elevated operating temperature. A joint may also serve as a seal, and be required to exclude gases or liquids.

Manufacturing conditions also impose constraints on joint design. Certain joining processes impose limits on the size of the assembly; electron beam welding, for instance, usually requires that the assembly fit within a vacuum chamber.

Figure 18.13   Joint geometries.

Figure 18.14   Standard modes of loading for joints.

On-site joining operations conducted out of doors require portable equipment, possibly with portable power supplies; an example is the hot-plate welding of PVC and PE pipelines. Portability, on the other hand, effectively means there is no limit on the size of the assembly—bridges, shipbuilding and aircraft make this self-evident.

An increasingly important consideration is the requirement for economic disassembly at the end of product life to facilitate recycling and reuse. Threaded fasteners can be disassembled and adhesives can be loosened with solvents or heat.

Physical limits in joining
Thermal joining processes—arc, laser and friction welding, for example—operate by imposing a local thermal cycle to the joint area to produce a bond in the liquid or solid state. Heat transfer and conduction set the physical limit to the size of joint. Here the intensity of heating is important. For translating heat sources of power Q moving at speed v along the joint line, the heat input per unit length, Q/v (J/m), is commonly used as a characteristic welding parameter. Peak temperatures and cooling rates with Q/v. Heat input is thus important in determining the extent of the 'heat-affected zone'—the region round the weld affected by the

thermal cycle imposed. The microstructural changes induced depend on the peak temperature and the cooling rate—especially in steels, which are commonly joined by welding. And local thermal cycles induce internal residual stresses. Because of their importance, microstructural changes during welding have been researched in depth and are now well understood; handbooks on the subject provide extensive documentation.

Process economics are dictated by the set-up time and the actual time for making the joint. For thermal processes, conduction of heat governs the operating speed with a given level of power, both in making the joint and in cooling of the component before it can be handled and moved. Some adhesives require prolonged curing times before they can take load, again slowing the throughput. And many processes take much longer to set up than to make the joint itself, need to avoid misalignment or lack of cleanliness in the joint region. Inventive workstation design can help to improve the economics by arranging that multiple set-up and product cooling areas run in parallel with a high-speed joining machine, a strategy used to enable the economic performance of capital-intensive processes like laser and electron beam welding.

18.6 Surface treatment (finishing) processes: attributes and origins

Material compatibility
Material–process compatibility for surface treatments is shown at the bottom of the matrix in Figure 18.2. As noted previously, surface finishing is more important for metals than for polymers.

The purpose of the treatment
The most discriminating finishing-specific attribute proves to be the purpose of applying the treatment. Table 18.3 illustrates the diversity of functions that

Table 18.3   The design functions provided by surface treatments

- Corrosion protection (aqueous)
- Corrosion protection (gases)
- Wear resistance
- Friction control
- Fatigue resistance
- Thermal conduction
- Thermal insulation
- Electrical insulation
- Magnetic properties
- Decoration
- Color
- Reflectivity

surface treatments can provide. Some protect, some enhance performance, still others are primarily aesthetic. All surface treatments add cost but the added value can be large. Protecting a component surface extends product life and increases the interval between maintenance cycles. Coatings on cutting tools enable faster cutting speeds and greater productivity. And surface hardening processes may enable the substrate alloy to be replaced with a cheaper material—for example, using a plain carbon steel with a hard carburized surface or a coating of hard titanium nitride (TiN), instead of using a more expensive alloy steel.

Secondary compatibilities

Material compatibility and design function are the first considerations in select-ing a finishing process. But there are others. Some surface treatments leave the dimensions, precision and roughness of the surface unchanged. Deposited coatings obviously change the dimensions a little, but may still leave a perfectly smooth surface. Others build up a relatively thick layer with a rough surface, requiring refinishing. Component geometry also influences the choice. 'Line-of-sight' deposition processes coat only the surface at which they are directed, leaving inaccessible areas uncoated; others, with what is called 'throwing power', coat flat, curved and re-entrant surfaces equally well. Many surface-treatment processes require heat. These can only be used on materials that can tolerate the rise in temperature. Some paints are applied cold, but many require a bake at up to 150°C. Heat treatments like carburizing or nitriding to give a hard surface layer require prolonged heating at temperatures up to 800°C. This thermal exposure of the substrate can change the underlying microstructure, not neces-sarily for the better. There are some innovative developments to match the thermal treatments required in both substrate and surface, achieving two treatments in one go. An example is the paint–bake cycle for heat-treatable aluminum alloy automotive panels, which cures and hardens the paint while simultaneously age hardening the sheet itself.

Physical limits to surface treatments

For thermal processes, the surface heating intensity and absorptivity, and the thermal properties of the substrate, dictate the rise time in temperature. For a stationary heat source of given power density, the temperature rises with the square root of time $\sqrt{t}$. This in turn determines the depth of surface layer that reaches the temperature at which the desired microstructural change takes place. Since diffusion of matter and heat follow the same mathematical equations, the depth of treatment in diffusion processes such as carburizing has the same dependence on time. For a spray-coating process delivering a given volumetric flow rate, the coating thickness increases linearly with time.

18.7 Estimating cost for shaping processes

Estimating process costs accurately—at the precision needed for competitive contract bidding, for example—is a specialized job. It is commonly based on

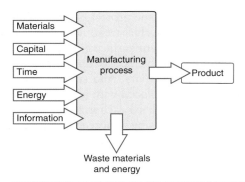

Figure 18.15   The inputs to a manufacturing process.

interpolation or extrapolation of known costs for similar, previous jobs, appropriately scaled. Our interest here is not in estimating cost for this purpose; it is to compare approximate costs of alternative process routes. This can be useful even when imprecise. We illustrate the method with a cost model for one of the most common choices: that of a batch shaping process. It requires certain *user-specified inputs*, such as local labor costs, as well as values for *cost-related attributes* of each process.

The cost model

The manufacture of a component consumes resources (Figure 18.15), each of which has an associated cost. The final cost is the sum of those of the resources it consumes. They are defined in Table 18.4. Thus, the cost of producing a component of mass m entails the cost C_m ($/kg) of the materials and consumable feed-stocks from which it is made. It involves the cost of dedicated *tooling*, C_t ($), and that of the capital *equipment*, C_c ($), in which the tooling will be used. It requires *time*, chargeable at an overhead rate $\dot{C}_{oh}$ (thus with units of $/h), in which we include the cost of labor, administration and general plant costs. It requires energy, which is sometimes charged against a process step if it is very energy intensive, but more usually is treated as part of the overhead and lumped

Table 18.4   Symbols, definitions and units in the cost model

Resource		Symbol	Unit
Materials:	Including consumables	C_m	$/kg
Capital:	Cost of tooling	C_t	$
	Cost of equipment	C_c	$
Time:	Overhead rate, including labor, Administration, rent, etc.	$\dot{C}_{oh}$	$/h
Energy:	Cost of energy	C_e	$/h
Information:	R & D or royalty payments	C_i	$/year

into $\dot{C}_{oh}$, as we shall do here. Finally, there is the cost of information, meaning that of research and development, royalty or license fees; this, too, we view as a cost per unit time and lump it into the overhead.

Consider now the manufacture of a component (the 'unit of output') weighing m kg, and made of a material costing C_m \$/kg. The first contribution to the unit cost is that of the material mC_m. Processes rarely use exactly the right amount of material but generate a scrap fraction f, which ends up in runners, risers, machining swarf and rejects. Some is recycled, but the material cost per unit needs to be magnified by the factor $1/(1-f)$ to account for the fraction that is lost. Hence the material contribution per unit is:

$$C_1 = \frac{mC_m}{(1-f)} \tag{18.2}$$

The cost C_t of a set of tooling—dies, molds, fixtures and jigs—is what is called a *dedicated cost*: one that must be wholly assigned to the production run of this single component. It is written off against the numerical size n of the production run. However, tooling wears out at a rate which depends on the number of items processed. We define the tool life n_t as the number of units that a set of tooling can make before it has to be replaced. We have to use one tooling set even if the production run is only one unit. But each time the tooling is replaced there is a step up in the total cost to be spread over the whole batch. To capture this with a smooth function, we multiply the cost of a tooling set C_t by $(1 + n/n_t)$. Thus, the tooling cost per unit takes the form:

$$C_2 = \frac{C_t}{n}\left(1 + \frac{n}{n_t}\right) \tag{18.3}$$

The *capital cost of equipment*, C_c, by contrast, is rarely dedicated. A given piece of equipment—a powder press, for example—can be used to make many different components by installing different die-sets or tooling. It is usual to convert the capital cost of *non-dedicated* equipment and the cost of borrowing the capital itself into an overhead by dividing it by a *capital write-off time*, t_{wo}, (5 years, say) over which it is to be recovered. The quantity C_c/t_{wo} is then a cost per hour—provided the equipment is used continuously. That is rarely the case, so the term is modified by dividing it by a *load factor*, L—the fraction of time for which the equipment is productive. This gives an effective hourly cost of the equipment, like a rental charge, even though it is your own piece of kit. The capital contribution to the cost per unit is then this hourly cost divided by the production rate/hour $\dot{n}$ at which units are produced:

$$C_3 = \frac{1}{\dot{n}}\left(\frac{C_c}{Lt_{wo}}\right) \tag{18.4}$$

Finally, there is the general background hourly *overhead rate* $\dot{C}_{oh}$ for labor, energy and so on. This is again converted to a cost per unit by dividing by the production rate $\dot{n}$ units per hour:

$$C_4 = \frac{\dot{C}_{oh}}{\dot{n}} \qquad (18.5)$$

The total shaping cost per part, C_s, is the sum of these four terms, C_1–C_4, taking the form:

$$C_s = \frac{mC_m}{(1-f)} + \frac{C_t}{n}\left(1 + \frac{n}{n_t}\right) + \frac{1}{\dot{n}}\left(\frac{C_c}{Lt_{wo}} + \dot{C}_{oh}\right) \qquad (18.6)$$

To emphasize the simple form of this equation, it can be written:

$$C_s = C_{material} + \frac{C_{dedicated}}{n} + \frac{\dot{C}_{capital} + \dot{C}_{overhead}}{\dot{n}} \qquad (18.7)$$

This equation shows that the cost has three essential contributions—a material cost per unit of production that is independent of batch size and rate, a dedicated cost per unit of production that varies as the reciprocal of the production volume ($1/n$), and a gross overhead per unit of production that varies as the reciprocal of the production rate ($1/\dot{n}$). The dedicated cost, the effective hourly rate of capital write-off and the production rate can all be defined by a representative range for each process; target batch size n, the overhead rate $\dot{C}_{oh}$, the load factor L and the capital write-off time t_{wo} must be defined by the user.

Figure 18.16 uses equation (18.6). It is a plot of cost, C_s, against batch size, n, comparing the cost of casting a small aluminum component by three alternative processes: sand casting, die casting and low-pressure casting. At small batch sizes the unit cost is dominated by the 'fixed' costs of tooling (the second term on the right of equation (18.6)). As the batch size n increases, the contribution of this to the unit cost falls (provided, of course, that the tooling has a life that is greater than n) until it flattens out at a value that is dominated by the 'variable' costs of material, labor and other overheads. Competing processes differ in tooling cost C_t, equipment cost C_c and production rate $\dot{n}$. Sand-casting equipment is cheap but slow. Die-casting equipment costs much more but is also much faster. Mold costs for low-pressure die casting are greater than for sand casting; those for high pressure die casting are higher still. The combination of all these factors for each process causes the C_s–n curves to cross, as shown in Figure 18.16.

The crossover means that the process that is cheapest depends on the batch size. This suggests the idea of an *economic batch size*—a range of batches for

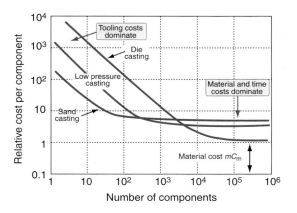

Figure 18.16   The cost of casting a small aluminum component by three competing processes.

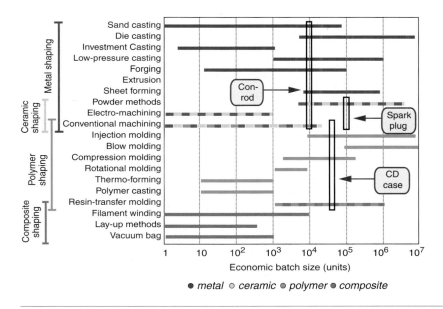

Figure 18.17   The process–economic batch size bar chart for shaping processes color coded by material.

which each process is likely to be the most competitive. Equation (18.6) allows the cost of competing processes to be compared if data for the parameters of the model are known. If they are not, the simpler economic batch size provides an alternative way of ranking. Figure 18.17 is a bar chart of this attribute, for the same processes and using the same color coding as the earlier charts.

Processes such as investment casting of metals and lay-up methods for composites have low tooling costs but are slow; they are economic when you want to make a small number of components but not when you want a large one. The reverse is true of the die casting of metals and the injection molding of polymers: they are fast, but the tooling is expensive.

18.8 Computer-aided process selection

Computer-aided selection

As with the material charts, the process charts of this chapter give an overview, but the number of processes that can be shown on any one of them is limited. Selection using them is practical when there are few constraints, but when there are many—as there usually are—checking that a given process meets them all is cumbersome, and the cost model cannot be presented in a useful way because of the need for user-defined parameters. All these problems are overcome in the CES implementation of the method.

Its database contains records for processes, organized into the families, classes and members shown in Figures 2.5–2.7 of Chapter 2. Each record contains attribute data for a process, with each attribute stored as a range spanning its usual values, and is linked to the records for the materials it can process. It also contains limited documentation in the form of text, images and references to sources of information about the process. The data are interrogated by the same search engine that is used for material selection, allowing superimposed stages like those suggested by Figure 18.18. On the left is a stage that limits the

Figure 18.18   Schematic of the screening steps in selecting process (here for shaping). First specify classes such as material and shape; then screen on numeric attributes, plotted graphically to the right as bar charts. In the CES software, target design requirements can also be entered through dialog boxes, shown in the center.

process choice to those able to handle a chosen material family, class or member (here, thermoplastics). In the center is a simple query interface for screening on one or more process attributes; the desired upper or lower limits for constrained attributes are entered and the search engine rejects all processes with attribute values that do not lie within the limits. On the right is a bar chart, constructed by the software, for any numeric attribute, (here, economic batch size). For screening, a selection box is superimposed on the bar chart with upper and lower edges that lie at the constrained values of the attribute. This retains the processes with values that penetrate the box, rejecting those that lie outside it.

The software includes the batch-process cost model described in the text. A dialog box allows the user to edit default values of the user-defined parameters L, t_{wo}, $\dot{C}_{oh}$, etc. The software then retrieves approximate values for the economic process attributes C_c, C_t, $\dot{n}$ from the database where they are stored as ranges. It allows the data to be presented in a number of ways, two of which are shown in Figures 18.19 and 18.20. The first is a plot of cost against batch size for a single process (here, injection molding), in the manner of the earlier Figure 18.16. The user-defined parameters are listed on it. The bandwidth derives from the ranges of the economic attributes: a simple shape, requiring only simple dies, lies near the lower edge; a more complex one, requiring multi-part dies, lies near the upper edge. Figure 18.20 shows an alternative presentation. Here the range of cost for making a chosen batch size (here, 10 000) of a component by a number of alternative processes is plotted as a bar chart. The user-defined parameters are again listed. Other selection stages can be applied in parallel with this one (as in Figure 18.18) applying constraints on material, shape, etc., causing some of the bars to drop out. The effect is to rank the surviving processes by cost.

Figure 18.19   The output of the cost model as implemented in the CES software, showing how unit cost changes with batch size. The range reflects the ranges in the input parameters to the model.

Figure 18.20   A bar chart made with the CES software showing unit cost of a set of processes for a specified component at a given batch size. The size of each bar reflects the range in the input parameters of the model.

18.9 Case studies

Here we have examples of the selection methodology in use. In each case, translation of the design requirements leads to a target list of process attributes for screening. Three shaping problems are approached using the charts of this chapter, with the cost model applied to the last of the three. These are followed by examples of selecting joining and surface treatment processes. For these we use the CES software; the charts show only the broad classes of joining and finishing processes, whereas the level of detail in the software is much greater. In each case, aspects of processing beyond the scope of simple screening are noted. These relate to defects, or material properties, or economic limitations—all of which can be addressed by seeking documentation.

Shaping a ceramic spark plug insulator

The anatomy of a spark plug is shown schematically in Figure 18.21. It is an assembly of components, one of which is the insulator. This is to be made of a ceramic, alumina, in an axisymmetric, hollow 3-D shape. The insulator is part of an assembly, fixing its dimensions. Given the material and dimensions, the expected mass can be estimated and the minimum section thickness identified. The insulator must seal within its casing, setting limits on precision and surface finish. Spark plugs are made in large numbers—the projected batch size is

Insulator

Body
Shell

Figure 18.21   A section through a spark plug. The insulator is cylindrical but not prismatic, making it 3-D hollow in the shape classification.

Table 18.5   Translation for shaping a spark plug insulator and potential process-sensitive defects

Function	• Spark plug insulator	
Objective	• Minimize cost	
Constraints	• Material: alumina	
	• Shape: 3-D hollow	
	• Estimated mass: 0.04–0.06 kg	Technical constraints
	• Minimum section: 1.5 mm	
	• Tolerance: $<\pm0.3$ mm	Quality constraints
	• Roughness: $<10\,\mu$m	
	• Batch size: 100 000	Economic constraint
Free variable	• Choice of shaping process and process operating conditions	
Defects	• Porosity	Influencing strength and toughness
	• Surface cracks	

100 000. Table 18.5 lists the constraints. The lower part of the table flags up key process-sensitive outcomes which cannot be checked by screening, but should be investigated by a documentation search for the processes that successfully pass the screening steps.

The constraints are plotted on the compatibility matrices and bar charts of Figures 18.2–18.4, 18.11 and 18.12. The material compatibility chart shows only three process groups for ceramics, and all three can handle the required shape. All three can cope with the mass and the section thickness. All three can also meet the quality constraints on roughness and tolerance, although powder

Table 18.6   Short-list of processes for shaping a spark plug insulator

Powder pressing and sintering
Powder injection molding

Table 18.7   Translation for shaping a CD case and potential process-sensitive defects

Function	• Case for compact disc	
Objective	• Minimize cost	
Constraints	• Material: polystyrene (thermoplastic) • Shape: solid 3-D • Estimated mass: 25–35 g • Minimum section: 1–2 mm	Technical constraints
	• Tolerance: 0.2 mm • Roughness: 1 μm	Quality constraints
	• Batch size: 50 000	Economic constraint
Free variable	• Choice of shaping process and process operating conditions	
Defects	• Cracks, • Trapped bubbles • Sink marks • Poor surface finish	Influencing strength and visual quality

methods are near their lower limit. The constraint on economic batch size (Figure 18.17) identifies powder methods as the preferred choice. Applying the same steps using the CES software (Exercise E18.8) identifies two specific variants of powder methods that meet the constraints, listed in Table 18.6. Powder injection molding is indeed the process used to make spark plug insulators.

A documentation search details good practice in die design, powder quality, die filling, and the cycle of pressure and temperature important for controlling porosity, finish and production rate.

Shaping a thermoplastic CD case

The unsatisfactory qualities of the 'jewel' cases of CDs were noted in Chapter 3 and Figure 3.8. The current material choice is polystyrene, but before considering a change of material we might wish to check our processing options for thermoplastics in this geometry. Table 18.7 shows the translation for the CD case. The dimensions and mass are set by the standardised size of CDs. The shape is specified as 'solid, 3-D' as it is not a flat panel but has features at the edges to make the hinges and tabs to hold the sleeve in place. These features also require a good tolerance. Low surface roughness is needed to preserve transparency. We assume a target batch size of 50 000.

Table 18.8   Short-list of processes for
shaping a polymer CD case

Injection molding
Compression molding

Figure 18.22   A steel connecting rod. The precision required for the bores and bore facing is much higher than for the rest of the body, requiring subsequent machining.

Refer again to the compatibility matrices and bar charts. Screening on the material (thermoplastic), shape, mass and minimum section leads to three possible processes: machining, injection molding and compression molding. Compression molding is just beyond its normal mass range. All three processes meet the targets for tolerance and roughness. The final bar chart shows that machining is not economic for the given batch size, but the other two are acceptable. Table 18.8 summarizes the surviving processes.

Good die design to include the protruding ridges and tabs on a CD case requires expert input. These re-entrant features exclude compression molding using uniaxial loading and open dies. Following up the avoidance of defects, a documentation search is principally a matter of good die design with correct location of the runners and risers.

Shaping a steel connecting rod
Figure 18.22 shows a schematic connecting rod ('con-rod') to be made of a medium carbon steel. The minimum section is around 8 mm, and an estimate of the volume gives a mass around 0.35 kg. Dimensional precision is important to ensure clearances at both the little and the big end. The con-rod carries cyclic loads with the consequent risk of fatigue crack initiation, so a low surface roughness overall is necessary (the bores will be finished by a subsequent machining operation). A modest batch size of 10 000 is required. Control of properties

Table 18.9   Translation for shaping a steel connecting rod and potential process-sensitive defects

Function	• Connecting rod ('con-rod')
Objective	• Minimize cost
Constraints	• Material: medium-carbon steel ⎫ • Shape: 3-D solid ⎬ Technical constraints • Estimated mass: 0.3–0.4 kg ⎪ • Minimum section: 8 mm ⎭
	• Tolerance: <0.25 mm (surface) ⎫ • Tolerance: <0.02 mm (bores) ⎬ Quality constraints • Roughness: <5 μm ⎭
	• Batch size: 10 000 Economic constraint
Free variable	• Choice of shaping process and process operating conditions
Defects	• Cracks, surface roughness ⎫ • Porosity ⎬ Influencing fatigue resistance • Residual stress ⎭

Table 18.10   Short-list of processes for shaping a steel connecting rod

Die casting
Forging + machining
Powder methods + machining

and of defects is dominated by the need to avoid fatigue failure. Table 18.9 summarizes.

These constraints are plotted on the matrices and charts. Screening on material and shape eliminates low-pressure casting, die casting, extrusion and sheet forming, but plenty of options survive. Sand casting is disqualified by the tolerance constraint. Investment casting, conventional machining, forging and powder methods remain. All achieve the target roughness, and surface tolerance, though not that required for the bores. However, by following the shaping process with a finishing process, the tolerance requirement can be reached. All of the finishing processes at the bottom of Figure 18.11 give better tolerance and roughness than specified— precision machining is most appropriate for finishing the bores and the faces surrounding them. Finally, the economic batch size chart suggests that forging and powder methods are suitable—machining (from solid) and investment casting are not economic for a batch size of 10 000. Table 18.10 summarizes.

This case study reveals that there is close competition for the manufacture of carbon steel components in the standard mid-range size and shape of a connecting rod. Surface finish and accuracy present some problems, but finishing by

Table 18.11 Process cost model input data for three shaping methods to make the con-rod

Parameters	Die casting	Forging	Powder methods
*Material, $mC_m/(1 - f)$	1	1	1
*Basic overhead, $\dot{C}_{oh}$ (per hour)	100	100	100
Capital write-off time, t_{wo} (years)	5	5	5
Load factor	0.5	0.5	0.5
*Dedicated tooling cost, C_t	17 500	125	1000
*Capital cost, C_c	50 000	110 000	1 000 000
Production rate, $\dot{n}$ (per hour)	50	50	2

* Costs normalized to $mC_m/(1 - f)$.

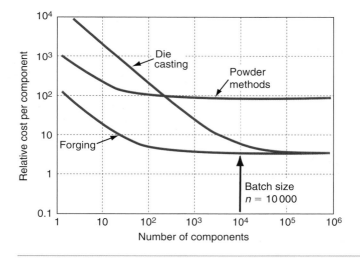

Figure 18.23 The cost model for three possible processes to manufacture the steel con-rod, using representative constant values for the model parameters. For the target batch size of 10 000, the cheapest option indicated is forging.

machining is not difficult for carbon steel. The final choice will be determined by detailed cost considerations and achieving the required fatigue strength. The properties will be sensitive to the composition of the steel and its heat treatment—something illustrated further in the next chapter.

Cost is considered further by applying the model of Section 18.7. Table 18.11 lists values for the parameters in the cost model for three processes: die casting, forging and powder methods. Assuming the same amount of material is used in each process, the costs are all normalized to the material cost. Figure 18.23 shows the computed relative cost per part against batch size for the three processes.

Figure 18.24   A section through a domestic radiator. The three pressed steel sections are joined by lap joints

For the batch size of 10 000 forging comes out the cheapest, followed closely by die casting. Powder is more expensive, principally because the slower production rate and high capital cost give a large overhead contribution per part. Since there is a spread of estimated cost (due to uncertainty in the inputs), and forging is likely to be followed by machining, the cost difference between forging and die casting is not significant—both remain on the list for investigation of the achievable properties.

Joining a steel radiator

Figure 18.24 shows a section through a domestic radiator made from corrugated pressed sheet steel. The task is to choose a joining process for the seams between the sheets. As always, the process must be compatible with the material (here, low-carbon steel sheet of thickness 1.5 mm). The lap joints carry only low loads in service but handling during installation may impose tension and shear. They must conduct heat, be watertight and able to tolerate temperatures up to 100°C. There is no need for the joints to be disassembled for recycling at the end of life, since the whole thing is steel. Quality constraints are not severe, but distortion and residual stress will make fit-up of adjacent joints difficult. Low-carbon steels are readily weldable, so it is unlikely that there is a risk of any welding process causing cracking or embrittlement problems—but this can be checked by seeking documentation. Table 18.12 summarizes the translation.

To tackle this selection we use the Cambridge Engineering Selector. It stores records and attributes for 52 joining processes. Applying the material compatibility check first, 32 processes are found to be suitable for joining low-carbon steel—those eliminated are principally polymer-specific processes. Further screening on joint geometry, mode of loading and section thickness reduces the list to 20 processes. The requirement to conduct heat is then the most discriminating—only 10 processes now pass. Water resistance and operating temperature do not

Table 18.12 Translation for joining a steel radiator and outcome (potential process-sensitive defects and properties

Function	• Domestic radiator	
Objective	• Minimize cost	
Constraints	• Material compatibility: low-carbon steel	
	• Joint geometry: lap joint	
	• Mode of loading: tension, shear (moderate)	Technical constraints
	• Sheet thickness: 1–2 mm	
	• Joint must conduct heat	
	• Joint must be watertight	
	• Service temperature >100°C	
	• Disassembly not required	Environmental constraint
Free variable	• Choice of shaping process and process operating conditions	
Defects	• Residual stress	Influencing fit-up and
	• Imperfect joint	service life

Table 18.13 Short-list of processes for joining a steel radiator

Brazing
Electron beam welding
Explosive welding
Metal inert gas arc welding (MIG)
Tungsten inert gas arc welding (TIG)
Laser beam welding
Manual metal arc welding (MMA)
Oxyacetylene welding
Riveting
Soldering

change the short-list further. The processes passing the screening stage are listed in Table 18.13.

Quality and economic criteria are difficult to apply as screening steps in selecting joining processes. At this point we seek documentation for the processes. This reveals that explosive welding requires special facilities and permits (hardly a surprise). Electron beam and laser welding require expensive equipment, so use of a shared facility would be necessary to make them economic. Resistance spot welding is screened out because it failed the requirement to be watertight. This is only necessary for the edge seams, so internal joint lines could be spot welded. This highlights the need for judgement in

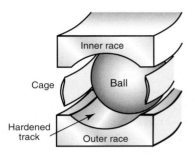

Figure 18.25   A section through a ball bearing race. The surface of the race is to be hardened to resist wear and fatigue crack initiation.

applying automated screening steps. Always ask: has an obvious (or existing) solution been eliminated? Why was this? Has something narrowly failed, so that a low-cost option could be accommodated with modest redesign? Could a second process step, like sealing the spot-welded joint with a mastic, correct a deficiency? The best way forward is to turn constraints on and off, and adjust numerical limits, exploring the options that appear.

Surface hardening a ball-bearing race
The balls of ball bearing races run in grooved tracks (Figure 18.25). As discussed in Chapter 11, the life of a ball race is limited by wear and by fatigue. Both are limited by using hard materials. Hard materials, however, are not tough, incurring the risk that shock loading or mishandling might cause the race to fracture. The solution is to use an alloy steel, which has excellent bulk properties, and to apply a separate surface treatment to increase the hardness where it matters. We therefore seek processes to surface harden alloy steels for wear and fatigue resistance. The precision of both balls and race is critical, so the process must not compromise the dimensions or the surface smoothness. Table 18.14 summarizes the translation.

The CES system contains records for 44 surface treatment processes. Many are compatible with alloy steels. More discriminating is the purpose of the treatment—to impart fatigue and wear resistance—reducing the list to eight. Imposing the requirement for a very smooth surface knocks out those that coat or deform the surface because these compromise the finish. The short-list of processes that survive the screening is given in Table 18.15. Adding the further constraint that the curved surface coverage must be *very good* leaves just the first two: carburizing and carbonitriding, and nitriding.

To get further we turn to documentation for these processes. The hardness of the surface and the depth of the hardened layer depend on process variables: the time and temperature of the treatment and the composition of the steel. And economics, of course, enters. Ball races are made in enormous batches and while their sizes vary, their geometry does not. This is where dedicated equipment, even if very expensive, is viable.

Table 18.14   Translation for surface hardening a ball-bearing race and potential process-sensitive defects

Function	• Ball-bearing race	
Objective	• Minimize cost	
Constraints	• Material: alloy steel • Function of treatment: hardening for fatigue and wear resistance • Curved surface coverage: good • Precision and surface finish not compromised	} Technical constraints
Free variable	• Choice of surface treatment process and process operating conditions	
Defects	• Cracks • Residual stress	} Influencing surface toughness and fatigue resistance

Table 18.15   Short-list of processes for hardening an alloy steel ball race

Carburizing and carbonitriding
Nitriding
Induction and flame hardening
Laser surface hardening and melting
Vapor metallizing

18.10 Summary and conclusions

Selecting a process follows a strategy much like that for materials: *translation* of requirements, *screening* of the options, *ranking* of the processes that survive screening, ending with a search of *documentation* for the most promising candidates. For processes, translation is straightforward—the requirements convert directly into the attributes of the process to be used in the screening step. Process attributes capture their technical capabilities and aspects of product quality. All process families require screening for compatibility with material to which they are to be applied, but after this the attributes and constraints are specific to the family. For shaping, component geometry, precision and finish are most discriminating. For joining it is joint shape and mode of loading. For surface treatments it is the purpose for which the treatment is applied (wear, corrosion, aesthetics, etc.). Ranking is based on cost if this can be modeled, as it can for batch processes. Cost models combine characteristic process attributes such as tooling and capital costs and the production rate, with user-specified parameters such as overhead rates and required batch size. The economic batch size provides a crude alternative to the cost model—a preliminary indication of the batch size for which the process is commonly found to be economic.

Translation of the design requirements frequently points to important aspects of processing that fall beyond a simple screening process. The avoidance of defects, achieving the required modification of material properties, and determining the process speed (and thus economics), all involve coupling between the process operating conditions, the material composition and design detail. This highlights the importance of the documentation stage in making the final choice of processes, and the need to refine the material and process specifications in parallel. The next chapter explores aspects of process-property interactions in more depth.

18.11 Further reading

Ashby, M.F. (2005) *Materials Selection in Mechanical Design*, 3rd edition, Butterworth-Heinemann, Oxford, UK, Chapter 4. ISBN 0-7506-6168-2. (*The source of the process selection methodology and cost model, with further case studies, giving a summary description of the key processes in each class.*)

ASM Handbook Series (1971–2004) Volume 4, *Heat Treatment*; Volume 5, *Surface Engineering*; Volume 6, *Welding, Brazing and Soldering*; Volume 7, *Powder Metal Technologies*; Volume 14, *Forming and Forging*; Volume 15, *Casting*; and Volume 16, *Machining*; ASM International, Metals Park, OH, USA. (*A comprehensive set of handbooks on processing, occasionally updated, and now available online at www.asminternational.org/hbk/index.jsp.*)

Bralla, J.G. (1998) *Design for Manufacturability Handbook*, 2nd edition, McGraw-Hill, New York, USA. ISBN 0-07-007139-X. (*Turgid reading, but a rich mine of information about manufacturing processes.*)

Campbell, J. (1991) *Casting*, Butterworth-Heinemann, Oxford, UK. ISBN 0-7506-1696-2. (*The fundamental science and technology of casting processes.*)

Houldcroft, P. (1990) *Which Process?*, Abington Publishing, Abington, Cambridge, UK. ISBN 1-85573-008-1. (*The title of this useful book is misleading—it deals only with a subset of joining process: the welding of steels. But here it is good, matching the process to the design requirements.*)

Kalpakjian, S. and Schmid, S.R. (2003) *Manufacturing Processes for Engineering Materials*, 4th edition, Prentice-Hall, Pearson Education, New Jersey, USA. ISBN 0-13-040871-9. (*A comprehensive and widely used text on material processing.*)

Lascoe, O.D. (1988) *Handbook of Fabrication Processes*, ASM International, Metals Park, Columbus, OH, USA. ISBN 0-87170-302-5. (*A reference source for fabrication processes.*)

Swift, K.G. and Booker, J.D. (1997) *Process Selection, from Design to Manufacture*, Arnold, London, UK. ISBN 0-340-69249-9. (*Details of 48 processes in a standard format, structured to guide process selection.*)

18.12 Exercises

Exercise E18.1   (a) Explain briefly why databases for initial process selection need to be sub-divided into generic process classes, while material selection can be conducted on a single database for all materials.

       (b) After initial screening of unsuitable processes, further refinement of the process selection usually requires detailed information about the materials being processed and features of the design. Explain why this is so, giving examples from the domains of shaping, joining and surface treatment.

Exercise E18.2   A manufacturing process is to be selected for an aluminum alloy piston. Its weight is between 0.8 and 1 kg, and its minimum thickness is 4–6 mm. The design specifies a precision of 0.5 mm and a surface finish in the range 2–5 μm. It is expected that the batch size will be around 1000 pistons. Use the process attribute charts earlier in the chapter to identify a subset of possible manufacturing routes, taking account of these requirements. Would the selection change if the batch size increased to 10 000?

Exercise E18.3   The choice of process for shaping a CD case was discussed in Section 18.5, but what about the CDs themselves? These are made of polycarbonate (another thermoplastic), but the principal difference from the case is precision. Reading a CD involves tracking the reflections of a laser from microscopic pits at the interface between two layers in the disc. These pits are typically 0.5 μm in size and spacing, and 1.5 μm apart; the tolerance must be better than 0.1 μm for the disc to work.

       First estimate (or measure) the mass and thickness of a CD. Use the charts to find a short-list of processes that can meet these requirements, and will be compatible with the material class. Can these processes routinely achieve the tolerance and finish needed? See if you can find out how they are made in practice.

18.13 Exploring design with CES

Exercise E18.4   Use the 'Browse' facility in CES to find:
- (a) The record for the shaping process injection molding, thermoplastics. What is its economic batch size? What does this term mean?
- (b) The machining process water-jet cutting (records for machining processes are contained in the Shaping data table). What are its typical uses?
- (c) The joining process friction-stir welding. Can it be used to join dissimilar materials?
- (d) The surface treatment process laser hardening. What are the three variants of this process?

Exercise E18.5   Use the 'Search' facility in CES to find:
- (a) Processes used for boat building.
- (b) Processes to make bottles.
- (c) Processes to make tail-light assemblies.
- (d) Processes for decoration.

Exercise E18.6   Use CES Level 3 to explore the selection of casting process for the products listed. First check compatibility with material and shape, and make reasonable estimates for the product dimensions (to assess mass and section thickness). Then include appropriate values for tolerance, roughness, and economic batch size.
- (i) Large cast iron water pipes.
- (ii) 10 000 Zn alloy toy cars (60 mm long).
- (iii) Small Ni–Co super-alloy (MAR-M432) gas turbine blades (best possible tolerance and finish).
- (iv) Large brass ship propeller.

Exercise E18.7   A small nylon fan is to be manufactured for a vacuum cleaner. The design requirements are summarized in the Table. Use CES Level 3 to identify the possible processes, making allowance if necessary for including a secondary finishing process. Suggest some aspects of the design that may merit investigation of supporting information on the selected processes.

Function	• Vacuum cleaner fan	
Objective	• Minimize cost	
Constraints	• Material: nylon • Shape: 3-D solid • Mass: 0.1–0.2 kg • Minimum section: 4 mm	Technical constraints
	• Tolerance: <0.5 mm • Roughness: <1 μm	Quality constraints
	• Batch size: 10 000	Economic constraint
Free variable	• Choice of shaping process	

Exercise E18.8   The selection of process for a connecting rod was discussed earlier in the chapter. Conduct the selection using CES Level 3, making reasonable estimates for any unspecified requirements. Explore the effect of changing to a 3 m con-rod for a ship, of approximate cross-section 10×10 cm, in a batch size of 10.

Exercise E18.9   Process selection for an aluminum piston was investigated in Exercise E18.2. Further investigation of the economics of gravity die casting and ceramic mold casting is suggested. Plot the cost against batch size for these processes, assuming a material cost of $2/kg and a piston mass of 1 kg. The overhead rate is $70/hour, the capital write-off time is 5 years and the load factor is 0.5. Which process is cheaper for a batch size of 1000? Assume that as the piston is simple in shape, it will fall near the bottom of each cost band.

Exercise E18.10   Two examples of selection of secondary processes were discussed in the chapter. The design requirements were summarized for joining processes for a radiator in Table 18.12 and for hardening a steel bearing race in Table 18.14. Use CES Level 3 to check the results obtained. For the radiator problem, use the 'Pass–Fail table' feature in CES to see if other processes could become options if the design requirements were modified.

18.14 Exploring the science with CES Elements

Exercise E18.11   Casting processes require that the metal be melted. Vapor methods like vapor metallizing require that the metal be vaporized. Casting requires energy: the latent heat of melting is an absolute lower limit (in fact it requires more than four times this). Vaporization requires the latent heat of vaporization, again as an absolute lower limit. Values for both are contained in the Elements database. Make a plot of one against the other. Using these lower limits find, approximately, how much more energy-intensive vapor methods are compared with those that simply melt.

Chapter 19
Follow the recipe: processing and properties

The Khafji rig disaster. (Image courtesy of
Thomas Brinsko with Bic Alliance Magazine.)

19.1 Introduction and synopsis

Some people are better at multi-tasking than others. The good ones should feel at home in manufacturing. Shaping, finishing and assembling components into products to meet the technical, economic and aesthetic expectations of consumers today requires the balancing of many priorities. Earlier chapters have introduced intrinsic properties—strength, resistivity and so on—that depend intimately on microstructure, and microstructure depends on processing. Microstructure is not accessible to observation during processing, so controlling it requires the ability to predict how a given process step will cause it to evolve or change. So manufacturers have their work cut out to turn materials reliably into good-quality products, while making themselves a decent profit. Manufacturing involves more than making materials into the right shapes and sticking them together; it is also responsible for producing properties on target.

There is a parallel with cooking. The recipe lists the ingredients and cooking instructions: how to mix, beat, heat, finish and present the dish. A good cook draws on experience (and creativity) to create new dishes with pleasing flavor, consistency and appearance; you might think of these as the attributes of the dish. But if the dish is not a success the first thing the cook might do is to cut right through it and examine what went wrong with its *microstructure*. Suppose, for instance, the dish is a fruit cake, then the distribution of porosity and fruit constitute aspects of its microstructure.

Microstructure is key to engineering properties too. Some of its components are very small—precipitate particles can consist of clusters of a few atoms. Others are larger—grains range in size from microns to millimeters. Here our goal is to provide an overview of the main microstructural features and how they evolve during processing, with the focus on *processing for properties*. The essence of the chapter is captured by the statement:

$$\text{Composition} + \text{Processing} \rightarrow \text{Microstructure} + \text{Properties}$$

Property charts were used in earlier chapters to illustrate examples of property changes imparted by composition and processing. Here we draw the discussion together and provide further examples from each of the material classes.

19.2 Microstructure of materials

The ability to tune properties is central to materials processing and design, but it brings with it the need for good process control. Failure to follow the 'recipe' leads, at best, to scrap and lost revenue and, at worst, to engineering failures (cover picture). We start by drawing together all of the microstructural features introduced in the book, by way of an overview.

Metals
Figure 19.1 summarizes the main microstructural features in metals. Starting at the bottom with atoms, we have crystalline packing (with the exception of the

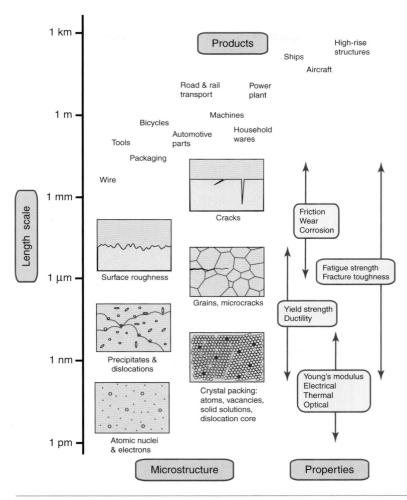

Figure 19.1 Microstructural features in metals, showing their length scale and the properties which they determine. Each interval on the length scale is a factor of 1000.

unusual amorphous metals)—responsible for elastic moduli and density. Atom-scale defects—vacancies and solute atoms—were introduced in Chapter 6. Thermal, electrical, optical and magnetic behavior are most directly influenced by this atomic scale of microstructure. Vacancies are responsible for diffusion so this, and the phenomena it causes (sintering, creep, heat treatments), also depend on atomic-scale structure. Strength, toughness and fatigue depend on structure at a slightly larger scale, that of dislocations and the obstacles to their motion sketched in Figure 19.1: precipitates and grain boundaries. Grains themselves usually fall into the 1–50 μm scale—a similar scale to the roughness on metal surfaces and of the porosity caused by the gases trapped in a metal when it is cast. Fatigue cracks start at grain scale, but grow to the dimensions of the component itself at fracture. Metal products themselves span a huge range: kitchen foil is around 10 μm thick, automotive panels a millimeter or so, ship propellers are several meters in diameter, while bridges and buildings reach the kilometer scale.

Ceramics and glasses

Ceramics are crystalline, glasses are amorphous. Figure 19.2 shows both near the bottom of our materials length scale. Most of the properties directly reflect the atomic layout and the intrinsically strong nature of the covalent or ionic

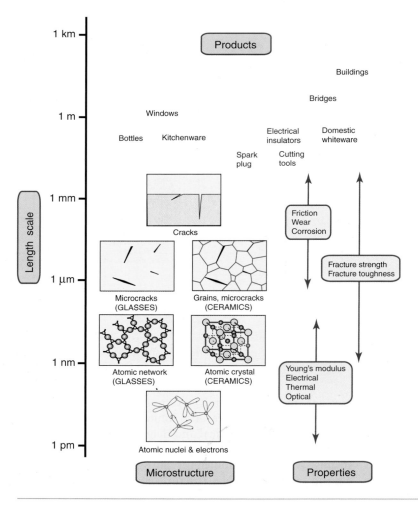

Figure 19.2   Microstructural features in ceramics and glasses, showing their length scale and the properties which they determine. Each interval on the length scale is a factor of 1000.

bonding—from elastic modulus to electrical insulation. The exceptions are strength and toughness. Cracks dominate failure in ceramics and glasses, so these are the key features—closely related to grain size (in ceramics) and to the surface finish.

Polymers and elastomers

Polymers and elastomers are inherently molecular rather than atomic. Figure 19.3 shows that there is wide diversity at this fine scale: polymer molecules can be amorphous, crystalline, cross-linked or aligned by drawing. Most properties again

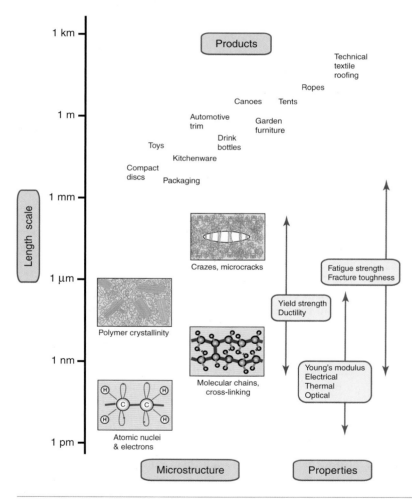

Figure 19.3   Microstructural features in polymers and elastomers, showing their length scale and the properties which they determine. Each interval on the length scale is a factor of 1000.

reflect behavior at this scale, with strength, toughness (and optical properties) bringing in larger features such as crazes and cracks. Polymer foams and fiber-reinforced polymer composites have additional length scales relating to their architecture—for example, pore size and fiber diameter. These and other 'hybrid' materials are discussed further in Section 19.6.

19.3 Microstructure evolution in processing

The role of shaping processes is to produce the right shape with the right microstructure. Achieving the first requires control of viscous flow or plasticity. Achieving the second means control of the nature and rate of structural evolution. Secondary processing (joining and surface treatment) may cause the structure to evolve further. Examples of microstructural evolution are discussed in later sections, but first some general principles on why and how it happens.

Phases, phase transformations and other structural change
A *phase* is defined as an arrangement of atoms with a definable structure and composition. Steam, water and ice are the vapor, liquid and solid phases of H_2O. Engineering alloys such as steel also melt and then vaporize if we heat them enough. More interestingly, they can change phase in the solid state, and it is this that gives us precipitation-hardened steels and non-ferrous alloys, made up of one phase, the precipitate, in a matrix of the other.

Phase transformations are changes of phase during material processing. They require a *driving force* and a *mechanism*. The first is determined by thermodynamics, the second by kinetic theory. The term driving force, somewhat confusingly, actually means a change in energy between the starting point (the initial phase or phases) and the end point (the final state). A ball released at the top of a ramp will roll down it because its potential energy at the bottom is lower than that at the top. Here the driving force is the potential energy. Phase changes are driven by differences in chemical free energy—primarily the internal bond energy (at least in crystalline materials). Solids melt on heating because being a liquid is the state of lowest free energy above the melting point. Chemical compounds can have lower free energy than solid solutions, so an alloy may decompose into a mixture of phases (both solutions and compounds) if, in so doing, the total free energy is reduced.

Microstructures can therefore only evolve from one state to another if it is energetically favorable to do so—thermodynamics points the way for change. But changes in microstructure need a *mechanism* for the atoms to rearrange from one structure to another. When a liquid solidifies, there is an interface between the liquid and the growing solid. This is where kinetic theory comes in—atoms transfer from liquid to solid by diffusion at the interface. Virtually all solid-state phase changes also involve diffusion to move the atoms around to enable new phases to form. Diffusion occurs at a rate that is strongly dependent on temperature—we calculated it in Chapter 13. The overall rate of structural evolution depends on both the magnitude of the driving force (no driving force, no evolution) and the kinetics of the diffusion mechanism by which it takes place. The rates of structural evolution are of great significance in processing. First they govern the length scale of the important microstructural features—grains, precipitates and so on—and thus the properties. But they also affect process economics, via the time it takes for the desired structural change to occur.

Solidification and precipitation involve changes in the phases present. There are other important types of structure evolution in processing that do not involve a phase change, but the principles are the same: there must be a reduction in internal energy, and a kinetic mechanism, for the change to occur. Concentrated solid solutions have higher free energy than dilute ones, so gradients in concentration tend to smooth out, if heated to give the solute atoms enough diffusive mobility. Recovery and recrystallization are mechanisms that change the grain structure of a crystal. The driving force comes from the stored elastic energy of the dislocations in a deformed crystal; the mechanism is diffusive transfer of atoms across the interface between the deformed and new grain structures.

With that brief introduction, we now illustrate the main changes in structure that take place in processing, in order to understand what can be manipulated to achieve the desired properties—or what can go wrong if we don't follow the recipe.

Solidification: metal casting

In casting, a liquid above its melting point is poured into a mold, where it cools by conduction. New solid forms by a process of *nucleation*: tiny crystals form in the melt, either spontaneously or (more usually) on the walls of the container or on foreign particles in the melt itself (Figure 19.4(a)). As noted above, the nuclei grow by diffusive attachment of atoms at the liquid–solid interface. Solidification is complete when crystals growing in opposing directions impinge on one another, forming grain boundaries—each original nucleus is the origin of a grain in the solid (Figure 19.4(b)). The initial grain size of the casting depends on the number of nuclei, since each one grows into a grain. The control of grain size, important in achieving optimal properties, is achieved by controlling nucleation through additions that stimulate it ('inoculants') and control of cooling rate. Growing a single crystal is a question of retaining just one nucleus, something that can be done by cooling the liquid slowly from one end so that the first nucleus to appear is made to grow along the entire length of the component. The rates of nucleation and growth also depend on the imposed cooling history and this is governed by heat flow. This is a multi-parameter problem: it depends on the thermal properties of the metal and mold, the contact between the two, the initial liquid temperature, the release of latent heat on solidification, and the size and shape of the casting. Here is an example of the design–process–material coupling discussed in Chapter 18: the geometry of the design combines with the type of casting process (mold material, metal temperature, etc.) and the choice of alloy to dictate the cooling rate; and the microstructural response to that cooling rate (and thus properties) depends on the alloy chemistry.

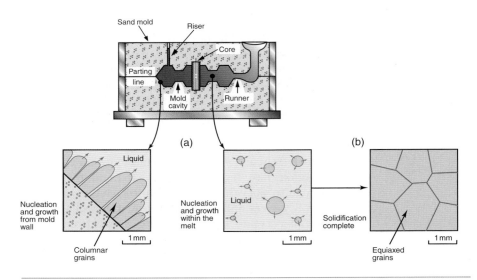

Figure 19.4   Solidification in metal casting. (a) Nucleation and growth of solid crystals in the melt. (b) Impingement of growing solid crystals forms the grains and grain boundaries.

A further complication in casting is that the liquid has a uniform concentration of solute but the solid does not. The solubility of impurities and solutes in the solid is less than that in the liquid, so solute is rejected ahead of a growing crystal. Concentration gradients are set up in the solid grains with higher solute and impurity levels in the last part to solidify (i.e. the grain boundaries). This *segregation* can be a source of problems such as embrittlement or corrosion sensitivity at the boundaries. Castings are often therefore held at high temperature for prolonged periods to enable some homogenization of these concentration gradients.

Polymer molding
Thermoplastics are molded as viscous liquids. Injection molding and extrusion dominate, but all molding processes impose flow that can orientate the molecules; if the molding is cooled fast enough the alignment is frozen in (Figure 19.5). If not, polymers mostly prefer to form an amorphous structure. In some polymers crystallinity may develop on slow cooling. All polymers shrink as the mold cools from the molding temperature to room temperature because of thermal contraction and the loss of free volume caused by crystallization. Allowance must be made for this when the mold is designed.

Deformation processing of metals
Most forming processes (rolling, forging, extrusion, drawing) are compressive rather than tensile. This enables large strains and changes of shape, without the problem of necking in tension. The grains change shape to accommodate the

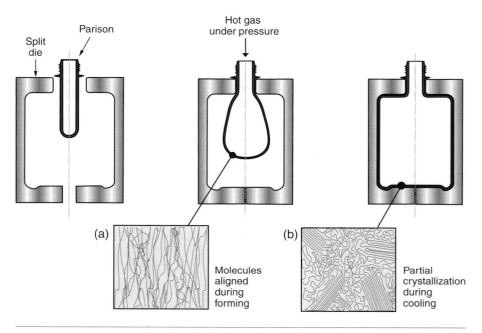

Figure 19.5   Evolution of molecular architecture in polymer molding. (a) Alignment of molecules during viscous flow in shaping. (b) Partial crystallization during cooling.

overall strain—this can give a bit of useful longitudinal strengthening, rather like a fiber composite (Figure 19.6(a)). Work hardening increases the dislocation density, useful for final properties but increasing the forming loads and limiting the possible strains. Raising the temperature to *anneal* the metal (soften with heat) overcomes the forming limits; it can be done by carrying out the forming process hot or by a separate heat treatment afterwards. Annealing involves *recovery* and *recrystallization*, introduced earlier. In recovery, dislocations interact and rearrange into organized patterns forming *sub-grains*, which are like tiny grains within grains (Figure 19.6(b)). More prolonged heating leads to *recrystallization* (Figure 19.6(c))—grain boundary migration over larger distances (by the atomic hopping mechanism at a boundary), sweeping up the dislocations and removing the work hardening completely. Both lower the yield stress, making it easier to work the material to shape (something blacksmiths have known for centuries though they had no idea of the microstructural changes). The scope for manipulating the grain structure during solid-state forming is a major distinction between the cast and wrought alloys. The grain sizes formed in wrought alloys are much smaller (typically 10–$100\,\mu\text{m}$). It's another coupled problem: grain structure depends on the alloy, process temperature, the rate of deformation, amount of strain and the annealing temperature. It is therefore difficult to get a

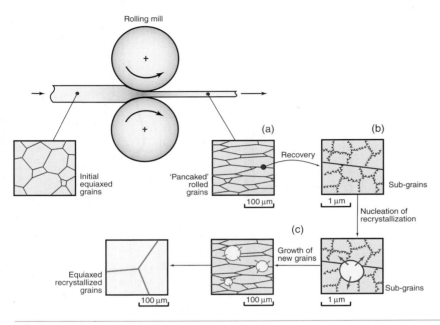

Figure 19.6   Grain structure evolution in the solid state by deformation and annealing.
(a) Grains pancake, following the change in shape of the component. (b) The
recovery mechanism: dislocations rearrange as sub-grains (shown on a large
scale). (c) The recrystallization mechanism: new grains form by migration of
boundaries from a few sub-grain nuclei, wiping out the dislocations in the
deformed microstructure.

uniform grain size in a formed part, as the strain, strain rate and temperature
vary with location.

Heat treatment of metals

Wrought metal products, and some castings and powder processed parts, are
often subjected to heat treatment. This serves various purposes. Annealing to
soften an alloy before further forming was one. Normalizing is another—slow
cooling from high temperature as the final step in making a component, the
object being to minimize the final residual stress, and to produce a microstructure
of lower strength but high toughness. Many alloys also undergo final heat treatments to enhance strength by precipitation hardening. The shaped component is
heated to high temperature, cooled at a controlled rate and usually reheated to
an intermediate temperature. This exploits the solid-state phase changes that
occur with temperature—for example, a common sequence is to solutionize
(i.e. form a solid solution at high temperature), quench to room temperature to
prevent an undesirable coarse structure forming and then precipitate fine-scale
phases in the subsequent reheat (Figure 19.7). Two heat treatments of this type
are particularly prevalent: the 'quench and temper' of steels, and 'age hardening'

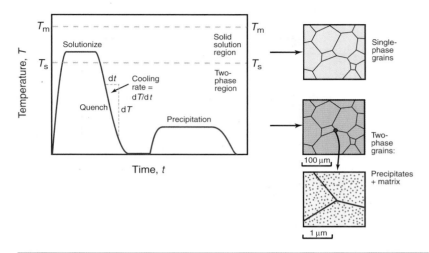

Figure 19.7   Schematic thermal profile for heat treatments that produce precipitation hardening, illustrating microstructural changes being induced: a solid solution at high temperature and precipitation of hardening particles (shown on an enlarged scale) at a lower temperature.

in aluminum alloys. But precipitation hardening is a generic mechanism in most alloy systems.

Powder processing

Ceramics, and many metals, can be shaped by filling a mold with loose powder and compacting it. The main microstructural evolution is the shrinkage of porosity during compaction, with each particle becoming a final grain. Powder compaction mechanisms are closely related to diffusional flow and creep (Figure 19.8). In purely thermal compaction ('sintering'), atoms diffuse along the particle boundaries to fill in the pores. Compaction is accelerated by imposing external pressure as well as temperature—'hot isostatic pressing' (or HIPing)—giving particle deformation by creep.

Joining processes

Thermal welding of metals involves heating and cooling. This may cause phase transformations (both melting and re-solidification, and solid–solid phase changes)—like a localized rapid casting or heat treatment process. Welds commonly have a surrounding 'heat-affected zone' (HAZ) with different properties to the rest of the component. The microstructural changes in the HAZ can soften or embrittle the material, or change the corrosion resistance. Figure 19.9 illustrates the contrasting responses to welding of low-carbon and low-alloy steels, and heat-treatable aluminum alloys, indicated by the profiles of hardness across the joint. Some mechanical joining processes (e.g. friction welding) also impose local changes in microstructure and properties, by deformation as well as heat.

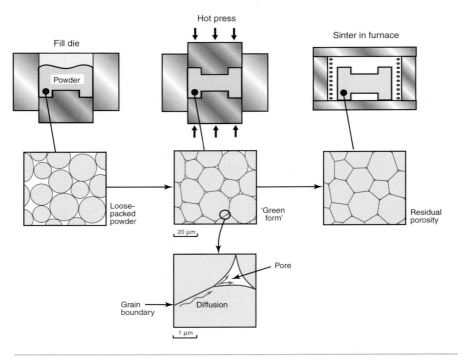

Figure 19.8   Mechanism of powder compaction in hot pressing and sintering.

Welding metallurgy is a big field of study—welds are often the 'Achilles heel' in design, being the critical locations that determine failure and allowable stresses. This may be in part because the material properties are damaged in some way at the joint (as in this example), but it may also be due to other side-effects of the process (residual stresses, or stress concentrations, or crevices, potentially causing fatigue and corrosion problems). Adhesive technologies are not trouble free—they still concentrate stress and contain defects, requiring good design and process control—but they have the advantage that they don't impose heat or deformation on the components being joined, leaving the microstructure and properties intact.

Surface engineering

Surface treatments exploit many different mechanisms and processes to change the surface microstructure and properties. Some simply add a new coating material, with its own microstructure and properties, leaving the substrate unchanged—the only problem is then making sure they stick. Others induce near-surface phase transformations by local heating and cooling. Figure 19.10 shows an example—laser hardening of steels, in which the final phase is much harder than the initial state. Direct diffusion of atoms into the surface is also feasible—but only for interstitial solute atoms at high temperature for diffusion to occur over a distance

Figure 19.9 A weld cross-section with corresponding thermal histories in the weld metal and heat-affected zone. Below are typical hardness profiles induced across welds in heat-treatable aluminum alloys, low-carbon steel and low-alloy steel.

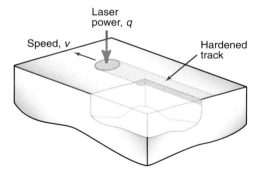

Figure 19.10 Laser hardening: a surface treatment process that modifies microstructure. The traversing laser beam induces a rapid thermal cycle, causing phase changes on both heating and cooling. The track below the path of the laser has a different final microstructure of high hardness.

of any significance. Chapter 13 revealed that, even near the melting point, there is a physical limit to diffusion distances ($\sqrt{Dt}$) of around 0.2 mm for a 1-hour hold, which is sufficient for surface treatments—and for bulk processing on a very small scale (as in semiconductor devices). This mechanism is used to increase the carbon content (and thus hardness) of steels—the surface treatment known as carburizing.

19.4 Processing for properties

There is a mantra a few pages back: *Composition + Processing → Microstructure + Properties*. Here is an example to flag up general issues about this interaction.

Aluminum bike frames

The material chosen for the bike frame, for its good stiffness and fatigue resistance at low weight (Chapters 5 and 10), is a heat-treatable aluminum alloy. Figure 19.11 illustrates the main steps in the process history. It makes a number of points.

Point 1: the chemistry of the chosen alloy is adjusted at the outset of processing, in the liquid state, and cannot be changed after casting. Metallic elements dissolve freely in one another in the liquid state, but it is very difficult to add them in the solid state; diffusion is much too slow.

Point 2: materials processing involves more than one step. Metal casting, powder processes and polymer molding are *near-net-shape processes*—the raw material is turned into the shape of the component in a single step, leaving only finishing operations (including heat treatment), so the number of steps is few. By contrast, wrought products (those that undergo some deformation processing) involve a longer chain of processes. The initial casting of wrought alloys is most economic on a large scale, so standard compositions are cast into large ingots as in Figure 19.11(a), which are transported to different factories for processing into different products. For the bike frame, the ingot is sliced into billets and extruded into tube. The figure shows, in (b), a temperature history for the material as it is shaped directly from a solid circular billet, around 200 mm in diameter, to a hollow tube, 30 mm in diameter and 3 mm thick. The extrusion is done hot to reduce the strength and increase the ductility. This enables the material to undergo the large plastic strains, allowing fast throughput at relatively low extrusion force. The quality of surface finish is determined at this stage.

Point 3: design focuses on the properties of the finished product, but some of these same properties (strength, ductility, etc.) are critical *during processing*. The design and the processing characteristics can be in conflict—strong alloys are more difficult to process and are thus more expensive. So in choosing materials for a component it is important to examine their suitability for processing as well as for performance in service. Returning to the extrusion step of the bicycle frame, we note that extrusion produces the required shape and it also enhances

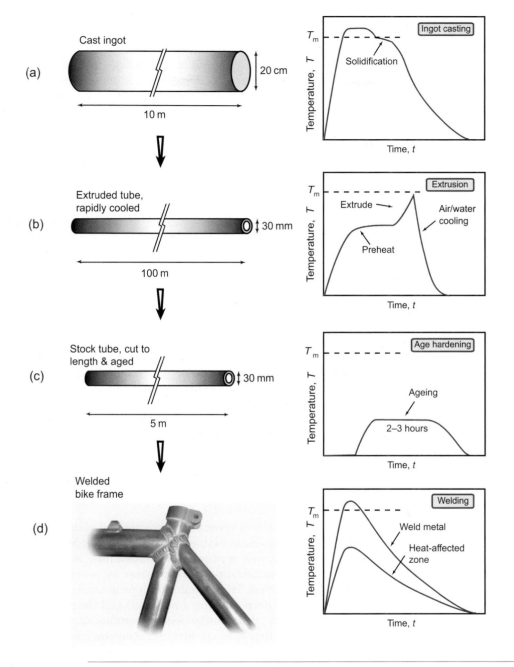

Figure 19.11   A schematic process history for the manufacture of a bike frame in heat-treatable aluminum alloy. (a) Casting of large ingot, cut into billets for extrusion. (b) Hot extrusion into tube, incorporating the solution heat treatment and quench. (c) Sections cut to length and age hardened. (d) Assembly into the frame by arc welding.

the microstructure. The hot deformation refines the as-cast grain structure and helps homogenize the alloy. Forced cooling of the tube as it emerges from the extrusion press quenches it, trapping solute in solution, thus obviating the need for a subsequent solution heat treatment prior to ageing. The tube is then cut to length and aged as in Figure 19.11(c), giving it its final strength. In a non-heat-treatable aluminum alloy, strength comes from solid solution and work hardening, not precipitation, so the extrusion is then done cold to maximize the work hardening, with no further heat treatment. Deformation and thermal histories are critical for controlling shape and final properties simultaneously, but depend in the detail on the particular alloy variant being processed.

Finally, the frame is assembled by arc welding. In the heat-affected zone the precipitates carefully produced by age hardening disappear back into solution. As the whole age-hardening cycle is difficult to repeat on the completed frame, the heat-affected zone round the weld ends up with different, inferior properties to the rest of the frame (Figure 19.9). Hence the last point in this short journey through a typical metal-processing operation is a reminder: watch out for joining steps (especially welding) that undo the good done by earlier process steps.

With this background, we now examine how diverse combinations of chemistry and processing are used to manipulate properties, in all material classes. The emphasis is on the interaction between processing and the 'ideal' properties stored in a property database, which can be achieved if all goes to plan. But the bicycle frame case study shows that we need to keep our eyes open in this game, and to be aware of why we often need to compromise between design objectives and manufacturing realities. The importance of the 'Seek Documentation' stage in material and process selection should not be underestimated!

19.5 Case studies

Metals—overview

It was clear in Chapter 18 that a favorable characteristic of metals is the diversity of manufacturing processes available to make and assemble components of complex shape. And the earlier sections have shown how, while shaping and joining components, the underlying microstructure is being manipulated to generate the properties needed. This includes both properties during processing (for example, ductility for deep drawing a beer can) and the properties needed in service (for example, strength and toughness for automotive alloys). Processing and service demands can often be in conflict, but there are cunning ways round this—e.g. shape it hot (and thus soft) and then heat treat it later (to raise the strength).

Steels (and other ferrous alloys) are the dominant engineering alloys, but similar principles apply to all the alloy systems, such as those based on Al, Cu, Ti, Ni, Zn, Mg and so on. Most alloy systems offer both cast and wrought variants, and many are also processed as powder. Usually either cast or wrought alloys dominate an alloy class. This reflects subtle differences in the ease of deformation of the underlying crystal structures in different elements. For example, Zn, Mg and Ti have the 'hexagonal close-packed (HCP)' crystal structure (see Chapter 4). This is inherently less ductile than the FCC or BCC crystal structures found in Al, Fe and Cu—so Zn, Mg and Ti are mostly cast, with only limited deformation processing being conducted, and usually then done hot (when the ductility is better). Ductile Al, Fe and Cu are mostly wrought, but all come as casting alloys too. Casting and wrought alloys in a given metal system tend to have quite different compositions. Good castability requires much higher levels of alloying additions than the relatively dilute wrought alloys (to lower the melting point). Casting leads to coarser microstructures and poorer strength and toughness than in wrought alloys (as illustrated for aluminum alloys in Chapter 8).

To illustrate the details of metals processing and property manipulation further, we now draw on examples from the dominant class: the ferrous alloys.

Ferrous alloys

For structural and mechanical applications, steels and other alloys based on iron dominate. They are intrinsically stiff, strong and tough, and mostly low cost. High density is a drawback for transport applications, allowing competition from light alloys, wood and composites. Figure 19.12 illustrates the diversity of applications for ferrous alloys. These reflect the many classes of alloy—different chemistries combined with different process histories. Examining each in turn highlights the key material property, composition and processing factors at work.

Figure 19.12   A selection of products made from alloys based on iron. (a) Cast iron: brake disk. (b) Low-carbon steel: I-beam. (c) Low-alloy steel: connecting rod. (d) Stainless steel: cutlery.

Cast iron brake disk.   Brake disks (Figure 19.12(a)) use sliding friction on the brake pads to decelerate a moving vehicle, generating a lot of heat in the process (Chapter 11). The key material properties are therefore hardness (strength) for wear resistance, good toughness and a high maximum service temperature. Low weight would be nice for a rotating part in a vehicle, but this is secondary. All ferrous alloys fit the bill—so why cast iron? Because casting is the cheapest way to make the moderately complicated shape of the disk, and the as-cast microstructure does not usually require further heat treatment to provide the hardness required. Cast irons typically contain 2–4% carbon (by weight) giving a lot of iron carbide—a hard compound giving excellent precipitation hardening. Cast iron may also contain free graphite, giving good machinability, the ability to be machined quickly with minimal or no lubricant—another important 'property during processing'.

As described in Section 19.3, the grain structure is formed during the shaping process and cannot be modified later. The key features to control are the size of the grains themselves (e.g. by using inoculants) and the distribution of impurities. Rather than trying to remove impurities from the melt, which is expensive, it is usually possible to render them harmless by giving them something else to react with, to form a solid compound distributed throughout the casting. Manganese is added to ferrous alloys to clean up sulfur, for instance. Tweaking the chemistry of castings to improve their properties has been a bit of a black art for centuries. Nowadays it is possible to use sophisticated software tools to model everything from the choice of composition to the way to pour the metal into the mold in order to minimize porosity and residual stress.

Plain carbon steel: I-beams, cars and cans.   If we had to single out one universally dominant material, we might well choose mild steel—iron containing 0.1–0.2% carbon. Almost all structural sections (Figure 19.12(b)), automotive alloys and steel packaging (beer and food cans) are made of mild steel (or a variant enhanced with a few other alloying additions). All of these applications are wrought—the alloy is deformed extensively to shape. The excellent ductility of plain carbon steel enables this, while the deformation process exploits their work hardening to give the required product strength. Ductility, toughness and decent strength are vital in a structural material—we would rather a bridge sagged a little rather than broke in two, and in a car crash our lives depend on the energy absorption of the front of the vehicle.

It was noted earlier that annealing is commonly used to control the grain structure in wrought alloys, including low-carbon steels. An important side-effect of deformation followed by annealing is the concept of crystal texture. The crystal planes in a casting are random—each nucleus forming in isolation within a melt. Not only does deformation 'pancake' the grains, it tends to align the crystallographic planes with respect to the axes of the deformation. Perhaps unexpectedly, recrystallization doesn't restore randomness—there remains a statistical distribution of orientations. This doesn't sound very important, but it is significant in working with sheet metal, since textured metals have yield properties which are *anisotropic* (i.e. different in different directions). This can be a problem if you want to make something smooth and circular out of a sheet, like a beverage container.

Alloy steels and heat treatment: cranks, tools and gears. The moving and contacting parts of machinery (Figure 19.12(c)) are subjected to very demanding conditions—high bulk stresses to transmit loads in bending and torsion (as in a crank or a drive shaft), high contact stresses where they slide or roll over one another (as in gears), and often reciprocating loads promoting fatigue failure (as in connecting rods). Strength with good toughness is everything. Plain carbon steels extend up to 0.8% carbon—this introduces more iron carbide into the microstructure, giving greater precipitation hardening. They do a reasonable job, but the density of iron is a problem. Make the steel stronger, and use less of it, and we can save weight. This is particularly true for fast-moving parts in engines, since the support structure can also be made lighter if the inertial loading is reduced (so-called 'secondary weight savings'). The trick is yet more alloying, and more processing.

A bewildering list of additions to carbon steel can be used to improve the strength—Mn, Ni, Cr, V, Mo and W are just the most important! Some contribute directly to the strength, giving a solid solution contribution (e.g. high-alloy tool steels, with up to 20% tungsten). More subtle though is the way quite modest additions (<5% in low-alloy steels) affect the alloy's response to heat treatment. The key process is 'quench and temper' (Figure 19.7). The steel is heated to a temperature at which the carbon and the alloy additions dissolve (850–1000°C) and then quenched rapidly to room temperature, usually in oil or water. Tempering is reheating of the steel to an intermediate temperature to precipitate a fine, uniform dispersion of iron carbide in iron in every single grain. The quench is critical—slow cooling produces a softer mixture of grains, some being pure iron and some containing a coarse dispersion of iron carbide. A bit of additional precipitation strength comes from the formation of alloy carbides, but this is not the main reason for alloying. Plain carbon steels can also be quenched and tempered. The problem is the severity of the quench needed to avoid the formation of the much less effective microstructure associated with slow cooling. Alloying effectively shuts this down, enabling slower quench rates to achieve the target microstructure. In consequence, bigger components can be heat treated this way—the cooling rate at the center being limited by heat conduction through the steel. The technical term for this is enhancing hardenability—it is another example of how material, process and design detail can interact in meeting the objectives of a design.

In passing, it is worthy of note that the intermediate microstructure (known as 'martensite') formed on quenching a hardenable steel is very hard, but also very brittle—like a ceramic. It is therefore useless as a bulk microstructure—tempering is essential to restore useful toughness. However, some surface treatments can produce a thin layer of martensite on a component—excellent for wear resistance on gears, bearings and so on. This is how laser hardening works

(Figure 19.10). In other respects, martensite is a bit of a hazard. It can cause cracking problems on quenching, due to stresses induced by differential thermal contraction (Chapter 12). And even worse, welding causes a thermal cycle similar to the first stage of heat treatment, so higher carbon and alloy steels are susceptible to inadvertent embrittlement if the cooling rate is too quick (Figure 19.9). Oil rigs and bridges have collapsed without warning as a result of this behavior.

Stainless steel: cutlery. The city of Sheffield in England made its name on stainless steel cutlery (Figure 19.12(d)), and today manufactures more steel (of all types) than ever before. The addition of substantial amounts of chromium (up to 20% by weight) imparts excellent corrosion resistance to iron, avoiding one of its more obvious failings: rust. It works because the chromium reacts more strongly with the surrounding oxygen, protecting the iron from attack (Chapter 17). Nickel is also usually added for other reasons—one being the preservation of the material's toughness at the very low cryogenic temperatures needed for the stainless steel pressure vessels used to store liquefied gases. Another is that both chromium and nickel provide solid solution hardening—this and work hardening (during rolling and forging) being the usual routes to strength in stainless steels, rather than heat treatment for precipitation strength.

Summary. In all of these applications, strength and toughness dominate the property profile. The manipulation of these properties was illustrated in Chapter 8 on a property chart for the main structural light alloys (aluminum). How do these varied ferrous alloys map out on this chart? Figure 19.13 shows a small selection of steels of the four types discussed above. This paints a remarkable picture. Starting with soft, tough pure iron (top left), we can manipulate it to produce more or less any combination of strength and toughness we like—increasing the strength by more than a factor of 20. Every final material (with the exception of as-quenched martensite) is comfortably above the typical fracture toughness threshold ($\approx 15 \, \text{MPa.m}^{1/2}$) for structural or mechanical application. The normalized condition is standard for structural use. The chart shows how the quench-and-temper treatment enhances the strength without damaging the toughness. This figure explains why steels are so important—no other material is so versatile, with literally hundreds of different steels and ferrous alloys available commercially.

Ceramics and glasses

Brittle materials have a problem. No matter what other advantages they have, they are susceptible to fracture, which may be traced back in many cases to the inherent defects built in during processing. Ceramic and glass processing for mechanical application are therefore about the control of defects. The microcracks in ceramics largely reflect the size of powders used in the first place, but good-quality processing is then needed to avoid porosity. This is surprisingly difficult to achieve, even in simple shapes. Loose powders contract by a factor of 2 or more when they are pressed and sintered together. Achieving uniform contraction of this magnitude is not easy, since different regions of the product

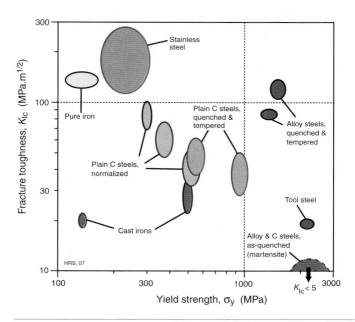

Figure 19.13   Fracture toughness—yield strength property chart for ferrous alloys. Composition and heat treatment enable a wide range of combinations of these properties.

densify first and then constrain the further compaction of the remainder. Glasses are easier to handle as they are melt processed. None the less, great care is taken over achieving a smooth finish—plate glass is mostly produced by flotation on a bath of molten tin. Ceramics and glasses are also very susceptible to thermal stresses during cooling. Being processed at high temperature, but inherently stiff and strong, stresses are not easily relieved by creep during cooling. In one instance, however, thermal stresses are exploited to good effect. Glass can be made more fracture resistant by deliberately generating a residual stress profile through-thickness by cooling the surfaces. This leads to beneficial compression at the surfaces and tension within—the surface being the likely location of the scratches and microcracks which cause fracture under tensile load.

Ceramic compositions tend to be largely discrete—each being a specific compound (alumina, silicon carbide, etc.). Some composite mixtures are processed together, such as tungsten carbide and cobalt, to make a hard, tough material for dies and cutting tools. In contrast, glass compositions can be tuned continuously over a wide range to achieve particular combinations of color and other optical properties (e.g. the graded compositions used in optical fibers). The adaptable spread of optical properties was evident in the property charts of Chapter 15. This is largely a matter of tweaking the chemistry rather than the process. More recent developments with glass processing are associated with coating technology—for scratch and chip resistance (in windscreens), or for self-cleaning and high reflectivity in high-rise buildings.

Polymers and elastomers

Polymers are versatile materials in many respects—easy to mold into complex shapes at low cost, easy to join together by snap fitting and easy to color. Bulk polymers provide a spread of Young's modulus, strength and toughness—all relatively low. Polymer chemistry, and the use of additives and fillers, give enormous freedom to adapt the bulk properties, though the ability to make large changes in strength is more limited than for metals. In contrast, polymers do offer scope for changing elastic properties, which metals do not.

Chapter 4 introduced the structure of polymers, and how chemistry is used to produce the thousands of variants available today. Long-chain molecules are produced by polymerization of a basic monomer unit of carbon, bonded with hydrogen, chlorine, oxygen and so on. Each monomer has a discrete chemistry—C_2H_4 makes polyethylene, for example. Unlike metal (and non-metal) atoms, monomers do not readily combine with one another. Only selected combinations will polymerize together, to make copolymers (with two monomers), terpolymers (with three) and so on. For example, ABS is a terpolymer, combining monomers of acrylonitrile, butadiene and styrene (hence the name). Blending polymers extend the range further, giving a mixture of more than one type of chain molecule—for example, ABS–PVC. And the molecular architecture itself can be manipulated. As described in Chapter 4, thermosets and elastomers are cross-linked, and some thermoplastics will partially crystallize. Furthermore, bulk polymers usually contain additives and fillers, for all sorts of reasons—for example, glass or ceramic particles or fibers (for stiffness, strength or color), flame retardants, plasticizers or UV protection.

Most of this has only a modest impact on processing of polymers—most processes work with most polymers. Additives change the viscosity, so the ease of injection molding will be influenced by the volume fraction of filler. Fusion welding only works for non-cross-linked thermoplastics, which can be re-melted. And, as noted earlier, crystallization is sensitive to temperature history—the rate of cooling of a molding will have some influence on the final stiffness and strength, as well as the dimensional precision. Processing can, however, play a significant role in determining the final molecular arrangement, and thus properties, if the molecules are aligned mechanically during the shaping process. Figure 19.5 showed the blow molding of a drinks bottle, from a pre-form closed tube called a parison. The stretching process of inflating the parison aligns the molecules in the bottle wall, enhancing stiffness and strength.

The most dramatic impact of processing on polymer properties is in making fibers. These are drawn from the melt, or for even better results, cold drawn. Now the molecular alignment is much more marked, so that the fiber properties exploit the covalent bonding along the chain. We can illustrate the resulting strength and stiffness on a chart—but to emphasize the lightweight performance with respect to other materials, Figure 19.14 shows the specific strength σ_y/ρ and specific stiffness E/ρ. Bulk polymers, in spite of all of the chemical trickery, only span about a decade in strength and stiffness, and are outperformed by metals—particularly on specific stiffness. Standard polymer fibers of nylon or

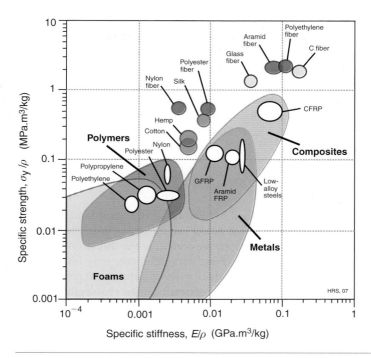

Figure 19.14 Specific strength—specific modulus property chart. Bulk polymers are stiffened and strengthened by drawing into fibres, competing with natural, carbon and glass fibres. Most products use the fibres in a composite, such as CFRP, or woven into a fabric.

polyester are stiffer and over 100 times stronger than the bulk polymer—they quickly overtake most natural fibers, such as cotton and hemp, but not silk, which is a particularly good natural fiber. However, aramid fibers (such as Kevlar), and special grades of polyethylene fibers, are up to 10 times stiffer and stronger than bulk polymer, with specific properties close or superior to carbon and glass fibers.

Fiber properties are all very impressive—the problem though is getting the fibers into a form that is useful to the engineer. One option is composites—aramid fibers are used in the same way as carbon and glass fibers, in a matrix of epoxy resin. As discussed in earlier chapters, the resulting composites shown in Figure 19.14 lie between those of the matrix and fiber, where they compete strongly with the best of the metals, such as low-alloy steels. Alternatively, we use polymer fibers in exactly the way that natural fibers have been used for millennia: twisted, tangled or woven to make rope, cables, fabrics and textiles. Fiber-based materials are a whole discipline of materials engineering in themselves, with many important applications. The geometric structure of ropes and textiles is essential to the way they work—it is this which gives them flexibility in bending, combined with enormous stiffness and strength in tension. Ropes can be stretched, coiled and knotted elastically due to their twisted and woven architecture. The loss of alignment of the fibers in making a rope or weave does mean that their load-carrying capacity is never as good as that of the fibers themselves in their pristine form. Nonetheless it is clear from the property chart that cables based on the best polyethylene fibers, for example, can potentially compete with conventional steel cabling. The drape of clothing exploits the differences in stiffness parallel to and at 45° to the weave (the 'bias').

19.6 Making hybrid materials

Hybrids were defined in Chapter 2 as a combination of materials from different classes—one of which could just be air (as in foams). Making a hybrid is often a processing challenge, so we conclude the discussion of processing for properties with some examples. First we might ask: why bother with hybrids at all? The answer is simple: hybrids can occupy spaces on property charts not occupied by monolithic materials. Sometimes design requirements can be in direct conflict, with no single material able to provide the property profile needed—for example, high strength (for which alloys are best) and high electrical conduction (for which pure metals are best). Hybrids allow innovative design solutions and improvements in performance, exploiting the individual properties of the component materials. Hybrids take a variety of forms—a simple classification is illustrated in Figure 19.15. Composites are the most straightforward—particles or fibers

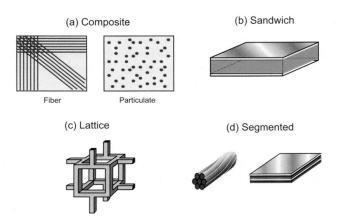

Figure 19.15   Classes of hybrid materials. (a) Fiber and particulate composites. (b) Sandwich panels. (c) Lattice structures, such as foams. (d) Segmented materials, such as ropes and laminates.

of one material embedded in another. Next are sandwich structures—skins of high-performance material on a lower grade core. Lattices (random or regular) include foams and other highly porous materials, while segmented materials are made of interlocking strands or layers of materials—ropes and cables are often hybrids of several types of material.

This classification is not 100% robust, since the boundaries get a bit fuzzy at the edges. For example, modern 'nanocomposites' are made by mixing nanometer-scale particulates into a molten metal and casting it—by Figure 19.15 this is a hybrid. However, heat treatment of steels produces an iron matrix with a fine dispersion of carbides—essentially the same microstructure, also produced for strength, yet we classify this as a metal. And at the other end of the scale, designing with a sandwich 'material' merges into designing a sandwich 'structure' (a ski, an aircraft floor panel or an industrial door). Essentially hybrids combine materials that are routinely available and used in their own right as bulk materials, with the bulk materials and the hybrid having definable material properties. In spite of some non-uniqueness in this definition, hybrid materials can all be thought of as some combination of component (bulk) materials plus configuration and scale. Configuration defines the proportions, geometric integration and shape of the components; scale defines the length scale on which they are mixed. The length scales of hybrids are mostly relatively coarse—fibers of order 1–10 μm, and the layers of laminates and sandwich panels of order 0.1–10 mm. Bulk material microstructures span everything from the atomic scale (nanometers) up to the micron scale (see Figures 19.1–19.3). However, since hybrids combine the properties of the component materials, their properties reflect microstructure all the way from atomic to product scale.

Making foams

Foams extend the range of stiffness and strength axes of the property charts downwards by several orders of magnitude. Some foams are made by trapping bubbles in a melt and solidifying the froth as a slab. Others are made by mixing in chemicals which react to form a gas, injecting it into a cavity and allowing it to inflate *in situ* as the gas bubbles are formed internally. The mechanical properties of an ideal foam depend systematically on the relative density (the fraction of solid), as shown in Chapter 4 for Young's modulus. In practice it is found that the properties are also sensitive to the variability in pore size. It is important in processing to control the size of the largest pores, which initiate premature failure. Metal foams and rigid polymer foams can be made to shape and then cut and assembled into other structured materials (often as the cores of sandwich materials themselves).

Making composites

Most polymers are in fact particulate composites—bulk polymer melts to which filler powders have been added before molding to shape. This is a hybrid which we opt to classify as a polymer variant—largely because the scale of the additives is very fine, almost at the scale of the polymer's own microstructure.

Metal–matrix composites, on the other hand, are regarded as distinct hybrids—SiC particles in aluminum alloys, for example—but they too are processed in exactly the same way as the matrix aluminum alloy. Long fiber-reinforced polymer composites—CFRP, GFRP and Kevlar (aramid FRP)—require dedicated manufacturing techniques. Their potential performance in lightweight design has been clearly illustrated in earlier chapters, but can the ideal properties be achieved routinely in practice? The answer depends on the complexity of the component geometry and the impact of joining. Composites are straightforward to produce in flat panels. 'Prepreg' layers of partly cured resin containing unidirectional fibers are stacked, usually aligned in several directions (e.g. 0, 90°, ±45°) to give reasonably uniform isotropy in the properties, and cured under pressure in an autoclave. Circular tubes are also easy to make by filament winding. More complex shapes are more difficult to laminate, particularly if the section changes in thickness. Joints are particularly troublesome. Fiber composites cannot be welded, while drilling holes for bolts and rivets damages the fibers and provides stress concentrations which can initiate cracks and delaminations around the hole. The best technique is adhesive joining, and the increasing use of composites has stimulated the development of improved adhesives. Careful joint design and manufacture is necessary to avoid premature failure in or around the joint.

Making laminates and sandwich panels

Most interest in laminates and sandwich panels reflects the potential to exploit fiber-reinforced composites in this form—a hybrid within a hybrid. Sandwich panels provide bending stiffness (and strength) at low weight. The lightweight core separates the skins, to maximize the second moment of area, without contributing directly to the bending resistance. Good-quality adhesive joints between skin and core are again critical for success. Comparison of sandwich panels with monolithic-shaped materials (such as I-beams) would normally be made on characteristics such as 'flexural rigidity' *EI*, as opposed to strict material properties. Laminates offer another hybrid solution. Various GFRP and aluminum laminates, such as GLARE and ARALL, have been developed for aerospace structures, as they offer a competitive alternative to purely aluminum or composite fabrication. And woods have had something of a renaissance in large-scale civil engineering in the form of laminates such as Glulam. The laminating process distributes the defects inherent in natural wood, reducing property variability, and enables much larger beam structures to be built. Once again, the glue between the layers in all laminates is a key part of the hybrid.

19.7 Summary and conclusions

Manufacturing processes are central to achieving the target properties in a material. Shaping processes such as casting, molding and forming always impose a thermal history and often a deformation history. These do more than make the shape—they also govern the evolution of the internal microstructure on which properties depend. Further processing at the surface can locally change the microstructure and properties there, e.g. for wear resistance. And joining components together can change things further—joints often being the design-limiting location, or the source of disaster. So processing offers many opportunities for innovation, together with responsibility for careful control. Metals show the greatest versatility—not surprising with the many different elements and alloy compositions available to work with, and their inherent castability and/or formability. One class alone—ferrous alloys—covers a wide domain of the key structural properties of strength and toughness. Other alloy systems use the same principles of modifying chemistry, forming and heat treatment to develop different property profiles.

Ceramics, glasses and polymers all have some scope for variation of composition and process. Blending, cross-linking and crystallinity provide modest ranges in the properties of polymers—including the elastic modulus. Most effective is the production of fibers, taking bulk polymers into completely new territory, well beyond the performance of natural fibers. Making use of these fiber properties in practical products needs imaginative hybrid construction, as composites, ropes, textiles and fabrics. Hybrids also extend to other forms—sandwich panels, foams and laminates. Processing again plays its part in the practicalities of using hybrids, with the development of high-performance adhesives receiving a boost from the drive to exploit their potential in design.

19.8 Further reading

Ashby, M.F. and Jones, D.R.H. (2005) *Engineering Materials II*, 3rd edition. ISBN 0-7506-6381-2 (*Popular treatment of material classes, and how processing affects microstructure and properties.*)

ASM Handbook Series (1971–2004) Volume 4, *Heat Treatment*; Volume 5, *Surface Engineering*; Volume 6, *Welding, Brazing and Soldering*; Volume 7, *Powder Metal Technologies*; Volume 14, *Forming and Forging*; Volume 15, *Casting*; and Volume 16, *Machining*; ASM International, Metals Park, OH, USA. (*A comprehensive set of handbooks on processing, occasionally updated, and now available online at www.asminternational.org/hbk/index.jsp.*)

Bralla, J.G. (1998) *Design for Manufacturability Handbook*, 2nd edition, McGraw-Hill, New York, USA. ISBN 0-07-007139-X. (*Turgid reading, but a rich mine of information about manufacturing processes.*)

Kalpakjian, S. and Schmid, S.R. (2003) *Manufacturing Processes for Engineering Materials*, 4th edition, Prentice-Hall, Pearson Education, New Jersey, USA. ISBN 0-13-040871-9. (*A comprehensive and widely used text on material processing.*)

Lascoe, O.D. (1988) *Handbook of Fabrication Processes*, ASM International, Metals Park, Columbus, OH, USA. ISBN 0-87170-302-5. (*A reference source for fabrication processes.*)

19.9 Exercises

Exercise E19.1 The table shows typical data for strength and fracture toughness of a selection of copper alloys, both cast and wrought. Sketch a property chart (on log scales) and plot the data. Use the chart to answer the following:

(a) How do the cast and wrought alloys compare on fracture toughness, at comparable strength?
(b) Rank the strengthening mechanisms (as indicated in the table) in order of effectiveness.
(c) Do the trends observed in (a) and (b) follow a similar pattern to aluminum alloys (Figure 8.15)?

Alloy	Process route	Main strengthening mechanisms	Yield strength (MPa)	Fracture toughness (MPa.m$^{1/2}$)
Pure Cu	Cast	None	35	105
Pure Cu	Hot rolled	Work	80	82
Bronze (10% Sn)	Cast	Solid solution	200	55
Brass (30% Zn)	Cast	Solid solution	90	80
Brass (30% Zn)	Wrought + annealed	Solid solution	100	75
Brass (30% Zn)	Wrought	Solid solution + work	400	35
Cu – 2% Be	Wrought + heat treated	Precipitation	1000	17

Exercise E19.2 The excellent specific properties of natural and artificial fibers were highlighted in Figure 19.14. It is also of interest to explore other property combinations which measure performance—for example, the maximum elastic stored energy. The table summarizes typical data for various fibers, together with some bulk polymers and steel for comparison.

Calculate the following performance indices for the materials given: (a) maximum elastic stored energy (per unit volume), σ_f^2/E; (b) maximum elastic stored energy (per unit mass), $\sigma_f^2/E\rho$.

Which material appears best on each criterion? Which criterion would be more important for climbing ropes? Why is it not practical for the properties of fibers to be exploited to their maximum potential?

	Hemp	Spider web silk	Bulk nylon	Aramid fiber	Polyester fiber	PE fiber	Nylon fiber	Alloy steel wire
Young's modulus (GPa)	8	11.0	2.5	124	13	2.85	3.9	210
Strength (MPa)	300	500	63	3930	784	1150	616	1330
Density (kg/m^3)	1490	1310	1090	1450	1390	950	1140	7800

19.10 Exploring design with CES

Exercise E19.3	Use CES Level 3 data to explore the properties of Cu–Ni alloys. Extract data for Young's modulus, yield stress, fracture toughness and electrical resistivity for annealed pure Cu and Ni, and a selection of alloys in between (e.g. 10%, 30%, 70% Ni). Sketch how the properties vary with Ni composition (between 0 and 100%).

(Note that in CES you can plot the composition of the dominant element on a chart axis—try plotting each property against the % Cu, selecting both Cu–Ni and Ni–Cu alloys to cover the full range, together with pure Cu and Ni.)

(a) Which properties follow an approximate linear rule of mixtures between the values for pure Cu and Ni?
(b) The alloys in CES are not binary Cu–Ni alloys. Which properties appear most strongly influenced by the other alloying additions?

Exercise E19.4	Use CES Level 3 to plot the strength and fracture toughness of Mg and its alloys. How do cast and wrought alloy variants compare?
Exercise E19.5	The records for polymers in CES include the % filler as a parameter. Plot each of the following properties against the % filler for PA, PP and epoxy, and explore how effective fillers are at enhancing the Young's modulus and strength of these polymers. How does the addition of filler affect the fracture toughness and the price/kg?

Chapter 20
Materials, processes and the environment

Power from the wind. (Image courtesy of Leica
Geosystems, Switzerland.)

20.1 Introduction and synopsis

The practice of engineering consumes vast quantities of materials and is dependent on a continuous supply of them. We start by surveying this consumption, emphasizing the materials used in the greatest quantities. Increasing population and living standards cause this consumption rate to grow—something it cannot do forever. Finding ways to use materials more efficiently is a prerequisite for a sustainable future.

There is a more immediate problem: present-day material usage already imposes stress on the environment in which we live. The environment has some capacity to cope with this, so that a certain level of impact can be absorbed without lasting damage. But it is clear that current human activities exceed this threshold with increasing frequency, diminishing the quality of the world in which we now live and threatening the well-being of future generations. *Design for the environment* is generally interpreted as the effort to adjust our present product design efforts to correct known, measurable, environmental degradation; the time-scale of this thinking is 10 years or so, an average product's expected life. *Design for sustainability* is the longer-term view: that of adaptation to a lifestyle that meets present needs without compromising the needs of future generations. The time-scale here is less clear—it is measured in decades or centuries—and the adaptation required is much greater.

20.2 Material consumption and its growth

Material consumption

Speaking globally, we consume roughly 10 billion (10^{10}) tonnes of engineering materials per year. Figure 20.1 gives a perspective: it is a bar chart of the consumption of the materials used in the greatest quantities. It has some interesting messages. On the extreme left, for calibration, are hydrocarbon fuels—oil and coal—of which we currently consume a colossal 9 billion tonnes per year. Next, moving to the right, are metals. The scale is logarithmic, making it appear that the consumption of steel (the first metal) is only a little greater than that of aluminum (the next); in reality, the consumption of steel exceeds, by a factor of 10, that of all other metals combined. Steel may lack the high-tech image that attaches to materials like titanium, carbon-fiber reinforced composites and (most recently) nano-materials, but make no mistake, its versatility, strength, toughness, low cost and wide availability are unmatched.

Polymers come next: 50 years ago their consumption was tiny; today the combined consumption of commodity polymers polyethylene (PE), polyvinyl chloride (PVC), polypropylene (PP) and polyethylene-terephthalate (PET) begins to approach that of steel.

The really big ones, though, are the materials of the construction industry. Steel is one of these, but the consumption of wood for construction purposes

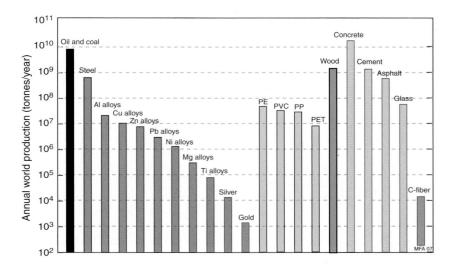

Figure 20.1   The consumption of hydrocarbons (left-hand column) and of engineering materials (the other columns).

exceeds that of steel even when measured in tonnes per year (as in the diagram), and since it is a factor of 10 lighter, if measured in m³/year, wood totally eclipses steel. Bigger still is the consumption of concrete, which exceeds that of all other materials combined. The other big ones are asphalt (roads) and glass.

The last column of all illustrates things to come: it shows today's consumption of carbon fiber. Just 20 years ago this material would not have crept onto the bottom of this chart. Today its consumption is approaching that of titanium and is growing fast.

The columns in this figure describe broad classes of materials, so—out of the 160 000 materials now available—they probably include 99.9% of all consumption when measured in tonnes. This is important when we come to consider the impact of materials on the environment, since impact scales with consumption.

The growth of consumption

Most materials are being consumed at a rate that is growing exponentially with time (Figure 20.2), simply because both population and living standards grow exponentially. One consequence of this is dramatized by the following statement: at a global growth rate of just 3% per year we will mine, process and dispose of more 'stuff' in the next 25 years than in the entire history of human engineering (see Exercises). If the current rate of consumption in tonnes per year is C then exponential growth means that

$$\frac{\mathrm{d}C}{\mathrm{d}t} = \frac{r}{100}C \qquad (20.1)$$

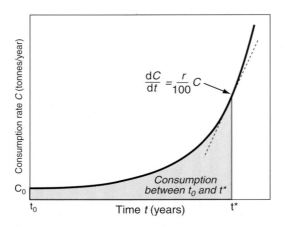

Figure 20.2   Exponential growth. Consumption rate C doubles in a time $t_d \approx 70/r$, where r is the annual growth rate.

where, for the generally small growth rates we deal with here (1–5% per year), r can be thought of as the percentage fractional rate of growth per year. Integrating over time gives

$$C = C_0 \, \exp\left\{\frac{r(t - t_0)}{100}\right\} \qquad (20.2)$$

where C_0 is the consumption rate at time $t = t_0$. The *doubling time* t_D of consumption rate is given by setting $C/C_0 = 2$ to give

$$t_D = \frac{100}{r} \log_e(2) \approx \frac{70}{r} \qquad (20.3)$$

After a period of stagnation, steel consumption is growing again, driven by growth in China; at 4% per year it doubles about every 18 years. Polymer consumption is rising at about 5% per year—it doubles every 14 years. During times of boom—the 1960s and 1970s, for instance—polymer production increased much faster than this, peaking at 18% per year (it doubled every 4 years).

The picture, then, is one of a global economy ever more dependent on a supply of materials, almost all drawn from non-renewable resources. To manage these in a sustainable way requires an understanding of the material life cycle. We turn to this next.

Figure 20.3   The material life cycle. Ore and feedstock are mined and processed to yield a material. This is manufactured into a product that is used and, at the end of its life, discarded or recycled. Energy and materials are consumed in each phase, generating waste heat and solid, liquid and gaseous emissions.

20.3 The material life cycle and criteria for assessment

Life-cycle assessment and energy
The materials life cycle is sketched in Figure 20.3. Ore and feedstock, drawn from the earth's resources, are processed to give materials; these are manufactured into products that are used and, at the end of their lives, discarded, a fraction perhaps entering a recycling loop, the rest committed to incineration or landfill. Energy and materials are consumed at each point in this cycle (we shall call them 'phases'), with an associated penalty of CO_2, SO_x, NO_x and other emissions—heat, and gaseous, liquid and solid waste, collectively called environmental '*stressors*'. These are assessed by the technique of *life-cycle analysis* (LCA). A rigorous LCA examines the life cycle of a product and assesses in detail the eco-impact created by one or more of its phases of life, cataloging and quantifying the stressors. This requires information for the life history of the product at a level of precision that is only available after the product has been manufactured and used. It is a tool for the evaluation and comparison of existing products, rather than one that guides the design of those that are new. A full LCA is time-consuming and expensive, and it cannot cope with the problem that 80% of the environmental burden of a product is determined in the early

stages of design, when many decisions are still fluid. This has led to the development of more approximate 'streamline' LCA methods that seek to combine acceptable cost with sufficient accuracy to guide decision-making, the choice of materials being one of these decisions. But even then there is a problem: a designer, seeking to cope with many interdependent decisions that any design involves, inevitably finds it hard to know how best to use data of this type. How are CO_2 and SO_x emissions to be balanced against resource depletion, toxicity or ease of recycling?

This perception has led to efforts to condense the eco-information about a material production into a single measure or *indicator*, normalizing and weighting each source of stress to give the designer a simple, numeric ranking. The use of a single-valued indicator is criticized by some. The grounds for criticism are that there is no agreement on normalization or weighting factors, and that the method is opaque since the indicator value has no simple physical significance. But on one point there is international agreement: the Kyoto Protocol of 1997 committed the developed nations that signed it to progressively reduce carbon emissions, meaning CO_2. At the national level the focus is more on reducing energy consumption, but since this and CO_2 production are closely related, they are nearly equivalent. Thus, there is a certain logic in basing design decisions on energy consumption or CO_2 generation; they carry more conviction than the use of a more obscure indicator. We shall follow this route, using energy as our measure. Before doing this, some definitions.

20.4 Definitions and measurement: embodied energy, process energy and end of life potential

Embodied energy H_m and CO_2 footprint

The *embodied energy* of a material is the energy that must be committed to create 1 kg of usable material—1 kg of steel stock, or of PET pellets, or of cement powder, for example—measured in MJ/kg. The CO_2 *footprint* is the associated release of CO_2, in kg/kg. It is tempting to try to estimate embodied energy via the thermodynamics of the processes involved—extracting aluminum from its oxide, for instance, requires the provision of the free energy of oxidation to liberate it. This much energy must be provided, it is true, but it is only the beginning. The thermodynamic efficiencies of processes are low, seldom reaching 50%. Only part of the output is usable—the scrap fraction ranges from a few percent to more than 10%. The feedstocks used in the extraction or production themselves carry embodied energy. Transport is involved. The production plant itself has to be lit, heated and serviced. And if it is a dedicated plant, one that is built for the sole purpose of making the material or product, there is an 'energy mortgage'—the energy consumed in building the plant in the first place.

Embodied energies are more properly assessed by *input–output analysis*. For example, for a material such as ingot iron, cement powder or PET granules, the embodied energy/kg is found by monitoring over a fixed period of time the total energy input to the production plant (including that smuggled in, so to speak, as embodied energy of feedstock) and dividing this by the quantity of usable material shipped out of the plant. The upper part of Figure 20.4 shows, much simplified, the inputs to a PET production facility: oil derivatives such as naphtha and other feedstock, direct power (which, if electric, is generated with a production efficiency of about 34%), and the energy of transporting the feedstock to the facility. The plant has an hourly output of usable PET granules. The embodied energy of the PET, $(H_m)_{PET}$, with usual units of MJ/kg, is then given by

$$(H_m)_{PET} = \frac{\sum \text{Energies entering plant per hour}}{\text{Mass of PET granules produced per hour}}$$

The *processing energy* H_p associated with a material is the energy, in MJ, used to shape, join and finish 1 kg of the material to create a component or product. Thus polymers, typically, are molded or extruded; metals are cast, forged or machined; ceramics are shaped by powder methods. A characteristic energy per kg is associated with each of these. Continuing with the PET example, the granules now become the input (after transportation) to a facility for blow-molding PET bottles for water, as shown in the lower part of Figure 20.4. There is no

Figure 20.4   An input–output diagram for PET production (giving the embodied energy/kg of PET) and for bottle production (giving the embodied energy/bottle).

need to list the inputs again—they are broadly the same, the PET itself bringing with it its embodied energy $(H_m)_{PET}$. The output of the analysis is the energy committed per bottle produced.

There are many more steps before the bottle reaches a consumer and is drunk: collection, filtration and monitoring of the water, transportation of water and bottles to bottling plant, labeling, delivery to central warehouse, distribution to retailers and refrigeration prior to sale. All have energy inputs, which, when totalled, give the energy cost of as simple a thing as a plastic bottle of cold water.

The *end-of-life potential* summarizes the possible utility of the material at life's end: the ability to be recycled back into the product from which it came, the lesser ability to be down-cycled into a lower-grade application, the ability to be biodegraded into usable compost, the ability to yield energy by controlled combustion and, failing all of these, the ability to be buried as landfill without contaminating the surrounding land then or in the future.

Recycling: ideals and realities

We buy, use and discard paper, packaging, cans, bottles, television sets, computers, furniture, tires, cars, even buildings. Why not retrieve the materials they contain and use them again? What could be simpler?

If you think that, think again. First, some facts. There are (simplifying again) two sorts of 'scrap', by which we mean material with recycle potential. In-house scrap is the off-cuts, ends and bits left in a material production facility when the usable material is shipped out. Here ideals are realized: almost 100% is recycled, meaning that it goes back into the primary production loop. But once a material is released into the outside world the picture changes. It is processed to make parts that may be small, very numerous and widely dispersed; it is assembled into products that contain many other materials; it may be painted, printed or plated; and its subsequent use contaminates it further. To reuse it, it must be collected (not always easy), separated from other materials, identified, decontaminated, chopped and processed. Collection is time-intensive and this makes it expensive. Imperfect separation causes problems: even a little copper or tin damages the properties of steel; residual iron embrittles aluminum; heavy metals (lead, cadmium, mercury) are unacceptable in many alloys; PVC contamination renders PET unusable, and dyes, water and almost any alien plastic renders a polymer unacceptable for its original demanding purpose, meaning that it can only be used in less demanding applications (a fate known as 'down-cycling').

Despite these difficulties, recycling can be economic, both in cash and energy terms. This is particularly so for metals: the energy commitment per kg for recycled aluminum is about one-tenth of that for virgin material; that for steel is about one-third. Some inevitable contamination is countered by addition of virgin material to dilute it. Metal recycling is both economic and makes important contributions to the saving of energy.

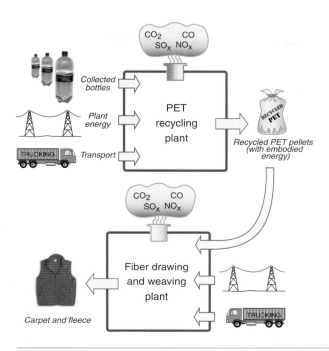

Figure 20.5   A simplified input–output diagram for the recycling of plastics to recover PET, and its use to make lower-grade products such as fleece.

Table 20.1   The energy-absorbing steps in recycling PET

1. Collection	7. Melting
2. Inspection	8. Filtration
3. Chopping	9. Pelletizing
4. Washing	10. Packaging
5. Flotation–separation	11. Plant heating, lighting
6. Drying	12. Transport

The picture for plastics is less rosy. The upper part of Figure 20.5 illustrates this for PET. Bottles are collected and delivered to the recycling plant as mixed plastic—predominantly PET, but with PE and PP bottles too. Table 20.1 lists the steps required to recycle the PET, each one consuming energy, with the results listed in Table 20.2. Some energy is saved, but not a lot—typically 50%.

Recycling of PET, then, can offer an energy saving. But is it economic? Time, in manufacture, is money. Collection, inspection, separation and drying are slow processes, and every minute adds dollars to the cost. Add to this the fact

Table 20.2   Embodied energy and market price of virgin and recycled plastics

Polymer	Embodied energy* (MJ/kg)		Price[†] ($/kg)	
	Virgin	*Recycled*	*Virgin*	*Recycled*
HDPE	82	40	1.9	0.9
PP	82	40	1.8	1.0
PET	85	55	2.0	1.1
PS	101	45	1.5	0.8
PVC	66	37	1.4	0.9

*Approximate values; see CES Edu 06 for details.
[†]Spot prices, December 2005.

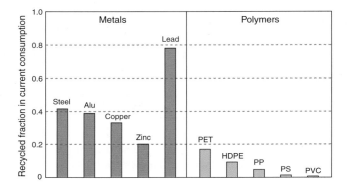

Figure 20.6   The fractional contribution of recycled material to current consumption. For metals, the contribution is large; for polymers, small (2005 data).

that the quality of recycled material is less good than the original, limiting its use to less demanding products, as suggested by the lower part of Figure 20.5—recycled PET cannot be used for bottles. Table 20.2 lists the current market price of granules of five commodity polymers in the virgin and the recycled states. If the recycled stuff were as good as new it would command the same price; in reality it commands little more than half. Thus, using today's technology, the cost of recycling plastics is high and the price they command is low, not a happy combination.

The consequences of this are brought out by Figure 20.6. It shows the current recycle fraction of commodity metals and plastics. The recycle fraction is the fraction of current supply that derives from recycling. For metals it is high: most of the lead, and almost half the steel and one-third of the aluminum we use today has been used at least once before. For plastics the only small success is PET, with a recycle fraction of about 18%, but for the rest the contribution is tiny, for many zero. Oil price inflation and restrictive legislation could change all this, but for the moment, that is how it is.

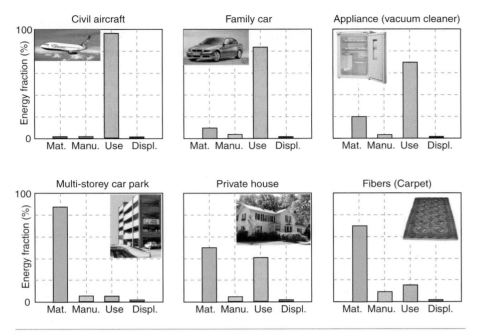

Figure 20.7   Approximate values for the energy consumed at each phase of Figure 20.3 for a range of products. The columns show the approximate embodied energy ('Mat.'), energy to manufacture ('Manu.'), use energy over design life ('Use') and energy for disposal ('Displ.').

The energy demands of products

With this background, we can proceed to look at the way products consume energy in each of the four life phases of Figure 20.3. The procedure is to tabulate the main components of the product together with their material and weight. The embodied energy and energy of processing associated with the product are estimated by multiplying the weights by the energies H_m and H_p, and summing. The use-energy of energy-using products may be estimated from information for the power, the duty cycle and the source from which the power is drawn. To this should be added the energy associated with maintenance and service over the useful life of the product. The energy of disposal is more difficult: some energy may be recovered by incineration, some saved by recycling, but, as already mentioned, there is also an energy cost associated with collection and disassembly. Transport costs can be estimated from the distance of transport and the energy/km.kg of the transport mode used.

Despite the uncertainty in some of the data, the outcome of this analysis is revealing. Figure 20.7 presents the evidence for a range of product groups. It has two significant features, with important implications. The product groups in the top row all consume energy as an unavoidable consequence of their use and for these the use-phase overwhelmingly dominates the life energy. The products in the bottom row depend less heavily on energy but are material intensive; for these it is the embodied energy of the material that dominates. If large changes are to be achieved, it is the dominant phase that must be the first target; when the differences are as great as those shown here, a reduction in the others makes little impact on the total, and the precision of the data for H_m and H_p is not the issue—an error of a factor of 2 changes the outcome very little. It is in the nature of those who conduct detailed LCA studies to wish to do so with precision, and better data is always a desirable goal. But engineers and designers need the ability to move forward without it, recognizing that precise judgements can be drawn from imprecise data.

20.5   Charts for embodied energy

Bar charts for embodied energy

Figures 20.8 and 20.9 show the embodied energy per kg and per m^3 for materials. When compared per unit mass, metals, particularly steels, appear as attractive choices, demanding much less energy than polymers. But when compared on a volume basis, the ranking changes and polymers lie lower than metals. The light

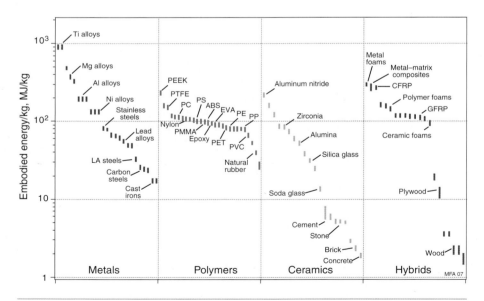

Figure 20.8   Bar chart of embodied energy of basic materials by weight. By this measure polymers are more energy intensive than many metals

alloys based on aluminum, magnesium and titanium are particularly demanding, with energies that are high by either measure. This prompts the question: what measure should we choose to make meaningful comparisons if we wish to minimize the embodied energy of a product? The answer is the same as the one we used with the objectives of minimizing mass or cost: it is to minimize embodied energy *per unit of function*. To do that we need the next two charts.

Property charts for embodied energy in structural design

Earlier chapters discussed the property trade-offs in problems of structural design. The function of the design might be, for example, to support a load without too much deflection, or without failure, while minimizing the mass. For this, modulus–density or strength–density were used. If the objective becomes minimizing the energy embodied in the material of the product while providing structural functionality, we need equivalent charts for these.

Figures 20.10 and 20.11 are a pair of materials selection charts for minimizing energy H_m per unit stiffness and strength. The first shows modulus E plotted against $H_m\rho$; the guidelines give the slopes for three of the commonest performance indices. The second shows strength σ_y plotted against $H_m\rho$; again, guidelines give the slopes. The two charts give survey data for minimum energy design. They are used in exactly the same way as the E–ρ and σ_y–ρ charts for minimum mass design.

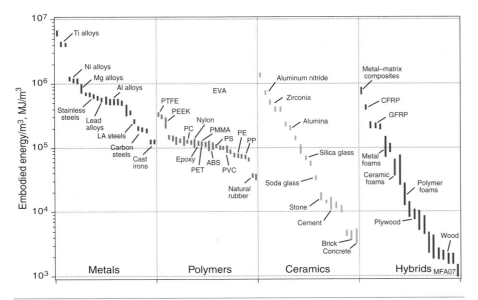

Figure 20.9   Bar chart of embodied energy of basic materials by volume. By this measure polymers are less energy intensive than any metal.

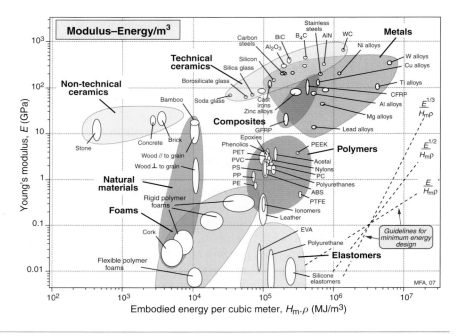

Figure 20.10   The modulus–embodied energy chart made with the CES software. It is the equivalent of the $E-\rho$ chart of Figure 4.6 and is used in the same way.

Figure 20.11   The strength–embodied energy chart made with the CES software. It is the equivalent of the $\sigma_y-\rho$ chart of Figure 6.6 and is used in the same way.

20.6 Design: selecting materials for eco-design

For selection of materials in environmentally responsible design we must first ask: which phase of the life cycle of the product under consideration makes the largest impact on the environment? The answer guides the effective use of the data in the way shown in Figure 20.12.

The material production phase

If material production consumes more energy than the other phases of life, it becomes the first target. Drink containers provide an example: they consume materials and energy during material extraction and container production, but, apart from transport and possible refrigeration, not thereafter. Here, selecting materials with low embodied energy and using less of them are the ways forward. Figure 20.7 made the point that large civil structures—buildings, bridges, roads—are material intensive. For these the embodied energy of the materials is the largest commitment. For this reason architects and civil engineers concern themselves with embodied energy as well as the thermal efficiency of their structures.

The product manufacture phase

The energy required to shape a material is usually much less than that to create it in the first place. Certainly it is important to save energy in production. But

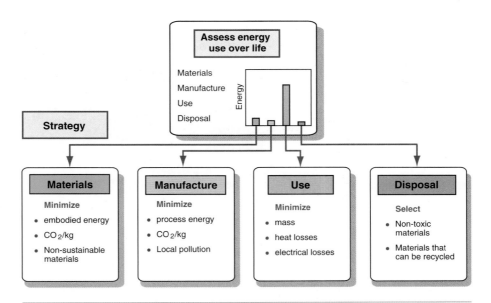

Figure 20.12   Rational use of the database starts with an analysis of the phase of life to be targeted. The decision then guides the method of selection to minimize the impact of the phase on the environment.

higher priority often attaches to the local impact of emissions and toxic waste during manufacture, and this depends crucially on local circumstances. Clean manufacture is the answer here.

The product use phase

The eco-impact of the use phase of energy-using products has nothing to do with the embodied energy of the materials themselves—indeed, minimizing this may frequently have the opposite effect on use energy. Use energy depends on mechanical, thermal and electrical efficiencies; it is minimized by maximizing these.

Fuel efficiency in transport systems (measured, say, by MJ/km) correlates closely with the mass of the vehicle itself; the objective then becomes that of minimizing mass. The evidence for this can be seen in Figure 20.13, showing the fuel consumption of some 4000 European models of car against their unladen mass, segregated by engine type (super-sport and luxury cars, shown as red symbols, are separated out—for these, fuel economy is not a design priority). The lines show linear fits through the data: the lowest, through the green symbols, for diesel-powered cars, the one above, through the blue symbols, for those with petrol engines. One hybrid model is included (yellow symbol). The correlation between fuel consumption and weight is clear. Here the solution is *minimum*

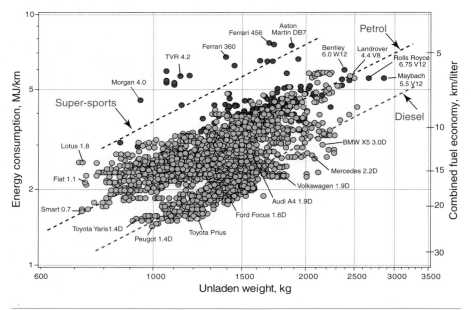

Figure 20.13   Energy consumption and fuel economy of 2005 model European cars, plotted against the unladen weight. The open red symbols are diesels, the open black are petrol driven and the full red symbols are for cars designed with performance, above all else, in mind. The broken lines are best fits to data for each type. Note the near-linear dependence of energy consumption on weight.

mass design, discussed extensively in earlier chapters; it is just as relevant to eco-design as to performance-driven design.

Energy efficiency in refrigeration or heating systems is achieved by minimizing the heat flux into or out of the system; the objective is then that of minimizing thermal conductivity or thermal inertia. Energy efficiency in electrical generation, transmission and conversion is maximized by minimizing the ohmic losses in the conductor; here the objective is to minimize electrical resistance while meeting necessary constraints on strength, cost, etc. Material selection to meet these objectives is well documented in other chapters and the texts listed under 'Further reading'.

The product disposal phase

The environmental consequences of the final phase of product life have many aspects. The ideal is summarized in the following guidelines:

- Avoid toxic materials such as heavy metals and organometallic compounds that, in landfill, cause long-term contamination of soil and groundwater.
- Examine the use of materials that cannot be recycled, since recycling can save both material and energy, but keep in mind the influence they have on the other phases of life.
- Seek to maximize recycling of materials for which this is possible, even though recycling may be difficult to achieve for the reasons already discussed.
- When recycling is impractical seek to recover energy by controlled combustion.
- Consider the use of materials that are biodegradable or photo-degradable, although these are ineffectual in landfill because the anaerobic conditions within them inhibit rather than promote degradation.

Implementing this requires information for toxicity, potential for recycling, controlled combustion and biodegradability. The CES software provides simple checks of each of these.

Case study: crash barriers

Barriers to protect driver and passengers of road vehicles are of two types: those that are static (the central divider of a freeway, for instance) and those that move (the fender of the vehicle itself) (Figure 20.14). The static type lines tens of thousands of miles of road. Once in place they consume no energy, create no CO_2 and last a long time. The dominant phases of their life in the sense of Figure 20.7 are those of material production and manufacture. The fender, by contrast, is part of the vehicle; it adds to its weight and thus to its fuel consumption. The dominant phase here is that of use. This means that, if eco-design is the objective, the criteria for selecting materials for the two sorts of barrier will differ.

The function of a barrier is to transfer load from the point of impact to the support structure, where reaction from the foundation or from crush elements in the vehicle support or absorb it. To do this the material of the barrier must have adequate strength, σ_y, and the ability to be shaped and joined cheaply, and (thinking of the disposal phase of life) recyclable. That for the car fender must

Figure 20.14 Two crash barriers, one static, the other—the fender—attached to something that moves. Different eco-criteria are needed for each.

meet these constraints with minimum mass, since this will reduce the use energy. As we know from Chapter 7, this means materials with high values of the index

$$M_1 = \frac{\sigma_y}{\rho}$$

where σ_y is the tensile strength and ρ is its density. For the static barrier embodied energy, not weight, is the problem. If we change the objective to that of *minimum embodied energy*, we require materials with large values of

$$M_2 = \frac{\sigma_y}{H_m \rho}$$

where H_m is the embodied energy per kg of material.

The chart of Figure 7.8 guides selection for the mobile barrier, where we seek strength at low weight. CFRPs excel by this criterion, but they are not recyclable. Heavier, but recyclable, are alloys of magnesium, titanium and aluminum. Ceramics are excluded both by their brittleness and the difficulty of shaping and joining them.

The chart of Figure 20.11 guides the selection for static barriers, where we seek strength at low embodied energy. The index M_2 is plotted in Figure 20.15. The chart shows that embodied energy per unit strength (leaving ceramics aside because of brittleness) is minimized by making the barrier from carbon steel, cast iron or wood; nothing else comes close.

Figure 20.15 The selection of materials for strength at minimum embodied energy. The best choices (rejecting ceramics because they are brittle) are cast iron, steel and wood.

Stiffness-limited design is treated in a similar way. Achieving it at minimum mass was the subject of Chapters 4 and 5. To do so at minimum embodied energy just requires that ρ is replaced by $H_m\rho$.

20.7 Summary and conclusions

Rational selection of materials to meet environmental objectives starts by identifying the phase of product life that causes greatest concern: production, manufacture, use or disposal. Dealing with all of these requires data not only for the obvious eco-attributes (energy, CO_2 and other emissions, toxicity, ability to be recycled, and the like) but also data for mechanical, thermal, electrical and chemical properties. Thus, if material production is the phase of concern, selection is based on minimizing the embodied energy or the associated emissions (CO_2 production, for example). But if it is the use phase that is of concern, selection is based instead on low weight, or excellence as a thermal insulator, or as an electrical conductor while meeting other constraints on stiffness, strength, cost, etc. The charts of this book give guidance in meeting these constraints and objectives. The CES databases provide data and tools that allow more sophisticated selection.

20.8 Appendix: some useful quantities

Energy contents of fuels

- Coal, lignite   15–19 MJ/kg
- Coal, anthracite   31–34 MJ/kg
- Oil   11.69 kWh/liter = 47.3 MJ/kg
- Gas   10.42 kWh/m^3
- LPG   13.7 kWh/liter = 46.5–49.6 MJ/kg

Approximate energy requirements of transport systems in MJ per tonne-km

- Sea freight   0.11
- Barge (river freight)   0.83
- Rail freight   0.86
- Truck   0.9–1.5, depending on size of truck (large are more economic)
- Air freight   8.3–15, depending on type and size of plane

Conversion factors

- 1 BthU = 1.06 kJ
- 1 kWh/kg (sometimes written kW/kg/h) = 3.6 MJ/kg
- A barrel of oil = 42 US gallons = 159 liters = 138 kg = 6210 MJ
- At $50 per barrel, a dollar buys 124 MJ

20.9 Further reading

Ashby, M.F. (2005) *Materials Selection in Mechanical Design*, 3rd edition, Butterworth-Heinemann, Oxford, UK. ISBN 0-7506-6168-2. *(A more advanced text developing the ideas presented here.)*

Brundlandt, D. (1987) *Report of the World Commission on the Environment and Development*, Oxford University Press, Oxford, UK. ISBN 0-19-282080-X. *(A much quoted report that introduced the need and potential difficulties of ensuring a sustainable future.)*

Dieter, G.E. (1991) *Engineering Design, A Materials and Processing Approach*, 2nd edition, McGraw-Hill, New York, USA. ISBN 0-07-100829-2. *(A well-balanced and respected text focusing on the place of materials and processing in technical design.)*

Goedkoop, M., Effting, S. and Collignon, M. (2000) *The Eco-indicator 99: A Damage Oriented Method for Life Cycle Impact Assessment, Manual for Designers* (14 April 2000), http://www.pre.nl. *(An introduction to eco-indicators, a technique for rolling all the damaging aspects of material production into a single number.)*

Graedel, T.E. (1998) *Streamlined Life-cycle Assessment*, Prentice-Hall, New Jersey, USA. ISBN 0-13-607425-1. *(An introduction to LCA methods and ways of streamlining them.)*

Kyoto Protocol (1997) United Nations, Framework Convention on Climate Change. Document FCCC/CP1997/7/ADD.1 (http://cop5.unfccc.de) O) 14040/14041/14042/ 14043, 'Environmental management—life cycle assessment' and subsections, Geneva, Switzerland. *(The international consensus on combating climate change.)*

20.10 Exercises

Exercise E20.1	What is meant by embodied energy per kilogram of a metal? Why does it differ from the thermodynamic energy of formation of the oxide, sulfide or silicate from which it was extracted?
Exercise E20.2	What is meant by the process energy per kilogram for casting a metal? Why does it differ from the latent heat of melting of the metal?
Exercise E20.3	Why is the recycling of metals more successful than that of polymers?
Exercise E20.4	The world consumption rate of CFRP is rising at 8% per year. How long does it take to double?
Exercise E20.5	Prove the statement made in the text that, 'at a global growth rate of just 3% per year we will mine, process and dispose of more 'stuff' in the next 25 years than in the entire history of human engineering'. For the purpose of your proof, assume consumption started with the dawn of the industrial revolution, 1750.
Exercise E20.6	Which phase of life would you expect to be the most energy intensive (in the sense of consuming fossil fuel) for the following products:

  - A toaster.
  - A two-car garage.
  - A bicycle.
  - A motorbike.
  - A wind turbine.
  - A ski lift.

  Indicate, in each case, your reasoning in one sentence.

Exercise E20.7	Show that the index for selecting materials for a strong panel, loaded in bending, with the minimum embodied energy content is:

$$M = \frac{\sigma_y}{H_m \rho}$$

Exercise E20.8 Use the chart $E–H_m\rho$ chart of Figure 20.10 to find the metal with a modulus E greater than 100 GPa and the lowest embodied energy per unit volume.

Exercise E20.9 Use the chart of $\sigma_y–H_m\rho$ (Figure 20.11) to find material for strong panels with minimum embodied energy per unit volume.

Exercise E20.10 Car fenders used to be made of steel. Most cars now have extruded aluminum or glass-reinforced polymer fenders. Both materials have a much higher embodied energy than steel. Take the mass of a steel bumper to be 20 kg and that of an aluminum one to be 14 kg; a bumper-set (two bumpers) weighs twice as much. Find an equation for the energy consumption in MJ/km as a function of mass for petrol engine cars using the data plotted in Figure 20.13.

(a) Work out how much energy is saved by changing the bumper-set of a 1500 kg car from steel to aluminum.

(b) Calculate whether, over an assumed life of 200 000 km, the switch from steel to aluminum has saved energy. You will find the embodied energies of steel and aluminum in the CES Level 2 database. Ignore the differences in energy in manufacturing the two bumpers—it is small. (The energy content of gasoline is 44 MJ/litre.)

(c) The switch from steel to aluminum increases the price of the car by $60. Using current pump prices for gasoline, work out whether, over the assumed life, it is cheaper to have the aluminum fender or the steel one.

20.11 Exploring design with CES

Exercise E20.11   Rank the three common commodity materials *Low carbon steel*, *Age hard-ening aluminum alloy* and *Polyethylene* by embodied energy/kg and embod-ied energy/m^3, using data drawn from Level 2 of the CES Edu database (use the means of the ranges given in the databases). Materials in products perform a primary function—providing stiffness, strength, heat transfer and the like. What is the appropriate measure of embodied energy for a given function?

Exercise E20.12   Plot a bar chart for the embodied energies of metals and compare it with one for polymers, on a 'per unit yield strength' basis, using CES. You will need to use the 'Advanced' facility in the axis-selection window to make the function:

$$\text{Energy per unit strength} = \frac{\text{Embodied energy} \times \text{Density}}{\text{Yield strength}}$$

Which materials are attractive by this measure?

Exercise E20.13   Drink containers co-exist that are made from a number of different materi-als. The masses of five competing container types, the material of which they are made and the specific energy content of each are listed in the table. Which container type carries the lowest overall embodied energy per unit of fluid contained?

Container type	Material	Mass (g)	Embodied energy (MJ/kg)
PET 400 ml bottle	PET	25	84
PE 1 liter milk bottle	High-density PE	38	80
Glass 750 ml bottle	Soda glass	325	14
Al 440 ml can	5000 series Al alloy	20	200
Steel 440 ml can	Plain carbon steel	45	23

Exercise E20.14   Iron is made by the reduction of iron oxide, Fe_2O_3, with carbon, aluminum by the electrochemical reduction of bauxite, basically Al_2O_3. The enthalpy of oxidation of iron to its oxide is 5.5 MJ/kg, that of aluminum to its oxide is 20.5 MJ/kg. Compare these with the embodied energies of cast iron and of carbon steel, and of aluminum, retrieved from the CES database (use means of the ranges given there). What conclusions do you draw?

Exercise E20.15   Calculate the energy to mold PET by assuming it to be equal to the energy required to heat PET from room temperature to its melting temperature, T_m. Compare this with the actual molding energy. You will find the molding energy, the specific heat and the melting temperature in the Level 2 record for PET in CES (use means of the ranges). Assume that the latent heat of melting is equal to that to raise the temperature from room temperature to the melting point. What conclusions do you draw?

Index

CES EduPack

Software support for your learning

CES EduPack is a software-based teaching toolkit that complements this book. It is used at hundreds of universities and colleges worldwide. With CES EduPack you can:

- **Get more from this text.** Complete supporting exercises as you read. Explore the world of materials in more depth. Together, the book and software offer a superb integrated learning resource for introductory materials courses and beyond.

- **Meet your learning needs.** CES EduPack is renowned for its outstanding materials selection capabilities. But the software offers much more to students of materials. You can 'drill down' into information that explains the science behind materials properties and processes. Different databases of these properties and processes are available to match varying needs, from first to final year. Special editions cover key engineering disciplines. CES EduPack supports you throughout the engineering curriculum.

- **Benefit from campus licensing.** Many departments use CES EduPack in multiple courses through a campus license, helping to make connections across the curriculum. This scheme also allows every student to install the software on their own computer.

CES EduPack is normally purchased by a college or university and then made available to students. Teachers - contact Granta Design (details below). We'll be happy to arrange an online demonstration of the software and to tell you about the lectures, projects, and exercises included in the tookit. Students - why not ask your teacher to contact us?!

MATERIAL INSPIRATION

www.grantadesign.com | info@grantadesign.com | UK/world +44 1223 518895

US toll free (800) 241-1546 | France 08 00 76 12 90 | Germany 0800 182 5026